Information Dynamics

NATO ASI Series

Advanced Science Institutes Series

A series presenting the results of activities sponsored by the NATO Science Committee, which aims at the dissemination of advanced scientific and technological knowledge, with a view to strengthening links between scientific communities.

The series is published by an international board of publishers in conjunction with the NATO Scientific Affairs Division

A	**Life Sciences**	Plenum Publishing Corporation
B	**Physics**	New York and London
C	**Mathematical and Physical Sciences**	Kluwer Academic Publishers
D	**Behavioral and Social Sciences**	Dordrecht, Boston, and London
E	**Applied Sciences**	
F	**Computer and Systems Sciences**	Springer-Verlag
G	**Ecological Sciences**	Berlin, Heidelberg, New York, London,
H	**Cell Biology**	Paris, Tokyo, Hong Kong, and Barcelona
I	**Global Environmental Change**	

Recent Volumes in this Series

Volume 250—The Application of Charge Density Research to Chemistry and Drug Design
edited by George A. Jeffrey and Juan F. Piniella

Volume 251—Granular Nanoelectronics
edited by David K. Ferry, John R. Barker, and Carlo Jacoboni

Volume 252—Laser Systems for Photobiology and Photomedicine
edited by A. N. Chester, S. Martellucci, and A. M. Scheggi

Volume 253—Condensed Systems of Low Dimensionality
edited by J. L. Beeby

Volume 254—Quantum Coherence in Mesoscopic Systems
edited by B. Kramer

Volume 255—Vacuum Structure in Intense Fields
edited by H. M. Fried and Berndt Muller

Volume 256—Information Dynamics
edited by Harald Atmanspacher and Herbert Scheingraber

Series B: Physics

Information Dynamics

Edited by

Harald Atmanspacher

and

Herbert Scheingraber

Max-Planck-Institut für extraterrestrische Physik
Garching, Germany

Plenum Press
New York and London
Published in cooperation with NATO Scientific Affairs Division

Proceedings of a NATO Advanced Study Institute on
Information Dynamics,
held June 15-26, 1990,
in Irsee/Kaufbeuren, Germany

Library of Congress Cataloging-in-Publication Data

NATO Advanced Study Institute on Information Dynamics (1990 : Irsee, Germany)
Information dynamics / edited by Harald Atmanspacher and Herbert Scheingraber.
p. cm. -- (NATO ASI series. Series B, Physics ; v. 256)
"Proceedings of a NATO Advanced Study Institute on Information Dynamics, held June 15-26, 1990, in Irsee/Kaufbeuren, Germany"--T.p. verso.
"Published in cooperation with NATO Scientific Affairs Division."
Includes bibliographical references and index.
ISBN 0-306-43912-3
1. Information theory--Congresses. 2. System theory--Congresses.
I. Atmanspacher, Harald. II. Scheingraber, Herbert. III. Title.
IV. Series.
Q350.N38 1990
003'.85--dc20 91-14448
CIP

A Division of Plenum Publishing Corporation
233 Spring Street, New York, N.Y. 10013

Printed in the United States of America

SPECIAL PROGRAM ON CHAOS, ORDER, AND PATTERNS

This book contains the proceedings of a NATO Advanced Research Workshop held within the program of activities of the NATO Special Program on Chaos, Order, and Patterns.

Volume 208—MEASURES OF COMPLEXITY AND CHAOS
edited by Neal B. Abraham, Alfonso M. Albano, Anthony Passamante, and Paul E. Rapp

Volume 225—NONLINEAR EVOLUTION OF SPATIO-TEMPORAL STRUCTURES IN DISSIPATIVE CONTINUOUS SYSTEMS
edited by F. H. Busse and L. Kramer

Volume 235—DISORDER AND FRACTURE
edited by J. C. Charmet, S. Roux, and E. Guyon

Volume 236—MICROSCOPIC SIMULATIONS OF COMPLEX FLOWS
edited by Michel Mareschal

Volume 240—GLOBAL CLIMATE AND ECOSYSTEM CHANGE
edited by Gordon J. MacDonald and Luigi Sertorio

Volume 243—DAVYDOV'S SOLITON REVISITED: Self-Trapping of Vibrational Energy in Protein
edited by Peter L. Christiansen and Alwyn C. Scott

Volume 244—NONLINEAR WAVE PROCESSES IN EXCITABLE MEDIA
edited by Arun V. Holden, Mario Markus, and Hans G. Othmer

Volume 245—DIFFERENTIAL GEOMETRIC METHODS IN THEORETICAL PHYSICS: Physics and Geometry
edited by Ling-Lie Chau and Werner Nahm

Volume 256—INFORMATION DYNAMICS
edited by Harald Atmanspacher and Herbert Scheingraber

PREFACE

Information and dynamics are key terms in many contemporary directions of research in numerous fields. Basic frameworks in this regard are information theory and the theory of dynamical systems. The origin of both areas essentially dates back into the 20s of this century. An excellent presentation of the history of information theory from its early days (e.g, Nyquist, Hartley) until to the 70s has been given by J.R. Pierce *et al.* in the *IEEE Transactions on Information Theory* IT–**19** (1973). System theory, the other branch mentioned above, also started with first publications in the 20s (e.g., Köhler, Lotka). A historical sketch can be found in L. v. Bertalanffy's *General Systems Theory* (Braziller, New York, 1968), the main body of which presents a proper introduction into the main concepts and applications.

At present, both branches, information theory and system theory, are hardly separable any more. Aspects of both appear in cybernetics, control theory, communication theory, computer science, game theory, cognitive science, and related fields. Combining knowledge of various individual scientific disciplines, they are particularly useful to deal with problems of basically interdisciplinary character.

The dynamical view toward systems, i.e., a specific emphasis on their temporal evolution, has been underlined by recent developments in the field of nonlinear dynamics. Although this field can be retraced to the work of Poincaré, it received its main impetus in the 70s when scientists fully recognized the relevance of nonlinear processes in dynamical systems. At that time, deterministic chaos and instabilities far from thermal equilibrium were discovered as extremely fruitful conceptual approaches to a kind of behavior which may – almost paradoxically – be characterized as *deterministic, but not determinable.* Here, "deterministic" is used to denote a temporal evolution which is mathematically determined by an equation system, and "not determinable" means that predictions about a future state of the system are not possible with arbitrary accuracy. As it is well known today the limited predictability in dynamical systems can be regarded as an intrinsic generation of information by a system in both space and time. This information flow motivates the notion of *information dynamics.*

Information dynamics was the title of an *Advanced Study Institute* (ASI) held at the Swabian conference center at Irsee monastery, a location southwest of Munich, from June 15 – 26, 1990. The ASI was sponsored by the NATO Scientific Affairs Division through its special program on *Order, Chaos, and Patterns*, by the Max Planck Institut für extraterrestrische Physik in Garching (FRG), and by the National Science Foundation of the USA. It was devoted to the impact of information theory and the theory of dynamical systems onto basic disciplines of physics and adjacent areas of scientific research.

This volume contains the lectures given at the meeting, which addressed mathematical concepts of uncertainty, information, entropy, Boolean and non–Boolean logic, information flow in nonlinear dynamical systems, chaotic behavior, symbolic dynamics, complexity, prediction, modelling, observation, perception, evolution, irreversibility and its quantum theoretical basis, stochastic and causal aspects of quantum theory, quantum chaos, and quantum measurement. The organization of the corresponding articles is guided by their relevance to classical and quantum systems, although this distinction can hardly be maintained in all cases. An epilogue, concluding the volume, relates the contents of individual articles to contemporary directions in the philosophy of science. In particular, it discusses the contextuality of information (and complexity) measures, which is relevant to any semantic or pragmatic aspect of information.

A number of letters reaching us subsequent to the meeting emphasized its warm interactive atmosphere besides its scientific value. We should like to forward corresponding thanks to all those who made this atmosphere possible, in particular to Valentin Demmel and Stefan Döbereiner, who did a lot of invisible organizational work, and to the staff of the center, who provided an excellent background concerning both, twentieth century comfort and monastically creative seclusion.

Garching, December 1990

Harald Atmanspacher
Herbert Scheingraber

CONTENTS

MEASURES AND PRINCIPLES OF UNCERTAINTY AND INFORMATION: RECENT DEVELOPMENTS

GEORGE J. KLIR

Systems Science Department
Thomas J. Watson School
State University of New York
Binghamton, NY 13901, USA

1. Introduction

Real-world problems are rarely free from uncertainty and, consequently, they require to make decisions. In his book, the British economist Shackle [1] explains this point:

> In a predestinate world, decision would be *illusory*; in a world of perfect foreknowledge, *empty*; in a world without natural order, *powerless*. Our intuitive attitude to life implies non-illusory, non-empty, non-powerless decision ... Since decision in this sense excludes both perfect foresight and anarchy in nature, it must be defined as choice in face of bounded uncertainty.

In my opinion, this quote captures well the importance of uncertainty in human affairs. Indeed, conscious decision making, in all its varieties, is perhaps the most fundamental capability of human beings. One of the key roles of science is to improve this crucial capability.

In order to understand properly the process of making decisions, we must understand the notion of uncertainty first. That requires, in turn, that the various facets of uncertainty be adequately conceptualized and formalized in appropriate mathematical languages.

In general, uncertainty in a problem situation emerges whenever information pertaining to the situation is deficient in some respect. It may be incomplete, imprecise, fragmentary, not fully reliable, vague, contradictory, or deficient in some other way. These various information deficiencies result in different types of uncertainty.

For three hundred years (from about the mid-seventeenth century, when the formal concept of numerical probability emerged, until the 1960s), uncertainty was conceived solely in terms of probability theory. This seemingly unique connection between uncertainty and probability is now challenged. The challenge comes from several mathematical theories, distinct from probability theory, which are demonstrably capable of characterizing situations under uncertainty. The most visible of these

Information Dynamics, Edited by H. Atmanspacher and
H. Scheingraber, Plenum Press, New York, 1991

theories, which began to emerge in the 1960s, are the theory of fuzzy sets (Zadeh [2]), the Dempster–Shafer theory (Shafer [3]), possibility theory (Zadeh [4], Dubois and Prade [5]), and the theory of fuzzy measures (Sugeno [6]).

When uncertainty was investigated within these new theories (mostly in the 1980s), it became clear that there are several distinct types of uncertainty. That is, it was realized that uncertainty is a multidimensional concept. Which of its dimensions are actually manifested in a description of a decision situation is determined by the mathematical theory employed. The multidimensional nature of uncertainty was obscured when uncertainty was conceived solely in terms of probability theory, in which it is manifested by only one of its dimensions.

Well justified measures of uncertainty of relevant types are now available not only in the classical set theory and probability theory, but also in the theory of fuzzy sets, possibility theory, and the Dempster–Shafer theory. The purpose of this chapter is to overview these measures and to describe three principles of uncertainty that are based upon them, a principle of minimum uncertainty, a principle of maximum uncertainty, and a principle of uncertainty invariance.

2. Terminology and Notation

In order to introduce terminology and notation for this chapter, I briefly overview in this section four mathematical theories capable of characterizing situations under uncertainty: probability theory, possibility theory, fuzzy set theory, and the Dempster–Shafer theory. The overview is not comprehensive; I assume that the reader is familiar with these theories at least on the level at which they are covered in the text by Klir and Folger [7].

Let X denote a universal set under consideration, assumed here to be finite for the sake of simplicity, and let $P(X)$ denote the power set of X. Then, the *Dempster–Shafer theory* is based upon a function

$$m: \ P(X) \rightarrow [0,1]$$

such that:

$$m(\emptyset) = 0 \quad \text{and} \quad \sum_{A \subseteq X} m(A) = 1$$

This function is called a *basic assignment*; the value $m(A)$ represents the degree of belief (based on relevant evidence) that a specific element of X belongs to set A, but not to any particular subset of A. Every set $A \in P(X)$ for which $m(A) \neq 0$ is called a *focal element*. The pair (F, m), where F denotes the set of all focal elements of m, is called a *body of evidence*.

Associated with each basic assignment m is a pair of measure, a *belief measure*, Bel, and a *plausibility measure*, Pl, which are determined for all sets $A \in P(X)$ by the equations:

$$\text{Bel (A)} = \sum_{B \subseteq A} m(B) \tag{1}$$

$$\text{Pl (A)} = \sum_{B \cap A \neq \emptyset} m(B) \tag{2}$$

These equations and the definition of the basic assignment form the core of the Dempster–Shafer theory. This theory was first formulated by Shafer in 1976 [3].

Belief and plausibility measures are connected by the equation

$$\mathrm{Pl}\ (\mathrm{A}) = 1 - \mathrm{Bel}\ (\bar{\mathrm{A}}) \tag{3}$$

for all $A \in P(X)$, where $\bar{A}$ denotes the complement of A. Furthermore,

$$\mathrm{Bel}\ (\mathrm{A}) \leq \mathrm{Pl}\ (\mathrm{A}) \tag{4}$$

for all $A \in P(X)$.

A belief measure (or plausibility measure) becomes a *probability measure*, Pr, when all focal elements are singletons. In this case, Pr (A) = Bel (A) = Pl (A) for all $A \in P(X)$, which follows immediately from Eqs. (1) and (2), and we obtain the *additivity property*

$$\mathrm{Pr}\ (\mathrm{A} \cup \mathrm{B}) = \mathrm{Pr}\ (\mathrm{A}) + \mathrm{Pr}\ (\mathrm{B}) - \mathrm{Pr}\ (\mathrm{A} \cap \mathrm{B}) \tag{5}$$

of probability measures. Any probability measure, Pr, on a finite set can be uniquely determined by a *probability distribution function*

$$p: \ X \rightarrow [0,1]$$

via the formula:

$$\mathrm{Pr}\ (\mathrm{A}) = \sum_{\mathrm{x} \in \mathrm{A}} \mathrm{p(x)} \tag{6}$$

From the standpoint of the Dempster–Shafer theory, clearly:

$$p(x) = m(\{x\})$$

When some focal elements are not singletons, Eq. (5) bifurcates into the inequalities:

$$\mathrm{Bel}\ (\mathrm{A} \cup \mathrm{B}) \geq \mathrm{Bel}\ (\mathrm{A}) + \mathrm{Bel}\ (\mathrm{B}) - \mathrm{Bel}\ (\mathrm{A} \cap \mathrm{B}), \tag{7}$$
$$\mathrm{Pl}\ (\mathrm{A} \cap \mathrm{B}) \leq \mathrm{Pl}\ (\mathrm{A}) + \mathrm{Pl}\ (\mathrm{B}) - \mathrm{Pl}\ (\mathrm{A} \cup \mathrm{B}) \tag{8}$$

When all focal elements are nested (ordered by set inclusion), the body of evidence is called *consonant*. In this case, we obtain special plausibility measures, which are called *possibility measures* (or *consonant plausibility measures*), and the corresponding belief measures, which are called *necessity measures*. A possibility measure, Pos, is conveniently (and uniquely) determined by a *possibility distribution function*

$$r: \ X \rightarrow [0,1]$$

via the formula

$$\mathrm{Pos}\ (\mathrm{A}) = \max_{\mathrm{x} \in \mathrm{A}} \mathrm{r(x)} \tag{9}$$

for all $A \in P(X)$. The corresponding necessity measure, Nec, is then determined for all $A \in P(X)$ by a formula equivalent to Eq. (3):

$$\mathrm{Nec}\ (\mathrm{A}) = 1 - \mathrm{Pos}\ (\bar{\mathrm{A}}) \tag{10}$$

A theory that deals with consonant bodies of evidence in terms of possibility and necessity measures is usually called a *possibility theory*. Possibility and necessity measures satisfy the equations:

$$\text{Pos}\,(\text{A} \cup \text{B}) = \max[\text{Pos}\,(\text{A}), \text{Pos}\,(\text{B})] \tag{11}$$

$$\text{Nec}\,(\text{A} \cap \text{B}) = \max[\text{Nec}\,(\text{A}), \text{Nec}\,(\text{B})] \tag{12}$$

Assume that $X = \{x_1, x_2, \ldots, x_n\}$ and let

$$A_1 \subset A_2 \subset \ldots \subset A_n$$

where $A_i = \{x_1, x_2, \ldots, x_i\}, i = 1, 2, \ldots, n$, be a complete sequence of nested subsets that contains all focal elements of a possibility measure Pos. That is, $m(A) = 0$ for each $A \notin \{A_1, A_2, \ldots, A_n\}$. Let $m_i = m(A_i)$ and $r_i = r(x_i)$ for all $i = 1, 2, \ldots, n$. Then, the n—tuples

$$m = (m_1, m_2, \ldots, m_n)$$

$$r = (r_1, r_2, \ldots, r_n)$$

fully characterize the basic assignment and the possibility distribution, respectively, by which the possibility measure Pos is defined. The possibility distribution r is ordered in the sense that $r_i \geq r_{i+1}$ for all $i = 1, 2, \ldots, n-1$. It is well known that

$$r_i = \sum_{k=i}^{n} m_k \tag{13}$$

$$m_i = r_i - r_{i+1} \tag{14}$$

for all $i = 1, 2, \ldots, n$, where $r_{n+1} = 0$ by convention (Klir and Folger [7]).

Possibility theory can be formulated not only in terms of consonant bodies of evidence within the Dempster–Shafer theory, but also in terms of fuzzy sets. It was introduced in this latter manner by Zadeh [4]. A *fuzzy set* is a set whose boundary is not sharp. That is, the change from nonmembership to membership in a fuzzy set is gradual rather than abrupt. This gradual change is expressed as a *membership grade function*, μ_A, of the form

$$\mu_A : X \to [0, 1]$$

where A is a label of the fuzzy set defined by this function within the universal set X. The value μ_A expresses the grade of membership of element x of X in the fuzzy set A or, in other words, the degree of compatibility of x with the concept represented by the fuzzy set. A fuzzy set A is called *normalized* when $\mu_A(x) = 1$ for at least one $x \in X$. If $\mu_A(x) \leq \mu_B(x)$ for all $x \in X$, then A is called a *fuzzy subset* of B.

An important concept associated with fuzzy sets is an α–cut. Given a fuzzy set A and a specific number $\alpha \in [0, 1]$, the *α–cut* A_α is a crisp (non–fuzzy) set:

$$A_\alpha = \{x \in X | \mu_A(x) \geq \alpha\} \tag{15}$$

The set of all elements of X for which $\mu_A(x) > 0$ is called a *support* of the fuzzy set A; it is usually denoted by supp (A).

For some applications, the concept of the fuzzy set can be extended in various ways. An important extension is to allow a more general form of the membership grade function

$$\mu_A : X \to L$$

where L denotes a partially ordered set (usually a lattice). Fuzzy sets defined by this more general function are called *L-fuzzy sets* (Goguen [8]). Simple but useful examples of L–fuzzy sets are *interval-valued fuzzy sets.* They are defined by membership grade functions of the form

$$\mu_A : X \rightarrow P([0,1])$$

where for each $x \in X$, $\mu_A(x)$ is a closed interval in $[0,1]$.

Given a regular fuzzy set A with membership grade function μ_A (the range of μ_A is $[0,1]$), Zadeh [4] defines a possibility distribution function r_A associated with A as numerically equal to μ_A, i.e.,

$$r_A(x) = \mu_A(x) \tag{16}$$

for all $x \in X$; then, he defines the corresponding possibility measure, Pos_A, by the equation

$$\mathrm{Pos}_A(B) = \max_{x \in B} r_A(x) \tag{17}$$

for all $B \in P(X)$. In this interpretation of possibility theory, focal elements correspond to distinct α–cuts A_α of the fuzzy set A. This follows from the property that $A_\alpha \subset A_\beta$ when $\alpha > \beta$.

As observed by Yager [9], the Dempster–Shafer theory can be fuzzified. In its fuzzified form, the basic assignment is a function

$$\widetilde{m} : \widetilde{P}(X) \rightarrow [0,1]$$

where $\widetilde{P}(X)$ denotes the set of all fuzzy subsets of X. This function must satisfy the same requirements for the extended domain $\widetilde{P}(X)$ as function m does for the domain $P(X)$. Plausibility and belief based upon $\widetilde{m}$ are expressed by the following generalized counterparts of Eqs. (1) and (2),

$$\widetilde{\mathrm{Bel}}(A) = \sum_{B \in F} \widetilde{m}(B) \left[1 - \max_{x \in X} \min(1 - \mu_A(x), \mu_B(x))\right] \tag{1'}$$

$$\widetilde{\mathrm{Pl}}(A) = \sum_{B \in F} \widetilde{m}(B) \left[\max_{x \in X} \min(\mu_A(x), \mu_B(x))\right] \tag{2'}$$

where $\mu_A(x)$ and $\mu_B(x)$ are degrees of membership of element x in fuzzy sets A and B, respectively, and F is the set of all focal elements (fuzzy sets) associated with $\widetilde{m}$.

3. Measures of Uncertainty

Assume that we can measure the amount of uncertainty associated with a decision situation. Assume further that this amount of uncertainty can be reduced by obtaining relevant information as a result of some action (observing a new fact, performing an experiment, finding an historical record). Then, the amount of information obtained by the action can be measured by the reduction of uncertainty that results from the action. In this sense, the amount of uncertainty (pertaining to a decision situation) and the amount of information (obtained by a relevant action) are intimately connected.

The nature of uncertainty depends on the mathematical theory within which decision situations are formalized. Each mathematical theory is capable of capturing only certain specific types of uncertainty. The more general the theory, the more types of uncertainty are captured by it. The classical mathematical theories for characterizing situations under uncertainty have been set theory and probability theory. Since the mid 1960s, a number of generalizations of these classical theories became available for formalizing the various types of uncertainty. In this section, I overview measures of uncertainty (and information) in the two classical theories and the three novel theories introduced in Sec.2 (possibility theory, fuzzy set theory, and the Dempster–Shafer theory). Let me begin with the classical set theory and probability theory.

When a decision situation is expressed in terms of a finite set A of possible alternatives that are left undecided, it is clear that the uncertainty assumes the form of nonspecificity. The more alternatives, the less specific the situation is; when only one alternative is possible, the situation is fully specific. A measure of this sort of uncertainty (and information) was derived by Hartley [10] in 1928. He showed that the function

$$I(A) = \log_2 |A| \tag{18}$$

where $|A|$ denotes the cardinality set of A, is the only meaningful measure (except for a multiplication constant) of the amount of information needed to characterize one element of the set, or, in other words, to eliminate the nonspecificity inherent in the set. Later, the uniqueness of this measure was proven axiomatically by Rényi [11].

A measure of probabilistic uncertainty (and the associated information) was established by Shannon [12]. This measure, whose basic form is

$$H\left(p(x)|x \in X\right) = -\sum_{x \in X} p(x) \log_2 p(x) \tag{19}$$

where $(p(x)|x \in X)$ denotes a probability distribution on X, is usually called the *Shannon entropy*. It is well justified, in a number of alternative ways, as a unique measure of uncertainty conceptualized in terms of probability theory (Klir and Folger [7]).

The Shannon entropy as well as the Hartley function measure uncertainty in units that are called *bits*. One bit of uncertainty is equivalent to the total ignorance regarding the truth or falsity of one proposition. The name "bit" is an abbreviation of "*b*inary digi*t*" (values 0 and 1 of a binary digit are often used for encoding truth values of a proposition). When defined on the same universal set X, the Shannon entropy and the Hartley function have the same range: $[0, \log_2 |X|]$.

What type of uncertainty does the Shannon entropy measure? We may easily conclude that it does not measure nonspecificity: the probabilities $p(x)$ are required to be real numbers in $[0, 1]$, and thus fully specific, and each of them focuses on a single (i.e., specific) alternative x. We observe, however, that each probability $p(x)$ in Eq. (19) expresses the degree of belief (based upon some evidence) that x is the true alternative in a given context. In probability theory, X (when finite) is always viewed as a set of exhaustive and mutually exclusive alternatives (outcomes, states, elementary events, basic propositions). Hence, the beliefs expressed by the probabilities in the distribution $(p(x)|x \in X)$ conflict with each other. That is, beliefs expressed in terms of a probability distribution are dissonant. The greater the lack of discrimination among the beliefs, the greater the dissonance. This is precisely how the Shannon entropy behaves. Hence, the Shannon entropy is a *measure of dissonance* in a probability distribution.

A natural generalization of the Hartley measure to possibility theory was discovered by Higashi and Klir [13]. They coined for it the name *U–uncertainty.* When a possibility distribution is expressed in terms of a normalized fuzzy set A, the U–uncertainty has the form

$$U(A) = \int_0^1 \log_2 |A_\alpha| d\alpha \tag{20}$$

where $|A_\alpha|$ denotes the cardinality of the α–cut of the fuzzy set A. The uniqueness of this function as a *possibilistic measure of nonspecificity* under appropriate requirements was proven by Klir and Mariano [14]. For fuzzy sets that are not normalized and for which $\max \mu_A(x) = a$, each value $U(A)$, given by Eq. (20), must be divided by a.

A generalization of the U–uncertainty in the Dempster–Shafer theory is expressed by a function V that, for any given body of evidence (F, m), assumes the form:

$$V(m) = \sum_{A \in F} m(A) \log_2 |A| \tag{21}$$

The uniqueness of this general measure of nonspecificity was proven by Ramer [15]. Its units of measurement are bits and its range is $[0, \log_2 |X|]$. When m corresponds to a possibility distribution represented by a fuzzy set A via Eq. (16), then $V(m) = U(A)$. Furthermore, the nonspecificity of a possibility measure can also be expressed by a convenient form

$$V(r) = \sum_{i=2}^{n} r_i \log_2 \frac{i}{i-1} \tag{22}$$

where $r = (r_1, r_2, \ldots, r_n)$ is a possibility distribution that corresponds to the basic assignment distribution via Eq. (13) and that is assumed to be normalized $(r_1 = 1)$ and ordered $(r_i \geq r_{i+1}$ for all $i = 1, 2, \ldots, n-1)$.

Since focal elements of probability measures are singletons, $V(m) = 0$ for every probability measure. That is, there is no nonspecificity in probability measures; all probability measures are fully specific, as already anticipated on intuitive grounds. Hence, *probability theory is not capable of conceptualizing nonspecificity*, one of the basic types of uncertainty.

Let me discuss now the meaning of the Shannon entropy within the Dempster–Shafer theory. Since the Dempster–Shafer theory is a generalization of probability theory, obtained by relaxing the additivity requirement, is it reasonable to expect that some appropriately generalized form of the Shannon entropy exists that is universally applicable within the broader theory? This question has already been answered in the affirmative by establishing that the function

$$E(m) = -\sum_{A \in F} m(A) \log_2 \mathrm{Pl}\ (\mathrm{A}) \tag{23}$$

is the generalized form of the Shannon entropy (Yager [16]). This function is usually called a *measure of dissonance.* Its units of measurement are bits and its range is $[0, \log_2 |X|]$.

What does function E actually measure? From Eq. (2) and the general property of basic assignments (satisfied for every $A \in P(X)$)

$$\sum_{B \cap A \neq \emptyset} m(B) + \sum_{B \cap A = \emptyset} m(B) = 1$$

we obtain:

$$E(m) = -\sum_{A \in F} m(A) \log_2 \left(1 - \sum_{B \cap A = \emptyset} m(B)\right) \tag{24}$$

The term

$$K = \sum_{B \cap A = \emptyset} m(B)$$

in this expression for $E(m)$ clearly represents the total conflict between the belief in A and other beliefs within a given body of evidence. It is obtained simply by adding the basic assignment values of all focal elements that are disjoint with A and, consequently, the beliefs allocated to them are in conflict with the belief focusing on A. The value of K ranges from 0 to 1. The function

$$-\log_2[1 - K]$$

which is employed in Eq. (24), is monotonically increasing with K; it extends the range from $[0, 1]$ to $[0, \infty)$. The choice of the logarithmic function is based on the same motivation as the choice of the logarithmic function in the Shannon entropy. Now, we can readily see that $E(m)$ defines the mean (expected) value of the conflict in beliefs associated with a given body of evidence (F, m); the name "measure of dissonance" is thus quite appropriate. This observation reinforces my previous argument that the Shannon entropy measures the degree of conflict (dissonance) among beliefs expressed by a probability distribution.

Since focal elements of possibility measures are nested, the plausibility of each focal element must be 1 (by Eq. (2)) and, consequently, $E(m) = 0$ when m defines a possibility measure. That is, consonant bodies of evidence (and the associated possibility and necessity measures) are free of dissonance (conflict in beliefs).

Lamata and Moral [17] proposed that the two uncertainty measures that coexist in the Dempster–Shafer theory be added to obtain a measure of *total uncertainty*, T, given by the formula:

$$T(m) = \sum_{A \in F} m(A) \log_2 \frac{|A|}{\mathrm{Pl}\,(\mathrm{A})} \tag{25}$$

The measurement units of T are again bits and, as proven by Ramer [18], the range of T is the same as the range of each of the two measures subsumed under it (V and E): $[0, \log_2 |X|]$.

The question of how to measure the degree of vagueness or fuzziness of a fuzzy set has been one of the basic issues of fuzzy set theory. Various measures of vagueness, more often called *measures of fuzziness*, have been proposed (Klir and Folger [7]). One way of measuring fuzziness, which was suggested by Yager [19] and further investigated by Higashi and Klir [20], is to view it as the lack of distinction between the fuzzy set and its complement. Clearly, the less a fuzzy set differs from its complement, the fuzzier it is. Using this approach, the measure of fuzziness depends on the complementation operator c and the Hamming distance. Then, assuming again a finite universal set X, f_c is given by the formula:

$$f_c(A) = |X| - \sum_{x \in X} |\mu_A(x) - c(\mu_A(x))| \tag{26}$$

Observe that the concept of the amount of fuzziness, measured by function f_c, has no applicability in classical probability theory: no vagueness is allowed in defining probabilities.

When the Dempster–Shafer theory is fuzzified, measures of the two relevant types of uncertainty, given by Eqs. (21) and (23), are still applicable provided that the entries in their formulae are properly interpreted: since focal elements A are now fuzzy sets, $|A|$ in Eq. (21) must be calculated by the formula

$$|A| = \sum_{x \in X} \mu_A(x) \tag{27}$$

which defines a simple (scalar) cardinality of fuzzy set A; in Eq. (23) values of Pl (A) must be calculated by Eq. (2').

In addition, it becomes also meaningful to measure fuzziness of a given fuzzified body of evidence. We may, for example, use the formula

$$F_c(\widetilde{m}) = \sum_{A \in F} \widetilde{m}(A) f_c(A) \tag{28}$$

where $F_c(\widetilde{m})$ denotes the fuzziness of $(F, \widetilde{m})$ and $f_c(A)$, which is determined by Eq. (26), denotes the fuzziness of the focal element (a fuzzy set) A.

The various measures of uncertainty and information introduced in this section are summarized in Table 1.

4. Principles of Uncertainty and Information

In this section, let me formulate three key principles of uncertainty, which, using the connection between uncertainty and information, can be also interpreted as principles of information. These principles are: a principle of minimum uncertainty, a principle of maximum uncertainty, and a principle of uncertainty invariance.

The *principle of minimum uncertainty* is an arbitration principle that helps us to select solutions in certain problems involving uncertainty. The principle requires that we accept only those solutions, from among all otherwise acceptable solutions, whose uncertainty (pertaining to the purpose involved) is minimal.

A major class of problems for which the principle of minimum uncertainty is applicable are *simplification problems*. When a system is simplified, it is usually unavoidable to lose some information contained in the system. The amount of information that is lost in this process results in the increase of an equal amount of relevant uncertainty. Examples of relevant uncertainties are predictive, retrodictive, or prescriptive uncertainty. A sound simplification of a given system should minimize the loss of relevant information (or the increase in relevant uncertainty) while achieving the required reduction of complexity. That is, we should accept only such simplifications of a given system at any desirable level of complexity for which the loss of relevant information (or the increase in relevant uncertainty) is minimal. When properly applied, the principle of minimum uncertainty guarantees that no information is wasted in the process of simplification.

There are many simplification strategies, which can perhaps be classified into three main classes:

- simplifications made by eliminating some entities from the system (variables, subsystems, etc.);

Table 1: Measures of uncertainty: a summary

Name	Formula	Meaning of Symbols
Hartley entropy (nonspecificity in classical set theory)	$I(A) = \log_2 \lvert A \rvert$	A: finite subset of universal set X $\lvert A \rvert$: cardinality of A
Shannon entropy (dissonance in probability theory)	$H(p) = -\sum_{i=1}^{n} p_i \log_2 p_i$	p: probability distribution $p = (p_1, p_2, ..., p_n)$
U–uncertainty (nonspecificity in possibility theory and fuzzy set theory)	$V(r) = \sum_{i=2}^{n} r_i \log_2 \frac{i}{i-1}$ $U(A) = \int_0^1 \log_2 \lvert A_\alpha \rvert d\alpha$	r: possibility distribution $r = (r_1, r_2, \ldots, r_n), r_{n+1} = 0,$ $r_i \geq r_{i+1}; i = 1, 2, \ldots, n-1$ A: fuzzy subset of X A_α: α–cuts of A
Dissonance (in Dempster–Shafer evidence theory)	$E(m) = -\sum_{A \in F} m(A) \log_2 \mathrm{Pl}(A)$	m: basic assignment Pl: plausibility function $\mathrm{Pl}(A) = \sum m(B),\ B \cap A \neq \emptyset$ F: set of focal elements
Nonspecificity (in Dempster–Shafer theory)	$V(m) = \sum_{A \in F} m(A) \log_2 \lvert A \rvert$	m: basic assingment $\lvert A \rvert$: cardinality of A F: set of focal elements
A representative measure of fuzziness	$f_c(A) = \lvert X \rvert - \sum_{x \in X} \lvert \mu_A(x) - c(\mu_A(x)) \rvert$	μ_A: membership grade function of fuzzy set A c: fuzzy complement $\lvert X \rvert$: cardinality of X
Fuzziness (in fuzzified Dempster–Shafer theory)	$F_c(\tilde{m}) = \sum_{A \in F} \tilde{m}(A) f_c(A)$	$\tilde{m}$: basic assingment on $\tilde{P}(X)$ A: fuzzy set on X F: set of focal elements $f_c(A)$: fuzziness of A

- simplifications made by aggregating some entities of the system (variables, states, etc.);
- simplifications made by breaking overall systems into appropriate subsystems.

Regardless of the strategy employed, the principle of minimum uncertainty is utilized in the same way. It is an arbiter which decides which simplifications to choose at any given level of complexity.

Another application of the principle of minimum uncertainty is the area of *conflict-resolution problems*. For example, when we integrate several overlapping models into one larger model, the models may be locally inconsistent. It is reasonable then to require that each of the models be appropriately adjusted in such a way that the overall model becomes consistent. It is obvious that some information contained in the given models is inevitably lost by these adjustments. This is not desirable. Hence, we should minimize this loss of information. That is, we should accept only those adjustments for which the total loss of information (or total increase of uncertainty) is minimal. The total loss of information may be expressed, for example, by the sum of all individual losses or by a weighted sum, if the given models are valued differently.

The second principle, the *principle of maximum uncertainty*, is essential for any problem that involves *ampliative reasoning*. This is reasoning in which conclusions are not entailed in the given premises. Using common sense, the principle may be expressed by the following requirement: in any ampliative inference, use all information available but make sure that no additional information is unwittingly added. That is, the principle requires that conclusions resulting from any ampliative inference maximize the relevant uncertainty within the constraints representing the premises. This principle guarantees that our ignorance be fully recognized when we try to enlarge our claims beyond the given premises and, at the same time, that all information contained in the premises be fully utilized. In other words, the principle guarantees that our conclusions are maximally noncommittal with regard to information not contained in the premises.

Ampliative reasoning is indispensable to science in a variety of ways. For example, whenever we utilize a scientific model for predictions, we employ ampliative reasoning. Similarly, when we want to estimate microstates from the knowledge of relevant macrostates and partial information regarding the microstates (as in image processing and many other problems), we must resort to ampliative reasoning. The problem of the identification of an overall system from some of its subsystems is another example that involves ampliative reasoning.

The principles of minimum and maximum uncertainty are well developed and tested within the classical information theory, where they are called the *principles of minimum and maximum entropy*. Perhaps the greatest skill in using these complementary principles in a broad spectrum of applications has been demonstrated by Christensen [21].

The *principle of maximum entropy*, which was founded, presumably, by Jaynes [22], has a history of more than thirty years (Kapur [23]). Its general formulation is: determine a probability distribution $(p(x)|x \in X)$ that maximizes the Shannon entropy (given by Eq. (19)) subject to given constraints $c_1, c_2, \ldots$, which express partial information about the unknown probability distribution, as well as general constraints (axioms) of probability theory. The most typical constraints employed in practical applications of the maximum entropy principle are mean (expected) values of one or more random variables or various marginal probability distributions of an unknown joint distribution.

Optimization problems that emerge from the maximum uncertainty principle outside the classical information theory have yet to be properly investigated and tested in praxis. These problems should be computationally rather simple in possibility theory since the possibility measure of uncertainty (the U–uncertainty) is a linear function. They will certainly be considerably more complex in the Dempster–Shafer theory, where two objective functions are involved, one linear (V) and one nonlinear (E).

Let me now turn to the third principle to be discussed here, the *principle of uncertainty invariance*. This principle, which was introduced recently by Klir [24], attempts to establish connections among representations of uncertainty and information in alternative mathematical theories. The principle requires that the amount of uncertainty (and information) be preserved when we transfer uncertainty formalized in one mathematical theory into an equivalent formalization in another theory. The most obvious examples are the transformation from a probabilistic formalization into an equivalent possibilistic formalization and the inverse transformation (Klir [24]). The principle of uncertainty invariance guarantees that no information is unwittingly added or eliminated solely by changing the mathematical framework by which a particular phenomenon is formalized.

The principle of uncertainty invariance, when properly developed, will allow us to use simultaneously several complementary theories for representing uncertainty and information. It will enable us to convert results obtained in one theory, most suitable for the purpose, into equivalent representations in the other theories. This will make all representations equally up-dated and ready for further use in an opportunistic fashion. That is, given a particular problem, we will use the theory most natural for dealing with it and, then, employ the uncertainty invariance principle for up-dating representations in the other theories.

5. Conclusions

Our repertory of mathematical theories by which we can characterize and deal with situations under uncertainty is now quite respectable. However, only the theories overviewed in Sec.2 are sufficiently developed in the sense that well justified measures of uncertainty have already been established in each of them.

While the less general theories (such as probability theory or possibility theory) are capable of characterizing only one type of uncertainty, the more general theories (such as the Dempster–Shafer theory) capture more than one type of uncertainty. This means that the more general a theory, the greater its flexibility to capture faithfully the various aspects of incompleteness or imperfection of information pertaining to a decision situation. As a rule, however, we pay for a greater generality by an increase in computational complexity. Furthermore, different theories (not necessarily comparable in terms of their generality) may be appropriate at different stages of a problem solving process or for different purposes at the same stage. As argued by Shackle [1, 25], for example, human reasoning about future events can be formalized more adequately by possibility theory rather then probability theory. Similarly, as argued by Cohen [26], possibility theory (and not probability theory) is the right mathematical framework for formalizing inductive reasoning. Possibility theory has also some attractive properties that are very useful in reconstructability analysis of systems (Klir *et al.* [27]). On the other hand, it is well known that probability theory is a natural tool for formalizing uncertainty in situations where evidence is based on outcomes of a series of independent random experiments.

The moral is that none of the theories is superior in all respects and for all purposes. That is, we need a flexible, multimodel representation of uncertainty. To utilize opportunistically advantages of the various theories of uncertainty, we need the capability of moving from one theory to another as appropriate. As explained in Sec.4, this is made possible by the principle of uncertainty and information invariance. Management of uncertainty within each theory is then handled by the complementary use of minimum and maximum uncertainty principles.

A turning point in our understanding of the concept of uncertainty was reached when it became clear that more than one type of uncertainty must be recognized within the theory of fuzzy sets (fuzziness and nonspecificity) and the Dempster–Shafer theory (nonspecificity and dissonance). This new insight into the concept of uncertainty was obtained by examining uncertainty emerging from mathematical frameworks more general than the two classical theories employed for characterizing uncertainty (classical set theory and probability theory).

The emergence of generalizations of existing mathematical theories is a significant current trend in mathematics, as exemplified by the change in emphasis from quantitative to qualitative, from functions to relations, from graphs to hypergraphs, from ordinary geometry (Euclidean as well as non–Euclidean) to fractal geometry, from ordinary automata to dynamic cellular automata, from classical analysis to a study of singularities (catastrophe theory), from ordinary artificial languages to developmental languages, from precise analysis to interval analysis, from classical logic to logic of inconsistency, from two–valued logic to multiple–valued logic, from single objective into multiple objective criteria optimization, and, as most relevant to the subject of this chapter, from probability measures to fuzzy measures and from classical set theory to fuzzy set theory. These generalizations stimulated primarily by advances in computer technology and modern systems thinking, have enriched not only our insights but also our capabilities for modelling the intricacies of the real world.

The relationship among uncertainty, complexity, and credibility of system models, which is of utmost importance to systems modelling, is not well understood as yet. We only know that uncertainty is a valuable commodity, which can be traded for a reduction of complexity or increase of credibility of models in the modelling business. Employing the results overviewed in this chapter, this trading can be made operational at a scale previously unsuspected. It is undeniable that major research must yet be undertaken not only to develop sound multidimensional principles of uncertainty in the novel mathematical theories, but also to learn how to use these theories in various application areas. After all, probability theory has been with us for over three centuries (Hacking [28]), while the new theories for conceptualizing uncertainty are a phenomenon of less than three decades.

References

1. G.L.S. Shackle. *Decision, Order, and Time in Human Affairs.* Cambridge University Press, Cambridge, 1961.
2. L.A. Zadeh. Fuzzy sets. *Information and Control* **8**, No.3, 338 - 353 (1965).
3. G. Shafer. *A Mathematical Theory of Evidence.* Princeton University Press, Princeton, 1976.
4. L.A. Zadeh. Fuzzy sets as a basis for a theory of possibility. *Fuzzy Sets and Systems* **1**, No.1, 3 - 28 (1978).
5. D. Dubois and H. Prade. *Possibility Theory.* Plenum Press, New York, 1988.
6. M. Sugeno. Fuzzy measures and fuzzy integrals: a survey. In *Fuzzy Automata and Decision Processes.* Eds. M.M. Gupta, G.N. Saridis, and B.R.Gaines. North Holland, Amsterdam/New York, 1977, pp. 89 - 102.
7. G.J. Klir and T.A. Folger. *Fuzzy Sets, Uncertainty, and Information.* Prentice Hall, Englewood Cliffs, 1988.
8. J.A. Goguen. L–fuzzy sets. *J. of Math. Analysis and Applications* **18**, 145 - 174 (1967).
9. R.R. Yager. Toward a general theory of reasoning with uncertainty: nonspecificity and fuzziness. *Int. J. of Intelligent Systems* **1**, No.1, 45 - 67 (1986).
10. R.V.L. Hartley. Transmission of information. *The Bell System Technical J.* **7**, 535 - 563 (1928).

11. A. Rényi. *Probability Theory.* North Holland, Amsterdam, 1970, Chapter IX: Introduction to Information Theory, pp. 540 - 616.

12. C.E. Shannnon. The mathematical theory of communication. *The Bell System Technical J.* **27**, 379 - 423, 623 - 656 (1948).

13. M. Higashi and G.J. Klir. Measures of uncertainty and information based on possibility distributions. *Int. J. of General Systems* **9**, No.1, 43 - 58 (1983).

14. G.J. Klir and M. Mariano. On the uniqueness of possibilistic measures of uncertainty and information. *Fuzzy Sets and Systems* **24**, No.2, 197 - 219 (1987).

15. A. Ramer. Uniqueness of information measure in the theory of evidence. *Fuzzy Sets and Systems* **24**, No.2, 183 - 196 (1987).

16. R.R. Yager. Entropy and specificity in a mathematical theory of evidence. *Int. J. of General Systems* **9**, No.4, 249 - 260 (1983).

17. M.T. Lamata and S. Moral. Measures of entropy in the theory of evidence. *Int. J. of General Systems* **14**, No.4, 297 - 305 (1988).

18. A. Ramer. Inequalities and generalized information measures. *Proceedings of NAFIPS '90, Toronto* (1990).

19. R.R. Yager. On the measure of fuzziness and negation. Part I: Membership in the unit interval. *Int. J. of General Systems* **5**, No.4, 221 - 229 (1979).

20. M. Higashi and G.J. Klir. On measures of fuzziness and fuzzy complements. *Int. J. of General Systems* **8**, No.3, 169 - 180 (1982).

21. R. Christensen. Entropy minimax multivariate statistical modeling. I. Theory. *Int. J. of General Systems* **11**, No.3, 231 - 277 (1985).
R. Christensen. Entropy minimax multivariate statistical modeling. I. Applications. *Int. J. of General Systems* **12**, No.3, 227 - 305 (1985).

22. E.T. Jaynes. *Papers on Probability, Statistics, and Statistical Physics.* Ed. R.D. Rosenkrantz. Reidel, Boston, 1983.

23. J.N. Kapur. Twenty–five years of maximum entropy principle. *J. Math. Phys. Sciences* **17**, No.2, 103 - 156 (1983).

24. G.J. Klir. A principle of uncertainty and information invariance. *Int. J. of General Systems* **17**, Nos.2/3, 249 - 275 (1990).

25. G.L.S. Shackle. *Imagination and the Nature of Choice.* Edinburgh University Press, Edinburgh, 1979.

26. L.J. Cohen. *The Implications of Induction.* Methuen, London, 1970.

27. G.J. Klir, B. Parviz, and M. Higashi. Relationship between true and estimated possibilistic systems and their reconstructions. *Int. J. of General Systems* **12**, No.4, 319 - 331 (1986).

28. I. Hacking. *The Emergence of Probability.* Cambridge University Press, Cambridge, 1975.

INFORMATION AND COMPLEXITY MEASURES IN DYNAMICAL SYSTEMS

PETER GRASSBERGER

Physics Department
University of Wuppertal
D-5600 Wuppertal 1, FRG

Abstract

These lectures deal with a number of information related concepts in dynamical systems. After a short general overview, emphasis is put on practical problems in numerical estimates of dynamical entropies and of attractor dimensions, and on the information dynamics of spatially extended systems. Regarding the latter, the possibility of measuring dimension densities and spatial information flow is discussed. In the last part, we investigate the relationship between chaos and (structured) complexity in chaotic dynamical systems, and in particular in systems showing emergent complexity.

1. Introduction

Emergence of structure and in particular of life, the most complex structure we know of, is an old conundrum in view of the second law of thermodynamics. It was observed already rather early that this seeming contradiction is resolved in principle by the observation that the second law applies to closed equilibrium systems only, while life is kept away from equilibrium by a constant flow of energy.

In spite of this, it still remains a challenge to understand in details the emergence of structure "out of nothing". It is here that toy models like maps, cellular automata, and simple coupled differential equations are essential: they show us that not all the complicatedness of real complex systems is necessary to observe interesting behaviour. Even with very few degrees of freedom, and with very simple basic laws, the resulting behaviour can be surprisingly rich. It is not yet clear whether the insight obtained in such models will be sufficient to understand the essentials of the complexity of life, but certainly the possibility to simulate such systems numerically has already led to a vast increase of our basic understanding.

In the present talk, I will shortly discuss this question. In particular, I will stress the point that it is not so much the energy flow, but rather an information flow which

Information Dynamics, Edited by H. Atmanspacher and
H. Scheingraber, Plenum Press, New York, 1991

is characteristic of open chaotic systems. Such systems have an infinite capacity of storing information. Since an observer never can deal with all of it, she has to concentrate on some "window", and effective randomness is observed when seemingly "irrelevant" information outside this window later becomes relevant. Complexity in the sense of meaningful structural richness arises in a further shift when this information, which up to this point had been at best a nuisance preventing simple predictable behaviour, acquires by itself an interest for the observer. One immediate consequence is that complexity will be viewed as a property not exclusively associated with the observed object, but rather as a subjective property reflecting the observer-object relationship. This is discussed in the next section.

While these concepts are easy to discuss in principle, real physical situations are corrupted by intrinsic or extrinsic noise, and practical considerations to overcome these problems are needed. Some of them are discussed in Sec.3.

Specific problems arise finally in spatially extended systems with infinitely many degrees of freedom. I shall only discuss the easiest case where the system is spatially uniform, without any frozen randomness and without large scale structures implied by other than collective instabilities. Thus, our considerations might apply to Bénard convection in very large aspect ratio cells, or to chaotic chemical pattern formation. It should not apply to fully developed turbulence, where the Galilean invariance and the conservation of mass imply the nonlinear derivative coupling responsible for the Kolmogorov cascade involving eddies of all sizes. Some of these problems are discussed in Sec.4.

In Sec.5, it is discussed in which sense chaotic systems can be complex as opposed to merely random. Several definitions of complexity are discussed there and applied to simple dynamical systems.

The last section, finally, contains a discussion and an outlook.

2. Emergence of Randomness in Chaotic Systems

In the simplest case, a dynamical system will be described by a set of coupled ordinary differential equations

$$\begin{aligned} \frac{dx_1}{dt} &= f_1(\vec{x}) \\ \frac{dx_2}{dt} &= f_2(\vec{x}) \\ \frac{dx_3}{dt} &= f_3(\vec{x}) \\ \ldots &= \ldots \end{aligned} \tag{1}$$

together with initial conditions:

$$\begin{aligned} x_1(0) &= a_1.b_1c_1\ldots \\ x_2(0) &= a_2.b_2c_2\ldots \\ x_3(0) &= a_3.b_3c_3\ldots \\ \ldots &= \ldots \end{aligned} \tag{2}$$

In realistic physical situations, the number of components will be equal to the number of degrees of freedom of a macroscopic system, and this will be a huge number (typically $\approx 10^{23}$). In a non-chaotic system not influenced by noise, it will be possible to find a small set of collective coordinates (center-of-mass coordinates, total momenta, ...) such that the macroscopic motion is effectively governed by them alone.

The traditional point of view is that noise comes in if such a clean separation is not possible. The set (1) of equations has to be cut off "vertically" (it is neither feasible nor desirable to describe a system by 10^{23} equations!), and the ignorance about the truncated degrees of freedom will appear as uncontrollable and random noise.

Deterministic chaos, on the other hand, is due to the necessary "horizontal" truncation in Eq. (2). It is impossible to specify initial conditions with infinite precision. In a regular system, this does not have serious consequences, since the ignorance about the cut off digits does not influence the significant digits. In chaotic systems, this is no longer true. There, there is a net flow of information (or, rather of ignorance) from less toward more significant digits [1], and as a consequence long-time behaviour is unpredictable. Notice only in very simple systems (and with suitably chosen codings!) this "flow" will resemble a directed laminar flow. In particular, the more significant digits will in general also influence the less significant ones. The directedness of the flow is not inherent to the dynamics, but due to the boundary conditions: we assume that the significant digits are known, while the unsignificant ones are unknown. Thus uncertainty can only flow in one direction.

The quantity measuring this information flow is the Kolmogorov-Sinai metric entropy. It and its close companion, the information dimension of the attractor, are defined as follows.

First of all we assume that there exists a measure (probability distribution) on the attractor which is invariant under time translation and which is stable against small perturbations. This is called the natural (or, if some further technical assumptions are fulfilled, the Bowen-Ruelle-Sinai [2, 3] measure μ. If this measure is unique, the system is called ergodic. All average values are in the following taken with respect to this measure.

If we have some partitioning of phase space into sets with weights $p_i = \int_i d\mu$, then the Shannon [4] entropy associated to it is defined as the average (with respect to μ) of $-\log p_i$:

$$H = -\sum_i p_i \log p_i = -\langle \log p \rangle \tag{3}$$

or, equivalently, as the average information needed to specify in which of the sets i falls a randomly chosen point (if informations are to be measured in bits, then $\log \equiv \log_2$).

Let us now consider an ϵ-tube $C(\epsilon, T)$ around some trajectory $\mathbf{x}(t)$ on the attractor of length T in time:

$$C(\epsilon, T) = \{\mathbf{y}(t) : \; |y_i(t) - x_i(t)| < \epsilon \;\; \forall i,\, 0 < t < T\} \tag{4}$$

Since the system is supposed to be deterministic, it will correspond to all starting points $\mathbf{y}(0)$ in some domain $B(\epsilon, T)$ (see Fig.1).

The average information needed to specify a trajectory which stays in this ϵ-tube, averaged over all $\mathbf{x}(0)$, can be decomposed into two parts: first we have to specify that $\mathbf{y}(0)$ is in an ϵ-ball around $\mathbf{x}(0)$, and then we have to restrict it further to $B(\epsilon, T)$ in order to guarantee that $\mathbf{y}(t)$ does not leave the tube. If the system is chaotic and has thus sensitive dependence on initial conditions, then we expect that the chance for a randomly chosen trajectory $\mathbf{y}(t)$ to leave the tube is independent of

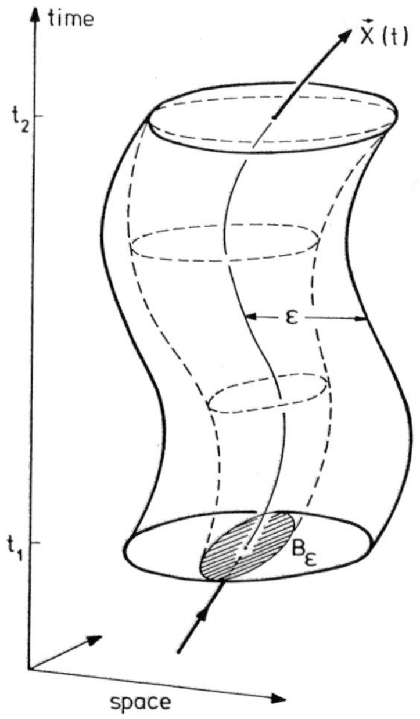

Figure 1: Typical ϵ–tube, and set $B(\epsilon, T)$ of initial values (from [9]).

time, and thus this second term should be proportional to t. Scale invariance suggests furthermore that the proportionality constant, called the *metric entropy* h, becomes independent of ϵ for $\epsilon \to 0$. On the other hand, the first term (needed to specify the initial point) should scale as $\log \frac{1}{\epsilon}$. If the measure were smooth in some R^d, then the proportionality constant would be just d: for each coordinate, one has to give $\approx \log \frac{1}{\epsilon}$ digits if one wants to specify a point in an ϵ-ball. In our case of fractal attractors, the proportionality constant defines just the *information dimension* D.

Summarizing the above, we expect thus the average information needed to specify that a trajectory is in an ϵ-tube to be

$$H(\epsilon, T) \sim hT - D \log \epsilon + \text{const.} \tag{5}$$

for $\epsilon \to 0$, $T \to \infty$. Notice that we are somewhat vague about the way in which the double limit is to be taken. In simple cases (essentially strictly hyperbolic systems, where all nearby points either converge or diverge asymptotically with power laws), Eq.(5) can be proven if either $\epsilon \to 0$ is taken first, or if $T \to \infty$ is taken first (in both cases the r.h.s simplifies) [2, 3]. But we conjecture that Eq.(5) holds also in the simultaneous limit $T \to \infty$, $T/\log \epsilon = \text{const}$.

Notice that the first term in Eq.(5) corresponds to the second law of thermodynamics: if we want to maintain our knowledge of the actual state of a chaotic system, we have to get more and more information. If we wouldn't observe the system as time progresses, we would know less and less about its state. If this state is a microstate, then this is just the increase of thermodynamic entropy.

On the other hand, the second term in Eq.(5) corresponds to the ability of dissipative systems to self–organize: if we started not on the attractor but on some larger region of phase space, the entropy first will decrease. Notice that this is a rather trivial aspect, and it illustrates that entropy is not a good measure of organization.

If we speak of organization, we should have in mind "structured" organization, and this is not measured by entropy.

Equation 5 has a further aspect. It shows that in chaotic systems knowledge of the past, since it is never perfect, does not allow perfect forecasts. More precisely, one can make arbitrarily precise forecasts, but the lack of information for making good forecasts is independent of the precision: one can never make forecasts with the same precision as the data on which the forecast is based.

Finally, let me mention the Lyapunov exponents which measure the exponential divergence of near-by trajectories. Let $B_\epsilon(\mathbf{x})$ be an ϵ-ball around $\mathbf{x}$. If ϵ is infinitesimal, then this ball will be transformed after a time t into an ellipsoid with semi-axes ϵ_i. We assume them to be ordered by magnitude, $\epsilon_1 \geq \epsilon_2 \geq \ldots$, and define:

$$\lambda_i = \lim_{t\to\infty} \langle \frac{1}{t} \log \epsilon_i \rangle \tag{6}$$

In other words, the λ_i's measure the (geometric average) growth of the length of vectors in tangent space.

There are close connections between dimension, metric entropy and Lyapunov exponents. While Pesin's [5] equality says that the entropy is just the sum of the positive Lyapunov exponents, the Kaplan-Yorke conjecture [6] says that the information dimension D is roughly such that the sum of the first D Lyapunov exponents is zero.

More precisely, we can associate a partial (information) dimension D_i to each direction i. It is essentially the information density in the digits of the i-th coordinate ξ_i, if we would use in each point of the attractor a local coordinate system $(\xi_1, \xi_2, \ldots)$ with the axes along the stable and unstable manifolds [7,8]. The i-th Lyapunov exponent is just the average velocity by which a perturbation in one of the digits of ξ_i propagates in digit space, i.e. the velocity of information (or rather ignorance) flow in digit space. The metric entropy is the total information flow rate in the variables along the stable directions [7,9]. For reversible systems, it is also the reversed flow along the unstable directions,

$$h = \sum_{\lambda_i > 0} \lambda_i D_i = - \sum_{\lambda_i < 0} \lambda_i D_i \tag{7}$$

During the last years, it has been very much stressed that typical strange attractors are multifractals [10]. This means that the measure is so non-uniform that averages like those in Eqs.(3) and (6) depend crucially on the fact that logarithms are taken first, and only then averages are formed (i.e., they are geometric means). Using different moments (e.g. arithmetic means) would give different results. For instance, Rènyi (or "generalized") entropies [11] are defined for any $q \neq 0$ by replacing Eq.(3) by:

$$H_q = (1-q)^{-1} \log \langle p^q \rangle \tag{8}$$

Again there are relations between these generalized entropies, dimensions, and Lyapunov exponents [8-10].

The interest in these generalizations is that on the one hand, different moments can allow simpler numerical or analytical treatments, as discussed in the next section. On the other hand, the whole spectrum of different Rènyi dimensions contains of course more information than the Shannon ($q \to 1$) case alone. But I think that the latter has been somewhat overemphasized. Even complete knowledge of all generalized dimensions and entropies does *not* specify an attractor completely (contrary to what has been sometimes claimed), and in most cases one needs *extremely* good data to get more information than already obtained, say, from the cases $q = 0, 1$, and 2.

The main result from this section is that there is a flow of information (or rather uncertainty) through the observables (the finite digits) in a dynamical system. On the one hand this is possible because dynamical systems are continuous, and nearly all real numbers need an infinite amount of information to specify. On the other hand, it only tells us about the *random* aspects of chaotic systems. Much more fascinating is that chaotic systems are in some very intuitive sense "complex", but for describing this complexity the above concepts are too crude.

3. Technical Problems in Estimating Dimensions, Entropies, and Lyapunov Exponents

3.1 *Lyapunov Exponents*

If the dynamical system is given analytically, in terms of maps or ordinary differential equations, then the easiest way is to compute Lyapunov exponents by iterating the system together with the equations for a sufficiently large set of tangent vectors which are rescaled and re-orthogonalized at regular time steps. From the Lyapunov exponents, entropies and dimensions are obtained via the Pesin or Kaplan-Yorke equations, respectively. If the system is only given implicitly by a time series, this is no longer so easy.

The first problem arises from the fact that usually only a small number of observables (typically one) is measured. Thus one has to "reconstruct" a sufficiently high dimensional phase space, in which the system has a chance to be deterministic. The simplest way is to just use "delay coordinates" [12, 13]:

$$\mathbf{x}_t = (x_t, x_{t-\tau}, x_{t-2\tau}, \ldots x_{t-(m-1)\tau}) \tag{9}$$

The number m is called the embedding dimension, and τ is the delay. We shall discuss good choices for them later.

One strategy could be again to first estimate Lyapunov exponents, and to obtain from them entropies and dimensions. Unfortunately, estimating Lyapunov exponents (except for the biggest ones) from a time series is not easy.

The cleanest method [14] (at least theoretically) fits a tangent map $\mathbf{T}_t$ at each point $\mathbf{x}_t$ of the time series, by looking at all points $\mathbf{x}_{t_k}$ in a suitably chosen neighbourhood and fitting the maps

$$\mathbf{x}_t \to \mathbf{x}_{t+1}, \qquad \mathbf{x}_{t_k} \to \mathbf{x}_{t_k+1} \tag{10}$$

by a linear function $\mathbf{x}' = \mathbf{T}_t \mathbf{x}$ (see Fig.2; notice that in delay coordinates, the non-trivial part of the map is indeed only $\mathbf{R}^m \to \mathbf{R}$, instead of $\mathbf{R}^m \to \mathbf{R}^m$ as suggested by Fig.2). The Lyapunov exponents are then obtained as usual from the eigenvalues of the products $\Pi_t \mathbf{T}_t$.

The main problem preventing the safe estimation of non-leading or even negative Lyapunov exponents is that the neighbourhoods needed in this method are not infinitesimal, and hence the correct map should not be linear. If the embedding dimension is too high (which is *a priori* hard to verify), then it is easy to see that non-linear terms introduce fake large Lyapunov exponents.

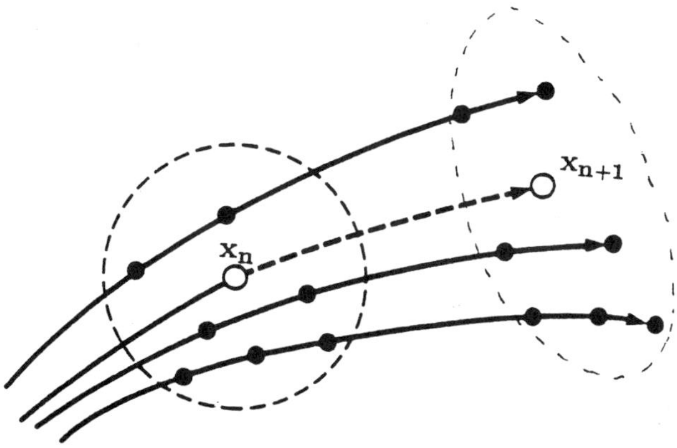

Figure 2: Evolution of the neighbourhood of a point $\mathbf{x}_n$ during one time step. In estimating Lyapunov exponents, one assumes $\mathbf{x}_{n+1}$ to be known and uses it in the fit. In forecasting, $\mathbf{x}_{n+1}$ is unknown and computed via the fitted linear map (adapted from [15]).

One possibility to avoid this problem could consist in fitting non-linear local maps (e.g., quadratic maps [15-17] or superpositions of radial basis functions [18]). This is of course much more cumbersome. But essentially the same maps can also be used for forecasting and noise reduction [15, 18-20], and the great interest in these problems has already led to more efficient algorithms [21], with still more efficient algorithms to be expected.

Even with linear local maps, some improvement could be obtained by carefully choosing the neighbourhoods in which the maps are fitted. Neighbourhoods isotropic with respect to the Euclidean metric in delay coordinate space are usually used, but are certainly not optimal. On the one hand, coordinates corresponding to times further back in the past should have less influence, on the other hand neighbourhoods should extend further into directions in which the invariant measure is sparse then into directions in which it is continuous. One way to formulate this is to say that the "natural" metric in the embedding space for Lyapunov exponent estimates and forecasting is probably *not* the Euclidean one.

Before leaving the subject of direct Lyapunov exponent estimates, I want to mention just one aspect of computational complexity reduction by efficient codes. A repeated problem in all methods discussed in this section is the search for nearest neighbours. If done naively, this takes a time $O(N^2)$, where N is the length of the time series. This can be reduced to $O(N \log N)$ by using trees [21-23]. In [24] (see also [25]) it was shown that even greater speed (and simpler routines!) can be obtained by using boxes ("radices") and linked lists.

3.2 *Dimensions and Entropies*

If one uses delay coordinates, the same routines can be used to estimate entropies and dimensions, since both need estimates of the invariant measure of typical (T, ϵ)-tubes. Dimensions are obtained from the scaling with ϵ, while entropies are obtained from the time scaling.

While the most natural choice of Shannon (instead of Rènyi) entropies corresponds to using geometric averages of these measures, it was observed in [26, 27] that the easiest estimate is for arithmetic averages. In order to estimate the average measure of a (T, ϵ)-tube, we have just to count the number of pairs (t, t') in the time series for which $||\mathbf{x}_t - \mathbf{x}_{t'}|| < \epsilon$, or (if we use the maximum norm):

$$|x_{t-k\tau} - x_{t'-k\tau}| < \epsilon \qquad \forall k = 0, 1, \ldots m-1 \tag{11}$$

The total length of the tube is given by $T = m\tau$. The number of such pairs divided by the total number $N(N-1)/2$ of pairs is called the correlation sum $C_N(T, \epsilon)$, its limit as $N \to \infty$ is called the correlation integral $C(T, \epsilon)$ (it is just the integral over the pair correlation, integrated from zero to ϵ). It is easy to see that the "correlation" dimensions and entropies estimated from the correlation integral via

$$\log C(T, \epsilon) \sim hT - D \log \epsilon \, + \, \text{const} \tag{12}$$

are just Rènyi dimensions or entropies, respectively, with $q = 2$.

The numerical advantages of the correlation method are on the one hand that it needs less storage and is easier to program (there is no need to store pointwise how many other points fall into neighbourhoods), and that there are no systematic finite-sample corrections (i.e., $C_N(T, \epsilon)$ is an unbiased estimator for $C(T, \epsilon)$). Neither of this would be true for estimators of Shannon entropies, or of Rènyi entropies with $q \neq 2$. The leading systematic corrections in these latter cases are discussed in [28].

There are a number of comments appropriate about implementing correlation dimension estimates:

(i) CPU times can be cut by several orders of magnitude by the fast neighbour search algorithms mentioned above. When this is done, dimension analyses with up to 10^6 points (and taking *all* close pairs!) can be done easily on work stations.

(ii) In the double sum over t and t', diagonal terms $t = t'$ should of course not be included. Moreover, since what one wants to measure is the clustering due to the geometry and to the properties of the invariant measure, one has to be careful not to be fooled by *dynamic* correlations between $\mathbf{x}_t$ and $\mathbf{x}_{t'}$. Thus, in the double sum also terms with $|t - t'| < t_{corr}$ should be excluded [29], where t_{corr} is the correlation time. It seems that a number of papers where authors found spuriously small attractor dimensions are obsolete due to a neglect of this. A popular alternative to excluding just the pairs with $|t - t'| < t_{corr}$ is to take only a subset of the time series where $|t_n - t_{n-1}| > t_{corr}$. While this does indeed avoid the above problem, it can have the drawback of substantially reducing the statistics, and is thus not recommended.

(iii) The time series has of course to be long enough to sample the attractor reasonably, and the system has to be stationary (if it is not stationary, the time series could be rendered so by differentiating or de-trending). The latter is not true, e.g., for systems undergoing random walks, and low measured *orbit* dimensions [30] should not be confused in such cases with small *attractor* dimensions. If the walks are unrestricted, then time sequences of arbitrary length will show small dimensions ($D =$ 2 for ordinary random walks). But if they are submitted to drift forces preventing escape from some bounded region, then a very long time sequence would show the expected large dimension.

(iv) Similarly, low measured correlation dimensions can be misleading in systems with strong intermittency or with several different dynamics. Take for instance wind speed measurements near a coast with strong winds during days and feable winds during the night. Essentially all pairs $(\mathbf{v}_t, \mathbf{v}_{t'})$ with small distance will result from night measurements, representing a region in phase space which on the scale of the global "attractor" is essentially a single point. Small effective correlation dimensions would in this case be trivial, and represent no sign of deterministic dynamics. In cases like this, it is essential to measure Shannon instead of Rènyi entropies.

(v) There exists a large literature on "optimal" embeddings, concentrating mostly on the choice of the delay τ and the embedding dimension m. It should be noted, however, that *universally* optimal embeddings don't exist. We mentioned already that for Lyapunov exponent estimates and for forecasting, non-isotropic metrics putting less weight on earlier delay coordinates might be optimal. These same metrics are presumably not optimal for dimension and entropy estimates. There, the maximum norm (see Eq.(11)) should be optimal since it gives no errors when a too small delay τ is used. It is often claimed that the optimal τ corresponds to the first zero of the mutual information [31]. Without further specification, this is not true if the maximum norm is used. There, using a smaller τ and keeping at the same time $T = m\tau$ fixed corresponds just to adding further constraints to Eq.(11). This can only improve the dimension estimate in the noiseless case, and could be detrimental only if the data are corrupted by shot noise which is not taken out by any smoothing. On the other hand, a criterion like that of Ref. [31] might be useful if one faces serious CPU time and/or storage limitations. Similarly, we should not expect the same embedding dimension to be optimal in Lyapunov exponent and in dimension estimates.

(vi) Alternative dimension estimators [32, 33] count the average radii of balls containing fixed numbers of neighbour points instead of counting the average number of neighbour points in balls of fixed radii. Though more cumbersome, they have several advantages. One of these is that systematic finite-sample corrections can be estimated exactly [32] for them. Applying this idea to discrete sequences, i.e., to symbolic dynamics, and taking the length of the longest common prefix as a distance measure between two sequences, these estimators have interesting connections to Ziv-Lempel codes [34, 35].

4. Information Dynamics of Spatially Extended Systems

In chaotic systems with few degrees of freedom, it is the infinite information storage capacity of each degree of freedom which is responsible for the unpredictability. In spatially extended systems, we should distinguish systems with continuous observables (partial differential equations or coupled map lattices, CML [36]) from systems with discrete variables (cellular automata, CA).

In the latter case, there is only a finite information storage capacity locally, and chaos can only arise if there is an information flow spatially. Kolmogorov-Sinai entropies of (deterministic) 1-dimensional CA are thus finite, and there exist inequalities between them, the entropy D per site in the invariant measure, and the average spreading velocities $\lambda_{right}, \lambda_{left}$ of disturbances [37]:

$$h \leq D(\lambda_{right} + \lambda_{left}) \tag{13}$$

These inequalities are analogous to the Pesin-Ledrappier-Young equation (7). They are *in*equalities in general since information in a CA does not only "flow" but may also be filtered and screened. In contrast, information processing in the case of smooth chaotic systems is essentially just a flow in the digit space of a suitably chosen coordinate and number system.

In generic d-dimensional CA, the metric entropy is neither constant nor extensive. Instead, the information needed per time step in order to maintain the information about the configuration of the CA within a volume V is bounded by the information which can flow in through the boundary of V:

$$h \leq D \oint_{\partial V} \lambda_{\mathbf{n}} \mathbf{n}.\mathbf{df} \tag{14}$$

In spatially extended but finite (volume V) systems with *continuous* variables, both the metric entropy and the attractor dimension will in general be extensive, leading to the concepts of dimension or entropy densities, respectively:

$$h = \eta V, \qquad D = \delta V \tag{15}$$

Combining these with the Kaplan-Yorke relation suggests that the Lyapunov exponents should scale as

$$\lambda_i = f(i/V) \tag{16}$$

While these relations might break down in strongly intermittent systems (where h and D are sub-extensive), there is elsewise ample evidence for them, both theoretically [38, 39] and numerically [40, 41].

In infinite spatially extended systems, we might expect naively a combination of Eqs.(14) and (15). The information loss per time unit about the state in an open volume should be due to the endogeneous information production ηV plus the flow through the boundaries given by Eq.(14) while the variables in the volume V should fill a phase space volume of dimension δV [42]. For 1-d systems, the latter could be measured by a Takens-Grassberger-Procaccia type correlation integral, but using spatial sequences $x_1(t), x_2(t), \ldots, x_L(t)$ at fixed time instead of time sequences $x_k(t_1), \ldots, x_k(t_m)$. Plotting the correlation sums on a log-log plot versus the embedding length L, one should find slopes equal to ηL+const.

While this was confirmed by early simulations on CML's [42], there are both numerical [43] and theoretical [44] arguments that this simple reasoning has to be modified. Indeed, we should expect the chaotic motion of the system outside V to act like some random noise on the system within V. Thus, the variables within V should fill a phase space volume of the full dimension nV, if there are n degrees of freedom per unit volume. Since the influence of far away regions should however be filtered, the corresponding "noise" should be of low amplitude, and the filled phase space volume should be very "thin" in the corresponding directions.

For 1-d systems, we expect thus that correlation sum estimates with spatial strings of length L as described above should give slopes $\sim \eta L$ if L is large and statistics is small, but slopes nL if L is small and statistics is large enough to see structures on small amplitude scales. Very high statistics results for coupled logistic maps,

$$x_n^{t+1} = 1 - 2[\overline{x}_n^t]^2, \qquad \overline{x}_n^t = (x_{n-1}^t + x_n^t + x_{n+1}^t)/3 \tag{17}$$

are shown in Figs.3 and 4. From these we see that this is essentially verified.

In addition to these generic features, coupled map lattices tend to show very strong intermittency effects. One such effect is seen in Fig.4 (namely the dip for $L > 12$ and

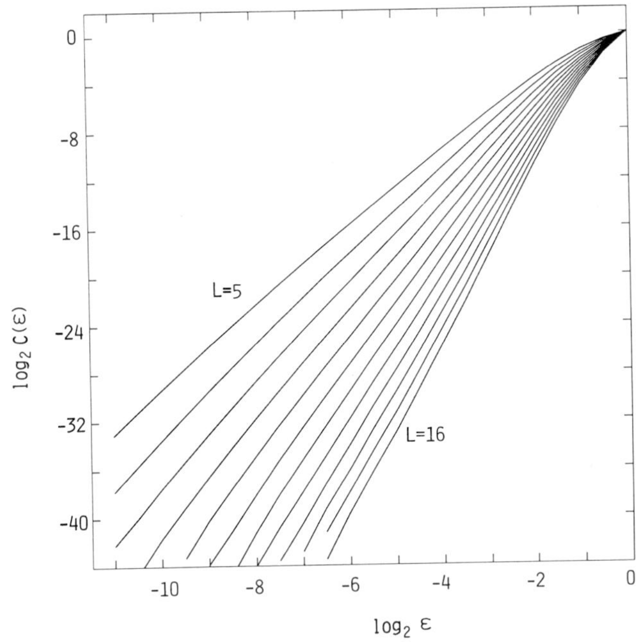

Figure 3: Correlation sum for spatial strings evolving under coupled logistic maps, Eq.(17). The fast algorithm of [24] was used to search near neighbours among altogether 3.4×10^{14} pairs. "Embedding dimensions" L vary from 5 (top curve) to 16 (bottom).

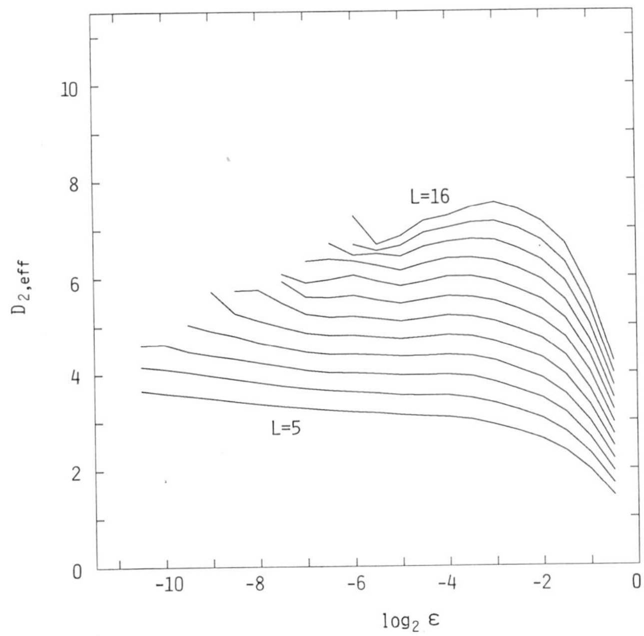

Figure 4: Local slopes obtained from Fig.3, giving estimates for the dimension of the invariant sets populated by spatial strings of length L. The structure at $L > 12$ and $\epsilon \approx 2^{-5}$ is presumably due to intermittency. For $L < 12$ and $\epsilon > 2^{-5}$, the results indicate that $\Delta D/\Delta L \approx 0.41$, but $D \to L$ for $\epsilon \to 0$.

$\epsilon \approx 10^{-5}$). Though they are very interesting, and certainly are one aspect where the *complexity* of chaotic systems is most clearly demonstrated, I have no time to discuss them here. I might just mention that, in contrast to previous investigations [45], we found in [46] that some of them are closely related to directed percolation.

It is an intriguing question whether (and in what sense) spatial information flow can be measured directly. In [47], the mutual correlation between space-time points in an inhomogeneous lattice of coupled ODE's was found to be maximal at distances $\Delta x \sim v\Delta t$, with constant "velocity" v. From this it was suggested that information flowed indeed with velocity v. That such an interpretation can be very problematic was found in [48]: in coupled logistic maps, the maximal mutual information occured at $v > 1$, which would correspond to superluminous information propagation. The problem is of course that there need not to be any direct causal connection between two space-time points. Instead, the observed correlations could be (and in general will be) due to their common past.

The only way to guarantee that information *propagates* seems to consist in externally disturbing the system at some point (x, t), and to observe the reaction at (x', t'). This leads to the concept of velocity-dependent Lyapunov exponents [49]. A certain problem is that it seems not easy to measure them in experimental systems, since one would have to compare two *exact* replicas of the system, with identical initial conditions except at point x.

5. Complexity

5.1 *Generalities*

The above discussion of extended chaotic systems has not changed our previous conclusion that the randomness characteristic of chaotic motion is due to a shift of information from "irrelevant" towards "relevant" (i.e. observed) degrees of freedom.

Besides being random, the most conspicuous property of most chaotic systems is that they are "complex" in an intuitive sense. Examples are pictures of Julia sets, Mandelbrot sets, homoclinic tangles in Hamiltonian systems, etc. None of these models seems *a priori* (i.e., in its formulation) to be complex, and their complexity seems "emergent" or "self–generated".

General problems with any definition of complexity of physical objects are discussed in [50-52]. One of the problems is that the most commonly considered "complexity measure", the Kolmogorov-Chaitin [53, 54] algorithmic complexity, is really a measure of randomness and not a measure of complexity in the naive intuitive sense. At least, this is the case when it is applied to the objects of interest to us (such as symbolic codings of chaotic trajectories) which *can* be random. When applied to well defined mathematical objects such as, e.g., the string of digits of π, the algorithmic complexity (defined essentially as the length of a minimal encoding of the object) *is* a good complexity measure.

The conclusion of [50-52] is that there exists indeed no *universal* and *objective* definition of complexity of physical objects. Instead, complexity is characteristic of a *subjective* object-observer relationship. It measures the difficulty (for the observer!) of some task (done *on* the object) which the observer considers as meaningful.

On the one hand, all complexity measures become in this way essentially complexities of *algorithms*, but these algorithms are of course not the objects themselves whose complexity one naively seems to deal with.

On the other hand, since the observer has to decide what is meaningful, complexity is closely related to *meaning* [55]. Emergent complexity seems to arise if seemingly irrelevant details become important and thus acquire "meaning" during the evolution. Notice that this shift is similar to the emergence of randomness, but it goes beyond it. While in the former the observer was forced to register previously unobserved details, now she is even forced to attach specific importance to them. An example of this shift is the extinction of dinosaurs. While this certainly would have been a random "accident" to an uninterested observer, it is of crucial importance to *us* and thus we consider the resulting structures in the biosphere as "complex".

Physical objects contain typically $> 10^{20}$ degrees of freedom. Thus most of their (microscopic) properties must be considered as meaningless and thus random. Indeed, when we speak of physical objects we actually think of *ensembles* where all irrelevant details are neglected. In this way, e.g., the above difficulty of the algorithmic complexity is resolved: while the description of an *individual* "random" pattern might be incompressible and thus difficult, the description of its *ensemble* is trivial, and thus this ensemble is not considered as complex [54].

Typical of extremely complex objects (such as living beings, ecologies, human societies, ...) is that this distinction between irrelevant and relevant features is not clear cut or is even impossible. While the distinction between microscopic and macroscopic degrees of freedom is trivial for most anorganic objects, the same distinction is virtually impossible for a living being.

A last general aspect of complex objects is that they allow a strong "data compression", but that neither the compression nor the "decompression" (i.e., the reconstruction of the object from its shortest description) is simple. Apart from the simplifications resulting from neglecting irrelevant details, the main reason may be that there are strong correlations (and thus large mutual information) between its parts. Another reason might be that there are correlations with its environment or with the observer himself. It is in the moment when the observer recognises these correlations and when he learns what is relevant, that he grasps the "meaning" of the object and is able to simplify its description.

5.2 *Application to Time Sequences*

In this subsection, we shall apply the above considerations to the behaviour of dynamical systems as manifested in their symbolic dynamics. We thus assume that we have a "good" mapping of phase space trajectories onto sequences over a discrete alphabet (in the following, we shall always assume a binary alphabet). We also assume that any complexity of the system is reflected in a complexity of the discrete sequences, so that complexity measures of the symbol sequences are also good measures for the dynamical system itself.

If the actual sequences are of importance, then storing and transmitting a compressed version is a meaningful task, and the algorithmic complexity (measuring the length of the most compressed encoding) is a relevant complexity measure. But even then, there are other tasks which are equally important, and whose difficulty is not measured by algorithmic complexity. For instance, the difficulty of *storing and transmitting* a code sequence is not correlated to the difficulty of *performing* the *encoding*

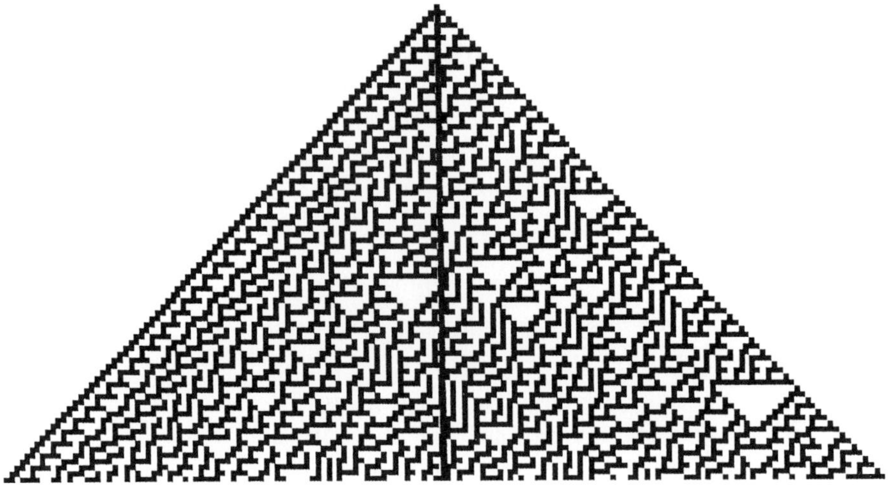

Figure 5: Space-time pattern generated by cellular automaton rule 86 [56], from an initial configuration having a single "1". The central column (marked grey) seems to be both statistically random and logically deep.

or *decoding*. Take for instance a completely random ("Bernoulli") sequence. Storing it is hard since no compression is possible. But coding is trivial, as it consists just of copying the original sequence. In contrast to this are the first 10^6 binary digits of π. There a compressed code consists of a program whose storage is easy (the program *is* short), but whose decoding may be hard (the program may *run* long).

Actually, the digits of π are a bad example, as very efficient programs for computing π exist. A better example [56] seems to be the 1-d cellular automaton which in Wolfram's notation is rule 86. This is a dynamical system over the space of doubly infinite binary sequences. The dynamics is discrete and "local", i.e., in each time step every symbol is replaced by a value computed from its old value and from its left and right neighbour. Starting with a string of 0's with a single "1", one obtains the space-time pattern shown in Fig.5. In this pattern, the sequence of values of the central site seems to be statistically random. Thus, if one wants to store a description of this sequence, one can either store them bit by bit, or one can store the rule and the initial condition. In the first case, decoding is trivial but storing is hard. In the second case, storing is trivial (both the rule and the initial condition can be described efficiently), but decoding is hard: as far as it is known, there exist no faster way to obtain N values than to simulate N time steps. This needs $O(N^2)$ operations.

Sequences which allow a compressed description and whose decoding is hard (the number of steps for decoding a single symbol takes a number of operations diverging with N) are called *logically deep* by C. Bennett [57]. His prime candidate for a logically deep system is life itself, since it is most likely that it started with a short "program", while evolution took very long to "decode" it.

Again another task related to a time sequence is forecasting. In relation with discrete sequences, we can distinguish two main problems. In the first, we predict only which symbol *can* come next. In the other, we also try to estimate *probabilities*, by either predicting which is the most likely next symbol or – preferably – by predicting the probabilities for every possible next symbol.

If we are not interested in predicting probabilities, we are treating the set of possible itineraries as a formal language. The difficulty in predicting possible next symbols is then exactly the complexity of the corresponding formal grammar. For *regular* grammars (grammars which can be verified by means of finite machines with finite storage), Wolfram [58] defined the "regular language complexity" (RLC) essentially as the memory of the smallest machine doing this job. An improvement is possible if probabilites are known (though not predicted!), as then this memory can be used more efficiently. In this case, a modified complexity measure (called "set complexity", SC) was introduced in [54] which is always a lower bound to the RLC. This measure can also be finite for non-regular languages, where the RLC is infinite by definition. A particular such case are binary sequences generated by the logistic map. As shown numerically in [59], the SC there is finite except at Feigenbaum and at tangent bifurcation points.

More interesting seems to be the problem where we also want to forecast probabilities. Notice that this is also closely related to coding. One way to optimally compress a sequence is to first forecast as good as possible the probabilities of the current next symbol, and then to encode only the information left over. In other words, when decoding any optimally compressed description of a symbol sequence, one has to forecast (implicitly or explicitly) probabilities, since otherwise the encoded information could not be sufficient for decompressing. The average minimal information about the past part of the sequence, needed to be stored for optimal forecasting, was called "forecasting complexity", FC, in [52]. It is essentially the average per symbol space complexity (in the sense of computation theory) of the decoding algorithm. Since decoding involves using this information (in addition to other operations), it follows tha FC is a lower bound to logical depth.

A lower bound to FC, in turn, is obtained from mutual information. Denote by H_n the (Shannon) information of blocks of n symbols:

$$H_n = - \sum_{s_1 \ldots s_n} p(s_1 \ldots s_n) \log p(s_1 \ldots s_n) \tag{18}$$

The additional information for a single new symbol, using the information about n previous symbols, is then:

$$h_n = H_{n+1} - H_n = - \sum_{s_1 \ldots s_{n+1}} p(s_1 \ldots s_{n+1}) \log p(s_{n+1} \mid s_1 \ldots s_n) \tag{19}$$

The difference

$$\delta h_n = h_{n-1} - h_n \tag{20}$$

is finally the amount by which the uncertainty of the $(n+1)^{st}$ symbol s_{n+1} decreases if the first symbol gets known in addition to all symbols between. It is thus the "useful" part of the information which has to be kept n time steps if an optimal forecasting of s_{n+1} is to be made. The total information stored at any time has thus to be larger than $\sum n\delta\, h_n$:

$$FC \geq \sum_{n=1}^{\infty} n\, \delta h_n \tag{21}$$

The right hand side was called "stored information" in [60], and "effective measure complexity", EMC, in [52]. The latter name is to indicate that EMC is effectively computable (in a statistical sense) for a given ensemble, while neither FC nor the logical depth are effectively computable. Simple algebra shows that it can also be written as the Legendre transform of H_n at $n \to \infty$:

$$EMC = \lim_{n\to\infty}[H_n - nh_n] \tag{22}$$

This quantity measures the increase in total message length if we break a long message into two parts, and do not take advantage in the encoding of their mutual information.

A very similar (though not identical) concept is the "sophistication" defined in [61]. There it is argued that any encoding can be considered as consisting of a "specific" part (the message proper) of length proportional to the sequence length, plus a general overhead describing the number of symbols, their probabilities, their correlations, etc. The sophistication is the length of the overhead of the coding with shortest total length. Apart from the fact that sophistication is defined within algorithmic information theory while the EMC is a probabilistic concept, the main difference between them is that in the EMC we include only the overhead due to mutual information between parts of the message. It is thus always smaller than the average sophistication.

In most situations of interest, the forecasting complexity is infinite. The reason is that in its definition we demanded the forecasts to be *optimal*. If, for instance, the exact probabilities to be forecasted are real numbers with some continuous distribution, then specifying such a probability will in general require an infinite number of digits and thus an infinite effort. Similarly, the sophistication will in general be infinite in any system depending on continuous parameters since the number of relevant digits of these parameters increases with the length of the sequence [62]. Except in very simple models [63], we will thus be forced to allow approximate forecasts. Improved forecasting complexity measures will then measure how the difficulty of such forecasts increase as the errors decrease [63, 64].

In contrast to this, the EMC is finite in many cases. Examples seem to be the symbol sequences of the logistic and the Henon map (for the former, we use the left-right coding, for the Henon map we use the partitioning proposed in [65]). In both cases, the block entropies and thus also the EMC seem to converge exponentially.

The EMC seems thus to be a more relevant complexity measure since it is large only if the sequences are complex in a naive sense and show very long range correlations. One such case might be natural languages. It is clear that orthographic/phonetic, syntactic and semantic constraints give rise to correlations and mutual information on very different time scales. The entropy of written English estimated from single-letter frequencies alone is ≈ 4.5 bits/character. Going to the longest feasible block lengths, this decreases to $h_n \approx 2$ bits/character with $n \approx 5$. And extrapolating to infinite block length, we find $h < 1$ bit/character [66, 35].

Finally, we want to mention bit sequences generated by the 1-d cellular automaton Nr.22. In the invariant state obtained by starting with random initial configurations, both spatial and temporal sequences seem to have entropies h_n (defined by Eq.(20)) which converge very slowly to zero [67]. If the extrapolations suggested by these simulations are correct, this would mean that these sequences are not random but highly complex. In a certain (very restricted!) way, this complexity is reminiscent of the complexity of life: on the one hand, the small entropy indicates that valid patterns are extremely restricted, implying a very high degree of (self-)organization. On the other hand, the slowness of the convergence of the h_n shows that the *absolute* number of valid patterns is still huge, and the restrictions are only seen when looking at very large patterns. Thus the organization is not at all trivial.

6. Conclusions

In these lectures, I have tried to cover the range from rather elementary considerations, via some technical remarks on time sequence analysis, to very speculative remarks on the complexity of self-organization. I have tried to argue that both chaos and complexity (in the subjective sense, not in the sense of algorithmic complexity) are due to unavoidable shifts of emphasis of the observer. Deterministic chaos is due to the shift of information in "unobservable" digits (and in infinitesimally low amplitude noise) towards the "window" of the observer. Complexity is due to the further shift from noise being a nuisance towards being the source of structures which acquire "meaning" for the observer. Both these shifts are *unavoidable* since the observer neither can store infinitely much information, nor can she consider infinitely many details as meaningful. Thus, though special definitions of complexity are necessarily subjective, its very existence is not.

Since complicated objects can be seen under many different aspects, they allow many different definitions of complexity. The two main aspects for symbol sequences were *performing coding/decoding* algorithms on the one hand, and *storing/transmitting* the codes on the other. Another aspect which maybe is most closely related to the subjective notion of complexity is the difficulty to *find* the optimal encoding, i.e., the problem to *understand* the system. I have not dealt with this problem since it seems very hard to formalize, and I do not know of any theoretical attempts to treat it mathematically.

In the course of these lectures, I have not yet said much about evolution. Evolution, or more precisely open-ended evolution, is a process which does not have any attractor or, in a probabilistic setting, any invariant measure. Of course, we want also to exclude situations without attractor where the trajectory just goes off towards infinity (like Brownian motion). We presumably also want to exclude pathological cases like the sets of parameter values of measure zero found in [68] for which invariant measures don't exist for the logistic map. It is also not clear what probability space we want to underly if we say that no invariant measure exists. Thus we should maybe replace the requirement that no attractor or no invariant measure exists by the requirement that a suitably defined complexity measure increases beyond any limit. A system close to this is the iterated prisoner's dilemma as formulated and simulated by Lindgren [69], where the complexity of the best strategies seems to increase forever. It is in my opinion one of the most challenging problems to find more deterministic or stochastic systems which show open-ended evolution in this sense.

References

1. R. Shaw. *Z. Naturforsch.* **36a**, 80 (1981).
2. D. Ruelle. *The Thermodynamic Formalism.* Addison–Wesley, Reading, Mass., 1978.
3. J.-P. Eckmann and D. Ruelle. *Rev. Mod. Phys.* **57**, 617 (1985).
4. C.E. Shannon and W. Weaver. *The Mathematical Theory of Communication.* Univ. of Illinois Press, 1949.

5. Ya.B. Pesin. *Russ. Math. Surveys* **32**, 55 (1977).
6. P. Frederickson, J.L. Kaplan, and J.A. Yorke. *J. Diff. Eqns.* **49**, 185 (1983).
7. F. Ledrappier and L.-S. Young. *Annals of Math.* **122**, 509 (1985).
8. P. Grassberger and I. Procaccia. *Physica* **13D**, 34 (1984).
9. P. Grassberger. Estimating the fractal dimensions and entropies of strange attractors. In *Chaos.* Ed. A.V. Holden. Manchester Univ. Press, 1986.
10. G. Paladin and A. Vulpiani. *Phys. Rep.* **156**, 147 (1987).
11. A. Rènyi. *Probability Theory.* Elsevier - North Holland, Amsterdam, 1970.
12. F. Takens. Detecting strange attractors in turbulence. In *Lecture Notes in Mathematics*, Vol. 898, p.366. Springer, Berlin, 1981.
13. N.H. Packard, J.P. Crutchfield, J.D. Farmer and R.S. Shaw. *Phys. Rev. Lett.* **45**, 712 (1980).
14. J.-P. Eckmann, S.O. Kamphorst, D. Ruelle and S. Ciliberto. *Phys. Rev.* **A 34**, 4971 (1986).
15. J.D. Farmer and J.J. Sidorowich. *Phys. Rev. Lett.* **59**, 845 (1987).
16. J.D. Farmer and J.J. Sidorowich. Exploiting chaos to predict the future and reduce noise. In *Evolution, Learning, and Cognition.* Ed. Y.C. Lee. World Scientific, Singapore, 1988.
17. P. Bryant, R. Brown and H.D.I. Abarbanel. *Phys. Rev. Lett.* **65**, 1523 (1990).
18. M. Casdagli. *Physica* **D 35**, (1989).
19. J. Crutchfield and B.S. McNamara. *Complex Systems* **1**, 417 (1987).
20. E.J. Kostelich and J.A. Yorke. *Phys. Rev.* **A 37**, 1649 (1988).
21. K. Stockbro, D.K. Umberger and J.A. Hertz. Preprint NORDITA - 90/28 S (1990).
22. S.M. Omohumdro. *Complex Systems* **1**, 273 (1987).
23. S. Bingham and M. Kot. *Phys. Lett.* **A 140**, 327 (1989).
24. P. Grassberger. *Phys. Lett.* **A148**, 63 (1990).
25. J. Theiler. *Phys. Rev.* **A 36**, 4456 (1987).
26. P. Grassberger and I. Procaccia. *Physica* **9D**, 189 (1983).
27. F. Takens. Invariants related to dimension and entropy. In *Atas do* 13° *Coloquio Brasileiro de Matematica* (1983).
28. P. Grassberger. *Phys. Lett.* **A 128**, 369 (1988).
29. J. Theiler. *Phys. Rev.* **A 34**, 2427 (1986).
30. A.R. Osborne and A. Provenzale. *Physica* **35D**, 357 (1989).
31. A.M. Fraser and H.L. Swinney. *Phys. Rev.* **A 33**, 1134 (1986).
32. P. Grassberger. *Phys. Lett.* **A 107**, 101 (1985).
33. R. Badii and A. Politi. *J. Stat. Phys.* **40**, 725 (1985).
34. J. Ziv and A. Lempel. *IEEE Trans. Inform. Theory* **24**, 530 (1978).
35. P. Grassberger. *IEEE Trans. Inform. Theory* **35**, 669 (1989).
36. K. Kaneko. *Physica* **23D**, 436 (1986).
37. S. Wolfram. *Physica* **10D**, 1 (1984).
38. D. Ruelle. *Commun. Math. Phys.* **87**, 287 (1982).

39. C. Foias, O.P. Manley, R. Temam and M.T. Treve. *Physica* **9D**, 157 (1983).
40. P. Manneville. In *Macroscopic Modeling of Turbulent Flows. Lecture Notes in Physics*, Vol. **230**, p.319. Springer, Berlin, 1985.
41. Y. Pomeau, A. Pumir and P. Pelce. *J. Stat. Phys.* **37**, 39 (1984).
42. P. Grassberger. *Physica Scripta* **40**, 346 (1989).
43. A. Politi, G. D'Alessandro, and A. Torcini. Firenze preprint, 1989.
44. L.A. Bunimovitch and Yu. Sinai. *Nonlinearity* **1**, 491 (1989).
45. H. Chate and P. Manneville. *Physica* **32D**, 409 (1988).
46. P. Grassberger and T. Schreiber. Wuppertal preprint WU B 90-9 (1990).
47. J.A. Vastano and H.L. Swinney. *Phys. Rev. Lett.* **60**, 1773 (1988).
48. T. Schreiber. Wuppertal preprint WU B 89-30 (1989).
49. R.J. Deissler and K. Kaneko. *Phys. Lett.* **A 119**, 397 (1987).
50. P. Grassberger. Randomness, information and complexity. In *Proceedings* 5th *Mexican Summer School on Stat. Mech. 1989.* Eds. R. Rechtman *et al.*
51. P. Grassberger. *Helv. Phys. Acta* **62**, 489 (1989).
52. P. Grassberger. *Int. J. Theor. Phys.* **25**, 907 (1986).
53. A.N. Kolmogorov. *Three Approaches to the Quantitative Definition of Information. Probl. of Inform. Th.* **1**, 3 (1965).
54. G.J. Chaitin. *Algorithmic Information Theory.* Cambridge Univ. Press, Cambridge, 1987.
55. H. Atlan. *Physica Scripta* **36**, 563 (1987).
56. S. Wolfram. *Adv. Appl. Math.* **7**, 123 (1986).
57. C.H. Bennett. In *Emerging Syntheses in Science.* Ed. D. Pines. Addison–Wesley, Reading Mass., 1985.
58. S. Wolfram. *Commun. Math. Phys.* **96**, 15 (1985).
59. P. Grassberger. *Z. Naturforsch.* **43a**, 671 (1988).
60. R. Shaw. *The Dripping Faucet as a Model Chaotic System.* Aerial Press, 1984.
61. M. Koppel and H. Atlan. Program length complexity, sophistication, and induction. Preprint 1987.
62. J. Rissanen. *IEEE Trans. Inform. Theory* **32**, 526 (1986).
63. D. Zambella and P. Grassberger. *Complex Systems* **2**, 269 (1988).
64. J.P. Crutchfield and K. Young. *Phys. Rev. Lett.* **63**, 105 (1989).
65. P. Grassberger and H. Kantz. *Phys. Lett.* **113A**, 235 (1985).
66. T.M. Cover and R.C. King. *IEEE Trans. Inform. Theory* **24**, 413 (1978).
67. P. Grassberger. *J. Stat. Phys.* **45**, 27 (1986).
68. G. Keller. Private communication.
69. K. Lindgren. NORDITA preprint (1990).

RECURSIVE SYMBOLIC ENCODING AND HIERARCHICAL DESCRIPTION OF COMPLEX SYSTEMS

R. BADII AND M. FINARDI

Paul Scherrer Institut
CH-5232 Villigen, Switzerland

G. BROGGI

Physik – Institut
Universität Zürich
CH-8057 Zürich, Switzerland

Abstract

We discuss the problem of quantifying complexity in the framework of a hierarchical modeling of physical systems which are suitably encoded by partitioning their phase-space. The dynamics underlying the formation of one-dimensional symbolic patterns is unfolded by using variable-length prefix-free codewords and described by means of allowed transitions on a "logic" tree. The derived model is employed to make predictions about the scaling behaviour of the system's observables at each level of resolution. The complexity of the system, relative to the unfolding scheme, is evaluated through a generalisation of the information gain by comparing prediction and observation.

1. Introduction

Many natural phenomena (motion of fluids, formation of neural networks, life) exhibit features that are clearly distinct from (space- or time-) periodicity and complete disorder. Highly organised structures appear spontaneously at different scales, without evident relations among the dynamical rules governing the evolution at each scale. Mathematical models like cellular automata, spin glasses and nonlinear dynamical systems also produce hierarchies of irregular (but structured) behaviour as the result of the repeated application of some basic scheme of sequential or parallel type. The characterisation of such systems cannot be obtained by evaluating entropy-like quantities, but rather requires the discovery of the nature and of the mutual relations

Information Dynamics, Edited by H. Atmanspacher and
H. Scheingraber, Plenum Press, New York, 1991

of the generating rules, steps that correspond to providing an interpretation and a description of the dynamics [1]. This investigation can be carried out efficiently by performing a coarse-graining of the phase-space X and by ordering its partition elements Δ_i in a hierarchical way, so that the generic domain Δ_i is labeled by a symbol s_i from an alphabet $A_0 = \{0, 1, \ldots, r-1\}$ $(r < \infty)$. Subdomains are in turn identified by concatenations $w_k = s_i s_j \ldots$ of symbols all beginning with the label s_i of the parent set Δ_i. A generic sequence $S = s_1 s_2 \ldots s_n$ represents a microstate of the system, characterised by the value of some local observable $Q(S)$, like the probability $P(S)$ of the sequence itself, or the sequence-length $|S|$, or the energy $E(S)$ for a spin configuration. Successive refinements of the original partition are represented by sets of increasingly longer sequences, which can be allocated on a "logic" tree. Every allowed path S corresponds to a succession of enlargements in a non-empty region of phase-space X. The structure of the latter is reflected in the set $\mathcal{L}$ (the "language") of all admissible sequences S. Given the knowledge of the previous $l-1$ levels of the tree, sequence-to-sequence transitions can be studied, and conditional expectations (such as $P[Q(S')|Q(S)]$) of the "thermodynamic" quantities $Q(S)$ [2] can be calculated and used to predict the structure of the system at the resolution level l. As a result, a description of the scaling dynamics [3] is obtained, in terms of a Markov model which governs the behaviour of $Q(S)$ along each allowed tree-path $S = s_1 s_2 \ldots s_n$, for $n \to \infty$. Systems which present, at increasingly finer levels of resolution, properties that cannot be accurately predicted from those measured at coarser scales are classified as complex; this notion of complexity is relative to the chosen unfolding scheme [1] and cannot therefore be measured by an absolute quantity. Systems governed by a finite number of rules (completely ordered or random-uncorrelated phenomena, regular languages [4]) are to be classified as simple [5]. Since the problem of characterising complexity has arisen in several disciplines (computer science, information theory, biology) with rather different aims, the corresponding measures (see Ref. [6] for a review) do not satisfy these two basic requirements, except for that of Ref. [5].

2. Decoding Symbolic Signals and Sequential Modeling

Our analysis starts by considering a stationary symbolic signal $\mathcal{S}_0 = s_1 s_2 \cdots$ of length $|\mathcal{S}_0| \gg 1$ (e.g., the output of a nonlinear dynamical system with known generating partition [7]) and by decomposing $\mathcal{S}_0$ into a succession of "primitive" words (sequences) w_1, w_2, ... which are assigned to the vertices of a tree. The l-th level $(l = 1, \ldots, \infty)$ will then contain by construction concatenations of l words. All branches leaving a generic vertex S point to the allowed continuations Sw_j, Sw_k, ... of sequence S. If the signal is aperiodic there exist "branching" vertices, on which strings that have several allowed extensions are allocated. Trees constructed in this way are equivalent to generalised Markov models (whose order depends on the average sequence-length per level) which describe the dynamics as a sequence of "events" w_i, occurring with measurable probabilities, according to unknown rules. The primitives should be chosen in such a way that the most compact description of the symbolic dynamics is obtained: their number may be either finite or not, depending on the nature of the signal. Physical systems may be thought to produce signals according to a code [8] which is unknown to the observer and is formed by variable-length words. This crucial aspect of our approach may be understood by considering, as an example, the discrete-time nonlinear dynamical system (roof map)

$$x_{k+1} = \begin{cases} a + 2(1-a)x_k & \text{if} x_k < 1/2 \ (s_k = 0) \\ 2(1-x_k) & \text{if} x_k \geq 1/2 \ (s_k = 1) \end{cases} \tag{1}$$

for $a = (3 - \sqrt{3})/4$, where a Markov partition [7] exists (the critical point $x = 1/2$ belongs to an unstable period-5 orbit). The unit interval can be divided into three subsets, labelled by the sequences 1 (interval $[1/2, 1]$), 01 (left preimage of element 1) and 001 (left preimage of 01). The symbolic signal contains all (random) combinations of these three strings, apart from the forbidden orbit 0011. The folding of phase-space is therefore incomplete, for any value of $a \neq 0$ (for $a = 0$ the trivial binary grammar is observed). Other examples are provided by the period-doubling (PD) accumulation point dynamics and by the golden mean quasiperiodic (QP) transition to chaos [7] which are most efficiently described by the infinite iteration of the (deterministic, parallel) transformations $(0, 1) \to (01, 10)$ and $(0, 1) \to (1, 01)$, respectively [9] (these generation schemes belong to the class of D0L grammars [10]).

The implementation of a computer algorithm for the automatic analysis of a signal implies a sequential scanning of it. If long-ranged "coherent" structures typical of parallel dynamics are present (as, e.g., in the case of Monte Carlo updates of spin chains or in cellular automata [11]), their detection is particularly difficult and requires consideration of increasingly long blocks of symbols. In general, even if the generation rule of the symbolic pattern is purely sequential, as in nonlinear dynamical systems, the optimal description of the actual dynamics is a mixture (endowed with stochastic ingredients) of sequential and parallel mechanisms. (Parallel rewriting schemes transform all or some symbols belonging to a string S into words and are called "grammars" [4, 10]; a sequential mechanism can be interpreted as a grammar in which only the last symbol is substituted). In our approach, the true dynamical behaviour is approximated by means of Markov trees, with an accuracy which grows with the level l: however, if the symbolic pattern is known to be generated by a grammar belonging to a well defined family [4], a more accurate description is obtained by constructing, instead, a "derivation" tree [4, 10] (however, the identification of the proper codewords may be, in general, still very difficult).

In the following, we introduce an efficient scheme for the unfolding of symbolic patterns generated by nonlinear dynamical systems [1]. We start by estimating the probability $P(S)$ of each sequence S, with length $|S| \in [1, n_{max}]$, as the frequency of occurrence of S in the signal $\mathcal{S}_0$ ($\sum_{|S|=n} P(S) = 1$, $\forall\ n$). We then define a (n_{min}-periodically-extendable) primitive as a substring w which can be periodically extended at least up to a length n_{min} ($1 < n_{min} \leq n_{max}$) and which does not contain a prefix with the same property. For example, 001 is a primitive if $(001)^{n_{min}/3}$ is allowed and $(0)^{n_{min}}$ is not: choosing n_{min} equal to n_{max} is equivalent to requiring ordinary periodicity (up to the maximum available length n_{max}). By means of this long-range topological property, regular structures in the signal are detected: the rule according to which the codewords w are chosen, however, depends on the nature of the problem, and may be based, e.g., on the probability $P(w)$ (as in the Huffman or Shannon-Fano codes [8]), or on the value $Q(w)$ of a generic observable. In the course of the analysis sequences of increasing length n are considered. If no single symbol ($n = 1$) satisfies this condition, all blocks of length 2 are examined, and so on, up to a cutoff length $n_{cut} \leq n_{min}$. If still no primitive is found, n_{min} is reduced by one and the whole process is iterated, until primitives of some length $n \leq n_{cut}$ are identified. All other allowed strings of length n are called "transient" orbits. The number of primitives in one-dimensional maps at generic parameter values (PD is an exception) is finite, whereas it may be infinite in higher-dimensional cases. The construction of the tree then proceeds by forming new sequences S' as concatenations of any admissible string S (including transient orbits) with a primitive. If sequence S' in turn exists (i.e., if $P(S') > 0$), it is allocated on the tree, at the position determined by its parental relations. Level 1 contains the primitives, whereas a virtual level 0 is attributed to the transient strings (see Ref. [1] for more details). If some level is not complete because the probability of sequences of length larger than n_{max} cannot be evaluated, a normalisation to one of the probabilities at the first level is carried out, and the rest of the tree is completed by using suitable predictors for $P(S)$. The errors introduced by this procedure are negligible if n_{max} is sufficiently large. The full-periodicity condition is the most useful one for the analysis of low-dimensional chaotic systems: in the example of Eq. (1), $w_1 = 1$, $w_2 = 01$ and $w_3 = 001$ (0 being transient). At PD or QP, instead, the appropriate regrouping is automatically obtained when n_{min} has decreased to 4 or 2, respectively, yielding the codewords $w_1 = 01$, $w_2 = 10$, $w_3 = 0010$, $w_4 = 1101$, ... (for PD) and $w_1 = 1$, $w_2 = 01$ (for QP). Notice that our unfolding procedure is applicable independently of the existence of a Markov partition.

3. Iterated Coding and Renormalisation

The hierarchical-logical analysis which has been hitherto described implies a coarse-graining of phase-space. In general, however, finite (and small) values of n_{min} do not allow to identify codewords which fully reflect the asymptotic properties of the system and the resulting description is not the most compact one. Therefore, it is necessary to resort to a higher-level modeling procedure, endowed with a parallel unfolding mechanism. This is obtained by assigning to the primitive words $w_1, w_2 \ldots$, identified in the analysis of the original signal $\mathcal{S}_0$, symbols from a new alphabet $A_1 = \{0, 1, \ldots\}$. The whole recoding procedure can be repeated on the transformed string $\mathcal{S}_1$ thus obtained, yielding a progressive coarse-graining of the symbolic signal (a process similar to observing phase-space with increased resolution). Obviously, the description of the image-signal $\mathcal{S}_k$ obtained after the k-th recoding step consists of the tree and of the code which keeps track of the previous block-renamings (relations between each symbol in the alphabet A_k and its pre-image string in $\mathcal{S}_0$). The recoding operation is the inverse of a step in a parallel generation scheme and is equivalent to a renormalisation-group transformation on the nonlinear map [9]. For self-similar parallel-generated languages [10], such as those of PD and QP, the trees obtained at each step are identical, i.e., an exact renormalisation is readily achieved. Conversely, for signals of purely sequential nature, there is no such possibility. Usually, however, it is possible and profitable to carry out some recoding steps, the number of which depends on the system: for the example in Eq. (1), the substitution $(w_1, w_2, w_3) \to (0, 1, 2)$ yields a tree in which the only forbidden concatenation is 20. Notice that in this case no further improvement is possible since the new symbols 0, 1 and 2 are obviously also primitive (the full-periodicity condition was required for the codewords in $\mathcal{S}_0$). A higher redundancy would be obtained using the four-element Markov partition (with the associated 9 prohibitions). The increased compactness of the description is achieved by exploiting the intrinsic asymmetries of the dynamics. That recoding yields more asymptotic estimates of the scaling observable can be evidenced by considering the probabilities $Q(S) = P(S)$ (for which $\sum_{\text{lev}l} P(S) = 1$, since to each complete level l corresponds, by construction, a covering of the whole phase-space). The recoding process lets all strings which previously occurred as overlaps between primitives (for the roof map, sequences 10, 100, etc.) simply disappear, so that overlap-free probabilities are computed and unnecessary strings are automatically neglected. An example of how the renaming reduces code redundancy is furnished by the analysis of PD, for which, although the number of primitives is infinite, the code consisting of the first two ($w_1 \equiv 01$ and $w_2 \equiv 10$) is complete and the signal can be rewritten in terms of them only. After renaming w_1 and w_2 as 0 and 1, and measuring the probabilities in the recoded signal, one has $P(0) + P(1) = 1$, whereas $P(w_1)+P(w_2) < 1$ in the original signal. Increased asymptoticity and reduced redundancy are essential properties when dealing with observables (e.g., nearest-neighbour distances $\delta(S)$ [3]) which do not scale as regularly as the probabilities $P(S)$ at PD.

4. Complexity

The detailed (microscopic, path-wise) description of the scaling behaviour of observable $Q(S)$ provided by the logic tree can also be condensed into a macroscopic indicator which measures the predicting power of the derived model. Let us consider, again for $Q(S) = P(S)$, a generic string S at the l-th level of resolution, and all its descendants at level $l + 1$, which are of the form $S' = Sw$, where w is a primitive. Concatenations Sw containing forbidden subwords are discarded *a priori*. If further strings are found to have probability $P(S') = 0$, topological prediction-errors occur, called "surprises". More generally, one proceeds to the comparison of the expected probability $P_0(S')$, assigned to each S' according to some prediction rule, with the actual value $P(S')$, previously measured. In the lowest-order approximation, the rule for the evaluation of P_0 may simply assume factorisation of the probabilities: $P_0(Sw) = P(S)P(w)$. An improved predictor with a memory extent of $l - 1$ primitives can be obtained by rewriting $S' = Sw$ as uVw (where u is another primitive and V a string of $l - 1$ primitives), and calculating $P_0(S')$ as $P_0(S') = P(uV)P(w|V) = P(uV)P(Vw)/P(V)$, where the conditional probability of the suffix w, given V, is considered. The metric complexity C_1 of the system, relative to the set of predictions P_0, is thus defined as the information gain

$$C_1 = \lim_{l\to\infty} \sum_{j=1}^{N(l)} P(S) \ln \frac{P(S)}{P_0(S)} \tag{2}$$

where the sum is over the $N(l)$ allowed sequences at level l. Metric complexity C_1 is a positive quantity, with the exception of the case of perfectly matching predictions ($P = P_0, \forall S$), in which it is identically zero. Analogous definitions can be given for any generic observable $Q(S)$, multiplicative along the path S. The finite-size ($l, |S| < \infty$) estimates provide a global reliability indicator for the associated predictive model, and are therefore meaningful. If the above described $(l-1)$-order predictor is used, the terms P/P_0 appearing in Eq. (2) are ratios between the values $\sigma_{l+1}(uVw) = P(uVw)/P(uV)$ and $\sigma_l(Vw) = P(Vw)/P(V)$ of a generalised scaling function σ [3] for the probabilities, evaluated at two consecutive levels of resolution. Therefore, C_1 has the meaning of a global measure of the convergence of the scaling function, and systems with regular scaling properties are simple, if analysed by using a sufficiently refined prediction rule. The factorisation assumption, for example, yields positive values of C_1 for most non-hyperbolic [7] attractors, since it disregards memory effects, whereas piecewise linear systems are metrically simple also within this lowest-order approach (the natural invariant measure [7] is non-singular). In this sense, C_1 is also a measure of nonlinearity. We have calculated the metric complexity C_1 of the well-known logistic, Lozi and Hénon maps at various parameter values, finding $C_1 < 10^{-2}$: hence, these systems exhibit scaling properties which are sufficiently regular to be estimated with good accuracy. As mentioned, the usage of more sophisticated estimators (i.e., essentially, the inclusion of memory effects) reduces the value of C_1 and improves therefore the convergence of any "thermodynamic" average [12] calculated over the tree structure.

Having as a goal a more complete characterisation of complexity, one can proceed to the generalisation of Eq. (2) and define a function C_q of a parameter q [1]. For $q = 0$, the topological complexity

$$C_0 = \lim_{l \to \infty} \lim_{n \to \infty} \ln \frac{N_0(l,n)}{N(l,n)} \tag{3}$$

is obtained, where $N_0(l,n)$ is the number of orbits predicted at level l, given the knowledge of all orbits of length $|S| < n$, and $N(l,n)$ the number of those with length $|S| \leq n$ allocated at the same level. Notice that Eq. (2) reduces to Eq. (3) if all existing and predicted sequences are separately equiprobable (i.e., $P(S) = 1/N$ and $P_0(S) = 1/N_0$, $\forall\ S$). The topological complexity is identically zero if all predicted orbits exist. Moreover, one- and two-dimensional maps can be easily shown to be topologically simple ($C_0 = 0$). In fact, the relative number $[N_0(l,n) - N(l,n)]/N(l,n)$ of prediction errors vanishes in the limit $(l,n) \to \infty$, since the number $N(n)$ of orbits of length n scales, for large n, as $\exp(K_0 n)$ (where K_0 is the topological entropy [7], whereas the quantity $N_0(n) - N(n)$, for an ordinary tree with fixed-length sequences, grows at most like $\exp(\gamma n)$, with $\gamma \leq K_0/2$ [13].

With our definitions (Eqs. (2) and (3)) numerical values are assigned to the complexity indicators only within the frame of a predictive method. At variance with all previous measures (with the exception of that of Ref. [5]) C_q is defined as a relative quantity, (trivially) identical to zero in the two limits of completely ordered (periodic, quasiperiodic) and random uncorrelated signals. It also satisfies the following "axiomatic" requirements [14]: (*i*) the complexity of a union of objects (patterns) $A^{(j)}$ does not exceed that of the most complex component ($C(\cup_j A^{(j)}) \leq \sup_j C(A^{(j)})$); (*ii*) the complexity of a Cartesian product $B = \otimes_j A^{(j)}$ of several signals is $C(B) = \sup_j C(A^{(j)})$ (at variance with Ref. [5], where C was defined as an extensive quantity).

5. Applications to Dynamical Systems

The unfolding procedure illustrated in Sections 2 and 3 provides a hierarchical model of the dynamics which uses the information stored in the tree and in the associated block-renaming translation table (code). This information is supplied to algorithms which estimate the asymptotic scaling behaviour of the system's observables (e.g., in the form of "thermodynamic" averages like $\Omega(z;q) = \sum z^{|S|} P^q(S)$) or reproduce time-signals having the same dynamical properties (tree-structure, transition probabilities) as the original one. This analysis can be applied to any system for which a generating partition can be either identified or closely approximated (e.g., by attributing different symbolic labels to all unstable periodic orbits up to a given order

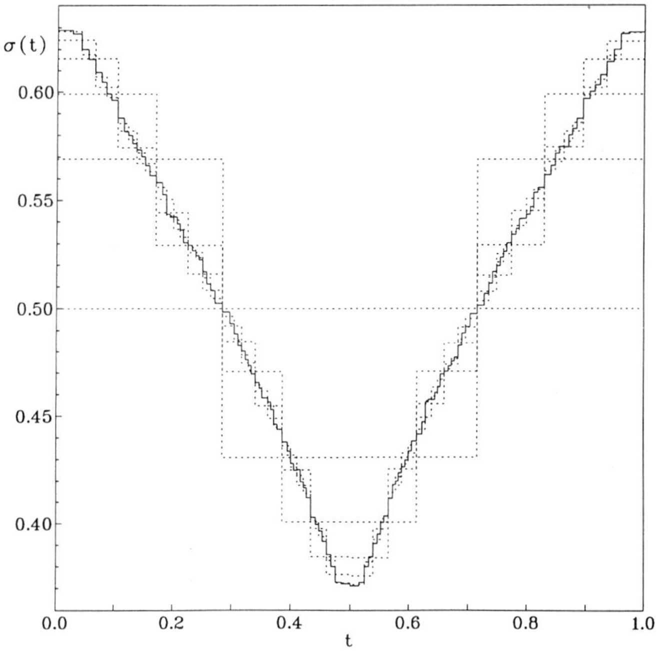

Figure 1: Approximations of the scaling function $\sigma(t)$ for the probabilities versus the ordering parameter t, evaluated on the first 7 levels of resolution for the Lorenz system at standard parameter values.

n_{max}). The main limitation is then represented by the size of the computer memory: with a binary alphabet, one is usually forced to choose $n_{max} \leq 22$.

In addition to the maps mentioned in Section 4, we have studied several flows, among which the well – known Lorenz system at standard parameter values [7]. Notwithstanding the absence of the two period-1 cycles, a complete binary tree description is appropriate, because prohibitions appear only for $|S| > 25 > n_{max}$. In order to show the convergence of the predictive models, we plot in Fig. 1 the corresponding probability scaling function $\sigma[t(S)]$ introduced in the preceding section. The ordering parameter $t(S)$ increases monotonically (from 0 to 1) with the binary values of the string S, read from right to left: $t(00) < t(10) < t(01) < t(11)$. The approximations σ_l obtained from the first 7 levels ($l = 1, \ldots, 7$) are drawn: within the statistical fluctuations, a fast convergence is observed for increasing l, indicating that Markov models of order as low as 5 or 6 already reproduce the symbolic dynamics with good accuracy. Correspondingly, the same convergence properties are exhibited by the thermodynamic averages, provided that the transfer operator technique [15] is implemented.

The results obtained for the asymptotic behaviour of the conditional probabilities (scaling function) show that the Markov models are able to generate signals with dynamical properties close to those of the original system. This corresponds to performing predictions of the time-dynamics in terms of a succession of deterministic paths (blocks of symbols) which appear at random in time, according to the measured transition probabilities [16]. A comparison between the power spectrum of the symbolic signal generated by the logistic map $x' = 1 - 1.85x^2$ (with primitives 1, 01 and 001) and those obtained from the level-1, -3 and -5 reconstructions is shown in

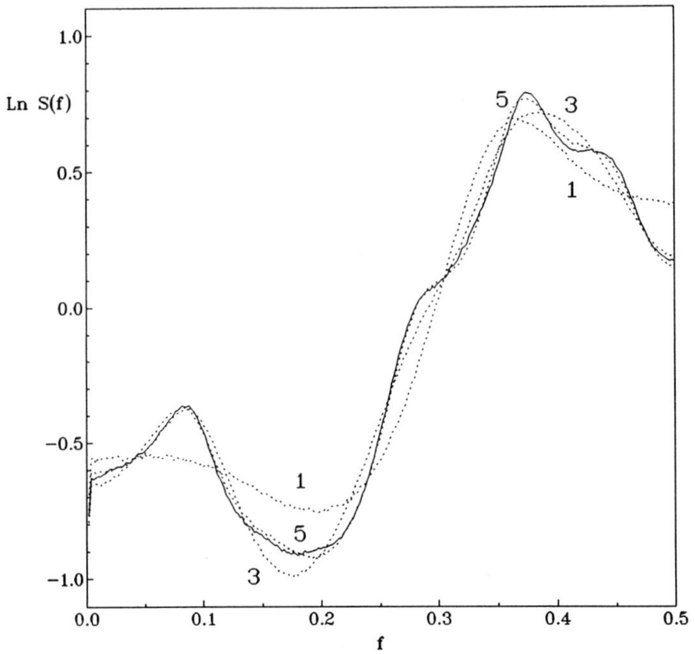

Figure 2: Comparison between the power spectrum $S(f)$ (vs. the frequency f) of the symbolic signal of the logistic map (thick line) and those corresponding to the successive hierarchical reconstructions obtained from levels 1, 3 and 5 of the logic tree (thin lines).

Fig. 2, as an illustration of the achieved accuracy. The structure of the power spectra appears to be dependent, in order of decreasing importance, on two ingredients which are invariant under smooth coordinate changes (the topology and the metric of the logic tree) and on one which is not (the values of the continuous variable x, not considered here).

The knowledge of the tree structure and of the recoding rules can be exploited, as mentioned, to construct global models (i.e., statistically "optimal", but not exact) [1]. The first one consists of a piecewise linear map that expands each of the phase-space elements of the l-th level approximation onto the image-elements. A refinement which takes into account also metric features is obtained by scaling the size of each subregion Δ according to the local expansion rates in such a way that the measured transition probabilities are reproduced. A further improvement requires consideration of nonlinear terms, which still satisfy the previously found topological structure, but which generate singularities in the invariant measure [7]: this corresponds to introducing nonhyperbolic ingredients in the model map and to including memory effects [1].

6. Conclusion

We have discussed the concept of complexity in the framework of a hierarchical modeling of physical systems, showing that complexity is "orthogonal" to the usual time-dynamics: the unpredictability of the future time evolution is a much less severe obstacle to the construction of reliable models than the unpredictability of the scaling dynamics, for the understanding of which a parallel unfolding mechanism must be included. Descriptions based on purely sequential (e.g., regular-language) interpretations [6, 17, 18], in fact, do not encompass the structure of higher-order languages such as those generated by formal grammars [4, 10].

References

1. R. Badii. In *Measures of Complexity and Chaos.* Eds. N.B. Abraham, A. Albano, T. Passamante, and P. Rapp. Plenum, New York, 1989.
   M.A. Sepúlveda and R. Badii. *ibid.*
   R. Badii. PSI Report **61**, 1. Villigen, Switzerland, 1990.
   R. Badii. Weizmann preprint, Rehovot, Israel, 1988.
2. D. Ruelle. *Thermodynamic Formalism*, Vol. 5 of Encyclopedia of Mathematics and its Applications. Addison-Wesley, Reading MA, 1978.
3. M.J. Feigenbaum. *J. Stat. Phys.* **52**, 527 (1988).
4. J.E. Hopcroft and J.D. Ullman. *Introduction to Automata Theory, Languages and Computation.* Addison-Wesley, Reading, MA, 1979.
5. S. Lloyd and H. Pagels. *Ann. Phys. (N.Y.)* **188**, 186 (1988).
6. P. Grassberger, Wuppertal preprint B 89-26 (1989).
7. J. Guckenheimer and P. Holmes. *Nonlinear Oscillations, Dynamical Systems and Bifurcations of Vector Fields.* Springer, New York, 1986.
8. R. Hamming. *Coding and Information Theory.* Prentice-Hall, Englewood Cliffs NJ, 1986.
9. I. Procaccia, S. Thomae, and C. Tresser. *Phys. Rev.* **A 35**, 1884 (1987).
10. G. Rozenberg and A. Salomaa. *The Mathematical Theory of L Systems.* Academic Press, London, 1980.
11. S. Wolfram. *Rev. Mod. Phys.* **55**, 601 (1983).
12. R. Badii. *Riv. Nuovo Cim.* **12**, N° 3, 1 (1989).
    R. Artuso, E. Aurell, and P. Cvitanović. Niels Bohr Institute preprints NBI-89-41 and NBI-89-42 (1989).
13. G. D'Alessandro and A. Politi. *Phys. Rev. Lett.* **64**, 1609 (1989).
14. R. Badii. PSI preprint, PSI-LUS-03 (1990).
15. M.J. Feigenbaum, M.H. Jensen, and I. Procaccia, *Phys. Rev. Lett.* **57**, 1503 (1986)
16. R. Badii, M. Finardi, and G. Broggi. To be published.
17. J.P. Crutchfield and K. Young. *Phys. Rev. Lett.* **63**, 105 (1989).
18. D. Auerbach and I. Procaccia. *Phys. Rev.* **A 41**, 6602 (1990).

RECONSTRUCTING LANGUAGE HIERARCHIES

James P. Crutchfield

Physics Department
University of California
Berkeley, CA 94720, USA
chaos%gojira @ lilac.bitnet

Abstract

Within an assumed language class optimal models can be estimated using Gibbsian statistical mechanics. But how are model classes themselves related? We consider the problem of moving from less to more computationally capable classes in the search for finite descriptions of unpredictable data series.

1. Complexity: The Essential Tension between Order and Chaos

Complexity arises at the onset of chaos. Natural systems that evolve with and learn from interaction with their immediate environment exhibit both structural order and dynamical chaos. Order is the foundation of communication between elements at any level of organization, whether that refers to a population of neurons, bees, or humans. For an organism order is the distillation of regularities abstracted from observations. An organism's very form is a functional manifestation of its ancestor's evolutionary and its own developmental memory.

A completely ordered universe, however, would be dead. Chaos is necessary for life. Behavioral diversity, to take an example, is fundamental to an organism's survival. No natural environment can be modeled in its entirety, though. Approximation becomes essential to any system with finite resources. Chaos, as we now understand it, is the dynamical mechanism by which nature develops constrained and useful randomness from finite resources. And from it follow diversity and the ability to anticipate the uncertain future.

There is a tendency, whose laws we dimly comprehend, for natural systems to balance order and chaos, to move to the interface between structure and uncertainty. The result is increased complexity. This often appears as a change in a system's

Information Dynamics, Edited by H. Atmanspacher and
H. Scheingraber, Plenum Press, New York, 1991

computational capability. The present state of evolutionary progress suggests that one need go even further and postulate a force that drives in time toward successively more sophisticated and qualitatively different computation. The evidence for this is immediate. We can look back to times in which there were no systems that attempted to model themselves, as we do now. This is certainly one of the outstanding puzzles: how can lifeless and disorganized matter exhibit such a drive? And the question goes to the heart of many disciplines, ranging from philosophy and cognitive science to evolutionary and developmental biology and particle astrophysics. The dynamics of chaos, the appearance of pattern and organization, and the complexity quantified by computation will be inseparable components in its resolution.

In the following we consider a restriction of this general problem to a tractable one that concerns modeling temporal sequences of measurements: i.e. inferring models of processes from their data series. The specificity gives a concrete picture of what we mean by a model, its class and language. Here a model will be a machine in the form of one of a variety of stochastic automata; its class, the list of architectural constraints; and its language, the full range of behavior it can describe. Our problem then becomes how to jump levels from a computationally less capable machine type to a more sophisticated one. Thus, we are not so much interested in the optimization of estimated models within any particular class, but rather the relation between levels. The relations consist of learning heuristics that indicate what needs to be innovated in order to define a new more powerful class.

2. ϵ – Machines

The central modeling abstraction is the ϵ-machine [1]. For a given, possibly infinite data stream this is the smallest deterministic machine at the least computationally powerful level that yields a finite description. ϵ here simply refers to the dependence on the nature of the data stream. In the past ϵ has denoted the measurement resolution with which a continuous state system is sampled. It is also consistent with the use in Shannon's dimension rate [2] and in the complexity theory of function spaces [3]. There are, of course, many other parameters that could be listed for a data stream; ϵ is a sufficient reminder.

There is no *a priori* reason to expect a hierarchy of finitely-describable models for nonlinear processes to extend ever upward or, for that matter, to terminate. The distinction between these alternatives is not unlike a Gödelian limitation. Computation theory indicates that the task of estimating optimal models becomes more difficult as the class capability increases. Nonetheless, we take ϵ-machines as a definition of what is humanly conceivable and humanly workable; and also as indicative of contemporary scientific method. From an operational viewpoint, for example, animal learning, scientific research, and biological evolution exhibit the property of compiling natural regularities into compact and usable structures. Physical degrees of freedom are reconfigured into analogical models of observed properties. These form predictive devices and a basis for generalization. The reconstruction of ϵ-machines gives a concrete approach to studying the common elements in these processes.

Table 1 summarizes the overall goal as a hierarchy of ever more sophisticated ϵ-machines. In the progression toward higher complexity less capable models form the representation basis for inference at the next higher level.

Table 1: Machine reconstruction hierarchy. Computational capability increases going down the table, i.e., with increasing level. See text for discussion.

Level	Grammar	Representation	State	Reconstruction Relation	Complexity	Symmetry (factored out)	Memory	Abstraction
0	subwords	Measurement sequence	observed symbol	$S = E / \sim_o$	Total storage	Choice of experiment E and observables O	Infinite one-dimensional symbol string	Information Observation Measurement
1	common subwords	Tree	tree node	$T = S / \sim_t$	#distinct words, Shannon information	time translation, subword invariance	Infinite tree of distinct subsequences	Probability Typicalness Ergodicity Stationarity
2	Definite	Subshift of finite type, Markov process	Markov state	$M = T / \sim_f$	# of states	finite subword conditioning	Finite set of states	Process
3	Regular	Semi-group	semi-group element	$G = T / \sim_p$	Semi-group complexity	predecessor conditioned future morph equivalence	Finite set of semi-group elements	Feedback Dynamics
4	Regular	Sofic system, Finite Automaton	dynamic state	$FA = G / \sim_m$ $FA = T / \sim_m$	Finitary Complexity	future-morph equivalence	Finite set of states and labelled transitions	Parity
5	Regular	String automaton	branching state	$SA = FA / \sim_c$	Branching Complexity	deterministic chain reduction	Finite sets of states and of finite transition strings	
6	Indexed Context Free	Register automaton	register and control state	$RA = SA / \sim_s$	Production Complexity	future state-morph equivalence	Infinite pushdown stack	Recursion
...	...	...	...	...	...	...	...	...
	Phrase structure	Bernoulli-Turing machine	tape state and control state	$BTM = S / \sim_{dr}$	Bernoulli -Kolmogorov	DTM computation and Randomness	Infinite tape	Universality

3. Reconstruction Hierarchy

At each level of the hierarchy, there is a corresponding model class or representation that determines the language. Each representation has a notion of current state and of transition to a successor state. The set of sequences that each can recognize and generate is indicated by the grammar type. And that in turn gives the structure of the language, i.e., the range of expression of the model class.

A representation is a set of assumptions about the underlying process that produced the data stream. These assumptions we refer to as symmetries, since in many cases they are easily represented by particular semigroups. The procedure of estimating a model within a class corresponds to factoring out the appropriate symmetries from the data stream. Formally, the "factoring out" reconstruction is given by an equivalence relation, generally denoted $\sim$.

The complexity $C(S|\mathcal{M})$ of a data stream S with respect to a given model class $\mathcal{M}$ is the representation's size [4, 5, 6]. And this typically is taken to be the number of inferred $\mathcal{M}$-states. Just as with Shannon information, which is a special case as discussed below, complexity is a fundamentally relative concept. It is important to note that if, at a given level, a process P producing a data stream S appears to consist of independent processes $P_1 \otimes P_2 \otimes \cdots$, then the complexity is additive:

$$C(S|\mathcal{M}) = \sum C(P_i|\mathcal{M}).$$

The models in the hierarchy are deterministic in the sense of computation theory. This means they deterministically recognize the measurement sequence from which they are reconstructed. In other words, the machines cannot make transitions to different successor states on the same input. Nonetheless, when used as models to generate new sequences, including some possibly never observed in the original data, they employ a source of randomness to select a uniquely labeled transition from a state. That source of guesses is referred to as a random oracle. In this sense the hierarchy is stochastic.

The computational capability of the classes increases going down the table, i.e., with increasing level. For example, Markov processes are less computationally capable than stochastic semigroups and finite automata. This trend to increasing capability is concomitant with the increasing specificity and strength of the assumed statistical and deterministic structure of the process that produced the data. In moving up the hierarchy new abstractions come into play as the appropriate assumptions about the underlying process. Thus, only at the tree level do the notions of stationarity and sequence probability first appear; and only at the semigroup level does intrinsic dynamics arise.

Above, we discussed the complexity in terms of factoring out symmetries from the data stream. While true, the reconstruction process as a whole is incremental. Lower levels provide a representation basis on top of which the next level is reconstructed. Moving stepwise up the hierarchy is an example of incremental learning. Regular patterns of lower level states become the states at the next higher level. In fact, the progression in the table represents, fairly closely, the inference method used in building up finitary and nonfinitary ϵ-machines; up to the level that is currently implemented: register production machines, the fifth level. The original question then is articulated in terms of the mathematical structure of incremental learning.

The first partial answer to the question of how to do incremental learning comes from considering the convergence properties of lower level representations. For example, as discussed below, the convergence of lower level complexity with increasingly accurate representation determines the complexity at the next higher level. For example, entropy convergence at the tree level determines the finite automaton complexity. And finite automata complexity convergence, in turn, affects the indexed context-free production complexity at the register automaton level.

The evolution to higher capability is simply described in statistical mechanical terms. There is a partition function for each class, $Z_\alpha = \sum_{\text{states}} e^{-\alpha \log p_{\text{state}}}$, that is the sum over that level's states. The complexity is the analog of the Helmholtz free energy, $C_\alpha = (1-\alpha)^{-1} \log Z_\alpha$. At each level one estimates an optimal model [7, 8, 9, 10]. This is taken as the minimal representation consistent with the data to within some error. It is found by minimizing the analog of the Gibbs free energy which relates the change in complexity to representation accuracy and prediction error.

Various labels in the table are given in terms of forward time reconstruction. These can be modified, however, for reverse time reconstruction by consistently replacing "future" with "past" and "predecessor" with "follower". The difference in the corresponding complexities is a measure of statistical irreversibility. The first level at which a useful (nonzero) measure of reversibility occurs is that of semigroup representation.

4. Data Streams and Processes

A data stream consists of a long measurement sequence $\mathbf{s} = \{s_0 s_1 s_2 \cdots : s_i \in \mathbf{A}\}$ whose elements are formal symbols in a finite set, the alphabet $\mathbf{A}$ of k symbols. Before discussing models it is natural to lay out the assumptions concerning what is being modeled, viz. a process that produces a symbol sequence. We are interested in *process* languages consisting of observable words and their substrings. One way to motivate this notion is to assume that there is some process P that emits an unending stream of symbols. We, the observers, have no control over how the process started and must simply take the data as it comes. If we have observed a particular word, then obviously we have observed its subwords. When we consider processes as machines, they can start in any state and can stop in any state. In contrast the machines we reconstruct have a single start state. This state represents the observer's condition of total ignorance of the process's state. The machine as a whole captures the temporal evolution of the observer's knowledge of the process as measurements are collected. The reconstruction problem is then how to infer properties of P at a specified level given an observed subset of the process's language $L(P)$. This is a central problem in the computational learning theory of inductive inference [11].

Given a data stream consisting of a long measurement sequence $\mathbf{s} = \{s_0 s_1 \cdots s_{N-1} : s_i \in \mathbf{A}\}$ we can ask how much storage it requires. Consider the lexicographic ordering of all such sequences. Since there are k^N possible sequences, it will take $N \log_2 k$ bits of storage to specify the ordinal number that uniquely identifies the particular sequence observed. There may be regularities (both deterministic: such as particular sequences being disallowed; and statistical: some sequences being more probable) that will allow for substantially less storage to be used than this upper bound.

5. Trees

For example, rather than explicitly listing the sequence, if we assume that similar sequences observed at different positions in the data stream represent the same process states, then there is a more compact storage method based on histograms. The associated data structure is a tree, each level L of which represents the number each length L sequence observed anywhere in the data stream.

Given this assumption, the second step in machine inference is the construction of a parse tree. A tree $T = \{\mathbf{n}, \mathbf{l}\}$ consists of nodes $\mathbf{n} = \{n_i\}$ and directed, labeled links $\mathbf{l} = \left\{l_i : \left(n \rightarrow_s n'\right),\ n, n' \in \mathbf{n},\ s \in \mathbf{A}\right\}$, connecting them in a hierarchical structure with no closed paths. An L-level subtree T_n^L is a tree that starts at node n and contains all nodes below n that can be reached within L links. The tree nodes are this representation's state; the links the state transitions. Trees, then, are machines with no feedback loops. There is no notion of recurrent sequences of tree states.

To construct a tree from a measurement sequence we simply parse the latter for all length D sequences and from this construct the tree with links up to level D that are labeled with individual symbols up to that time. We refer to length D subsequences $s^D = \left\{s_i \cdots s_{i+D-1} : s_j = (\mathbf{s})_j\right\}$ as D-cylinders. Hence a D level tree has a length D path corresponding to each distinct observed D-cylinder. The picture here is that a particular D-cylinder is a name for that bundle of the underlying process's orbits each of which visited the sequence of measurement partition elements indexed by the D-cylinder. The basic assumption in building a tree is that symbol sequences observed at different times in the data stream approximate the **same** process state. Nonstationary processes are examples for which this assumption fails.

Every node n in the tree is associated with the sequence or word ω_n that leads to it starting from the top tree node n_λ, the tree state of total ignorance. Probabilistic structure is added to the tree T by recording for each node the number $N(\omega_n)$ of occurrences of the associated D-cylinder ω_n relative to the observed total number $N(l)$ of length $l = |\omega_n|$, $p_n^T = N^{-1}(l)N(\omega_n)$ where $n_\lambda \rightarrow_{\omega_n} n$. This provides estimates of the node probabilities $\vec{p}_n = \{p_n : n \in \mathbf{n}\}$. And this in turn gives a hierarchical approximation of the measure in sequence space $\Pi_{\text{Time}}\mathbf{A}$. Tree representations of data streams are closely related to the hierarchical algorithm used for estimating dynamical entropies [12, 13].

6. Entropy

In fact, the notion of entropy as used in information theory is intimately related to tree data structures. For if we ask how much storage is required for a tree representation of a data stream, Shannon's entropy emerges as the most natural estimator for the growth rate of the number of nodes. This follows from consideration of the fundamental relationship between entropy, combinatorics, and probability [14]. Rather than specifying the particular data sequence as one out of k^N, it is identified via two parts: a histogram of counts and the ordinal number within the ensemble indexed by that histogram. For the given alphabet, we must specify k counts $\{v_i : i = 1, \ldots, k\}$ for each symbol and each count may be as large as N. This takes $k \log_2 N$ bits. Then the observed sequence is specified as one out of the $N!(v_i! \cdots v_k!)^{-1}$ possible sequences described by the ensemble. The total required storage for large N is then dominated by $-N \sum_{i=1}^{k} \frac{v_i}{N} \log \frac{v_i}{N}$. This is clearly related to Shannon's entropy. But that connection follows only if we then assume that probabilities exist and can be determined by the relative frequency estimator, $p_i = N^{-1} v_i$. The analysis up to this point has made no such assumption.

We have considered the histogram of individual symbols. It can be easily extended to account for L-sequences and so it applies to each level of a tree representation. Within the class of tree models the notions of typical subsequence and of probability first come into play, moving up the hierarchy. The analysis shows in what sense entropy is dual to tree models of data streams and how it is the appropriate measure of a tree representation's average size. The tree is built using a sliding window to move through the data stream. It captures in this way the distinct sequences and summarizes their occurrence at different points in the data stream via a count or probability. Entropy measures the number of distinct sequences. That number increases if there is branching as one moves down the tree and forward in time.

Assuming that probabilities are appropriate descriptions, then the total Shannon entropy [2] of length L sequences is:

$$H_{\text{Shannon}}(L) = - \sum_{n \in N, |\omega_n| = L} p_n \log_2 p_n$$

The total Hartley entropy is given simply by the total number of distinct sequences independent of their (positive) probability $H_{\text{Hartley}}(L) = \log_2 N(L)$. If the probability distribution is uniform on the nonzero probability cylinders then these two entropies are equal. Any difference is thus a measure of deviation of the cylinder distribution from uniformity.

The latter observation leads to a parametrized generalization of the entropy introduced by Renyi. This we put into a statistical mechanical formalism by defining a partition function for the tree. The α-order total Renyi entropy [15], or "Helmholtz free information", of the measurement sequence up to L-cylinders is $H_\alpha(L) = (1-\alpha)^{-1} \log Z_\alpha(L)$, where the tree partition function is

$$Z_\alpha(L) = - \sum_{n \in N, |\omega_n| = L} e^{\alpha \log p_n}$$

with the probabilities p_n defined on the tree nodes. The Shannon entropy corresponds to the $\alpha = 1$ case, with the application of L'Hopital's rule, and the Hartley entropy to the $\alpha = 0$, with the appropriate definition of the zeroth power of a variable as its indicator function.

The average branching rate in the tree measures the growth rate of the number of new sequences of increasing length. And as such it is a measure of unpredictability in that a periodic process will at some length give rise to no more new cylinders and a random one will. The Renyi specific entropy, i.e., entropy per measurement, is approximated [12] from the L-cylinder distribution by $h_\alpha(L) = L^{-1} H_\alpha(L)$ and is given asymptotically by $h_\alpha = \lim_{L\to\infty} h_\alpha(L)$. The growth rate of total Shannon entropy is often referred to in information theory as the source entropy and in dynamical systems as the metric entropy. It is given by $h_\mu = \lim_{L\to\infty} H_{\text{Shannon}}(L)$, The corresponding Hartley entropy growth rate is called the topological entropy $h_0 = \lim_{L\to\infty} H_{\text{Hartley}}(L)$.

These entropy growth rates can be also given by:

$$h_\alpha = \lim_{L\to\infty} \{H\alpha(L) - H_\alpha(L-1)\}$$

This form suggests a recursive algorithm for estimating them based on the difference in the cylinder distribution between successive levels in the tree. Consider the following information gain form for the metric entropy:

$$h_\mu = \lim_{L\to\infty} \sum_{s \in \mathbf{A}, |\omega| = L-1} p_{\omega s} \log_2 \frac{p_{\omega s}}{p_\omega}$$

The entropy rate is then the asymptotic information gain of the L-depth tree with respect to the $(L-1)$-depth tree representation. If a tree representation is good, then the information gain, or entropy rate, at some depth will vanish. This indicates that no further information need be stored to represent the process. This happens for a periodic process for trees deeper than the period. If the process is chaotic, with positive entropy, then the information contained in the tree representation will grow exponentially fast with modeling longer subsequences in the data stream. This indicates that the tree representation is inadequate and suggests that the modeler innovate a new class of representations. In the present case of temporal processes, the innovation is the notion feedback dynamics. As we will see many chaotic processes have a finite representation at this next level, whereas they have infinite tree representations.

7. Dynamic States

Let us assume then that the process is aperiodic, that is, that the information in the tree structure grows without bound. Then we must move up to the next level of representation whose central notions are feedback and dynamics. At this point we deviate a bit from following the increasing representation complexity given in table 1 by jumping ahead to what is essentially a more capable model class: finite automata.

We ask at this level if there is some dynamic process underlying the estimated tree structure. The reconstruction goal is to infer recurrent states. As a measurement is made, does the observer know what "state" the underlying process is in? The only window onto the process's state is through the intermediary of the chosen observables and measurement method. Isolated measurements do not necessarily correspond to direct detection of the process's state. They are only its indirect reflections.

At this computational level ϵ-machines are represented by a class of labeled, directed multigraphs, or l-digraphs [16]. They are related to the Shannon graphs of information theory [2], to Weiss's Sofic systems in symbolic dynamics [17], to discrete finite automata in computation theory [18], and to regular languages in Chomsky's hierarchy [19]. Here we are concerned with stochastic versions of these. Their topological structure is described by an l-digraph $G = \{\mathbf{V}, \mathbf{E},\}$ that consists of vertices $\mathbf{V} = \{v\}$ and directed edges $\mathbf{E} = \{e\}$ connecting them, each of the latter is labeled by a symbol $s \in \mathbf{A}$. An edge is a triplet consisting of an ordered pair of states $\mathbf{E} \subseteq \mathbf{V} \times \mathbf{V} \times \mathbf{A}$ and a symbol. An edge $e = \left(v \rightarrow_s v'\right)$ is interpreted as a transition from state v to state v' on symbol s.

To reconstruct a topological ϵ-machine we define an equivalence relation, subtree similarity, denoted $\sim_m$, on the nodes of a depth D tree T by the condition that the L-subtrees are identical: $n \sim n'$ if and only if $T_n^L = T_{n'}^L$. Naturally, we require that $L < D$ and in practice we take $2L = D$. Subtree equivalence means that the link structure is identical. We refer to the archetypal subtree link structure for each class as a "morph". This equivalence relation induces on T, and so on the measurement sequence $\mathbf{s}$, a set of equivalence classes $\left\{\mathbf{C}_m^L : m = 1, 2, 3, \ldots\right\}$ given by:

$$\mathbf{C}_l^L = \left\{n \in \mathbf{n} : n \in \mathbf{C}_l^L \text{ and } n' \in \mathbf{C}_l^L \text{ iff } n \sim n'\right\}$$

An l-digraph G_L is then constructed by associating a vertex to each tree node L-level equivalence class; that is, $\mathbf{V} = \left\{\mathbf{C}_m^L : m = 0, \ldots, V-1\right\}$.

If the number of inferred states is finite, $\| \left\{\mathbf{C}_m^L\right\} \| < \infty$, then the machine is finitary. The process associated with the machine states is a finite-order Markov chain, although the process associated with the symbol alphabet, i.e., the transitions, need not be finite Markovian [20].

But this boundedness need not always be the case. It simply defines the finitary ϵ-machines. An infinite number of states can be inferred. This is what is found, in fact, at the onset of chaos via period-doubling [1, 21, 22], quasiperiodicity [21, 22], or intermittency.

9. Transitions

Two vertices v_k and v_l are connected by a directed edge $e = (v_k \rightarrow v_l)$ if the transition exists in T between nodes in the equivalence classes: $n \rightarrow n' : n \in \mathbf{C}_k^L, n' \in \mathbf{C}_l^L$. The corresponding edge is labeled by the symbol associated with the tree links connecting the tree nodes in the two equivalence classes:

$$\mathbf{E} = \left\{e = (v_k \rightarrow_s v_l) : v_k \rightarrow_s v_l \text{ iff } n \rightarrow_s n'; n \in \mathbf{C}_k^L, n' \in \mathbf{C}_l^L, s \in \mathbf{A}\right\}$$

10. Occam's Razor and Explanation

In this way, ϵ-machine reconstruction deduces from the diversity of individual patterns in the data stream "generalized states": the morphs, associated with the graph vertices, that are optimal for forecasting. If an observer knows the process's state then the average uncertainty is the minimum possible, that is, the metric entropy or, equivalently, the process's entropy rate. From a structural viewpoint, the topological ϵ-machines so reconstructed capture the essential computational aspects of the data stream by virtue of the following instantiation of Occam's razor: Topological reconstruction of G_L produces the minimal and unique machine recognizing the language and the generalized states specified up to L-cylinders by the measurement sequence. This follows from the future morph equivalence relation which is the analog of right equivalence in finite automata minimization [18]. This type of equivalencing goes back to Huffman's DFA reduction [23] and was formalized by Nerode and Moore [18]. Although similar in principle, topological machine reconstruction differs from those approaches in that (i) they are not concerned with data series, rather they start with a given machine and minimize it and (ii) they are concerned with machines that are necessarily finite. Finally, we are ultimately interested in stochastic versions of reconstruction: including ways to estimate probabilities for the topologically reconstructed machines and for machines with stochastically similar states.

Minimality guarantees that the model contains no structure and no more properties than the process. For example, there is a 128 state DFA that accepts all binary strings. But we wish to interpret the number of states as a measure of the amount of the process's memory. The 128 state DFA is consistent with the data, but indicates that the process has seven bits of storage. But the process clearly does not. It produces the most unpredictable, ideally random sequences possible. The minimal representation has a single state and so no information storage.

From the Bayesian viewpoint minimality is important since the minimal model consistent with the data maximizes, with respect to all other nonminimal, but consistent models, the posterior probability that the model could have produced that data $\Pr(S|M)$. Thus, via Bayes' theorem, minimal machines are the most likely explanation: $\Pr(M|S)$ is maximized.

If the minimal machine is found to be independent of window length L then the storage requirement is $\mathcal{O}(|\mathbf{V}| + |\mathbf{E}|) = \mathcal{O}(V(1 + h_0))$, which is independent of the length of the data stream.

11. Stochastic Machine from Tree

If we wish to interpret the tree and so also the machine as representing the average increase in the observer's knowledge with successively longer sequences, then the machine state and transition probabilities need to be estimated. The transition probability from a state $v \in \mathbf{V}$ is found by looking at the branching probabilities at the tree node labeled with the shortest path to the state. Due to the sliding window construction of the tree this gives estimates that use the highest count statistics since the tree node is the closest to the tree's top. Nodes further down in the tree in the state's equivalence class will have relatively diluted statistics. The increased counts in turn reduce the fluctuations and so the estimation error. So the transition probability from a state on symbol $s \in \mathbf{A}$ is given by:

$$p(v \rightarrow_s v') = \frac{p_{n_{\omega s}}}{p_{n_\omega}}, \quad \text{where } n_\lambda \rightarrow_\omega n_\omega,\ n_\omega \in C_v,\ |\omega| \text{ smallest}$$

The state probabilities are estimated from the state-partitioning of tree nodes at morph depth level L. Recall that the probability distribution, and the total number of counts for that matter, on each level is normalized. Going down the tree then dilutes the given constant number of counts into more tree nodes. The deepest level at the tree nodes that have been classified gives the most refined partition. At that level the probability distribution has relaxed as close as possible to its asymptotic value. The state probabilities are then estimated via

$$p_v(L) = \|C_v^L\|^{-1} \sum_{n \in C_v^L} p_n$$

where C_v^L is the set of tree nodes at level L in the equivalence class of v.

In the case that the underlying process is k-order Markovian or k-periodic, then this method gives the appropriate estimates of the correct stochastic structure when $L > k$. Increasing the length of the data stream simply improves the probability estimates' accuracy. We have skipped over the details required to estimate the stochastic structure of the full class of DFAs or, equivalently, Sofic systems. But the gist of the estimation process is made clear by presenting the Markovian case.

12. Finitary Complexity

With this representation for the data, the question becomes how an ϵ-machine captures a process's dynamics. Many of the important properties of these stochastic automata models, again, are given concisely using a statistical mechanical formalism that describes the coarse-grained scaling structure of orbit space. The statistical structure of an ϵ-machine is given by a parametrized stochastic connection matrix $T_\alpha = \sum_{s \in \mathbf{A}} T_\alpha^{(s)}$, that is the sum over each symbol of the state transition matrices $T_\alpha^{(s)} = \left\{ e^{\alpha \log p(v_i | v_j; s)} \right\}$. The α-order finitary complexity is the free energy $C_\alpha(L) = (1-\alpha)^{-1} \log Z_\alpha^v(L)$ where the machine partition function is $Z_\alpha(L) = \sum_{v \in \mathbf{V}} e^{-\alpha \log p_v}$ and the probabilities p_v are defined on the ϵ-machine's vertices $v \in \mathbf{V}$. The finitary complexity is a measure of an ϵ-machine's information processing capacity in terms of the amount of information stored in the morphs. It is directly related to the mutual information of the past and future semi-infinite sequences and to the convergence [24, 25, 26] of the entropy estimates $h_\alpha(L)$. It can be interpreted, then, as a measure of the amount of mathematical work necessary to produce a fluctuation from asymptotic statistics. The units for complexity measures are bits of information. However, at this level we see that the complexity begins to more strongly reflect the degree of computational capability and so we refer to the units as Turings, rather than bits. At this low level the difference between bits and Turings is not as dramatic as at higher levels where each unit of machine structure is clearly associated with sophisticated computation.

The entropies and complexities are dual in the sense that the former is determined by the principal eigenvalue λ_α of T_α, $h_\alpha = (1-\alpha)^{-1} \log_2 \lambda_\alpha$ and the latter by the associated left eigenvector of T_α, $\vec{p}_\alpha = \{p_v^\alpha : v \in \mathbf{V}\}$ that gives the asymptotic vertex probabilities. A complexity based on the asymptotic edge probabilities

$\vec{p}_e = \{p_e : e \in \mathbf{E}\}$ can also be defined by $C_\alpha^e = (1-\alpha)^{-1} \log \sum_{e \in \mathbf{E}} p_e^\alpha$, where $\vec{p}_e$ is given by the left eigenvector of the ϵ-machine's edge graph. The transition complexity C_1^e is simply related to the entropy and graph complexity by $C_1^e = C_1 + h_1$. There are, thus, only two independent quantities for a finitary ϵ-machine [27].

The two limits for α warrant explicit discussion. For the first, topological case ($\alpha = 0$), T_0 is the l-digraph's connection matrix. The Renyi entropy $h_0 = \log \lambda_0$ is the topological entropy h. And the finitary complexity is $C_0 = \log |\mathbf{V}|$. This is $C(\mathbf{s}|\text{DFA})$: the size of the minimal DFA description, or "program", required to *produce* sequences in the observed ensemble of which $\mathbf{s}$ is a member. This topological complexity counts all of the reconstructed states. It is similar to the regular language complexity developed for cellular automaton spatial patterns [28]. The DFAs in that case were constructed from known equations of motion and an assumed neighborhood template. In the second, metric case ($\alpha = 1$), h_α becomes the metric entropy $h_\mu = \lim_{\alpha \to 1} h_\alpha = -\frac{d\lambda_\alpha}{d\alpha}$. The metric complexity $C_\mu = \lim_{\alpha \to 1} C_\alpha = -\sum_{v \in \mathbf{V}} p_v \log p_v$ is the Shannon information contained in the morphs. These measures have been discussed before as the "set complexity" version of the regular language complexity [26]. Following the preceding remarks, the metric entropy is also given directly in terms of the stochastic connection matrix:

$$h_\mu = \sum_{v \in \mathbf{V}} p_v \sum_{v' \in \mathbf{V}, s \in \mathbf{A}} p(v|v';s) \log p(v|v';s)$$

13. Knowledge Relaxation

ϵ-machines are representations of an observer's model of a process. Starting from the state of total ignorance about the process, successive steps through the machine correspond to a refinement of the observer's knowledge based on observations. The average increase is given by a diffusion of information throughout the given model. In the case of trees, we have a flow of probability downwards, in increasing time, toward the leaves. This is a unidirectional diffusion of information on an ultrametric structure [29]. The ultrametric distance on the tree is sequence length or, more simply, time itself. The tree and machine transition probabilities, especially those connected with transient states, govern the relaxation process of the observer gaining more information about the process with longer measurement sequences.

A measure of information relaxation on finitary machines is given by the length-dependent finitary complexity $C_\mu(t) = H(p_\mathbf{V}(t))$ where $H(P)$ is the Shannon entropy of the distribution P and $p_\mathbf{V}(t)$ is the probability distribution at time t beginning with the initial distribution $p_\mathbf{V}(0) = (1, 0, 0, \ldots)$ concentrated on the start state. The latter distribution represents the observer's state of total ignorance of the process's state, i.e., before any measurements have been made, and correspondingly $C_\mu(0) = 0$. The length-dependent complexity $C_\mu(t)$ is simply (the negative of) the Boltzmann H-function in the present setting. And we have the analogous result to the H-theorem for stochastic ϵ-machines: $C_\mu(t)$ is monotonically increasing. Furthermore, the observer has the maximal amount of information about the process, i.e., the observer's knowledge is in equilibrium with the process, when $C_\mu(t+1)) - C_\mu(t)$ vanishes for $t > t_{\text{lock}}$. Generally, we have $C_\mu(t) \to_{t \to \infty} C_\mu$. That is, the length-dependent complexity limits on the metric complexity.

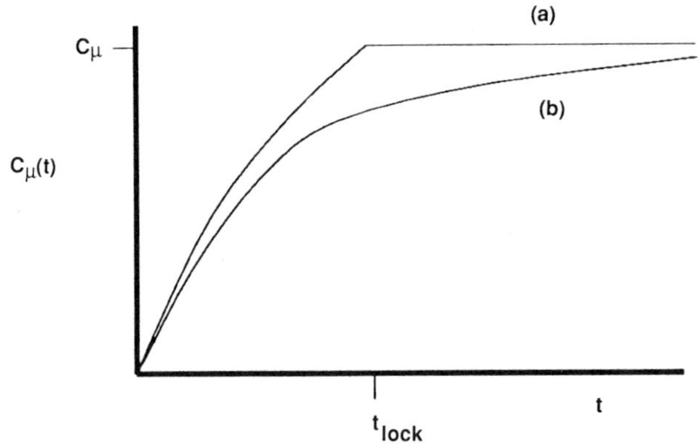

Figure 1: Temporal convergence of length-dependent complexity $C_\mu(t)$ for (a) periodic and Markovian (subshifts of finite type) processes and (b) for strictly Sofic systems.

For finitary machines there are two convergence behaviors for $C_\mu(t)$: finite time convergence for periodic and Markovian (subshift of finite type: SSFT) processes and asymptotic convergence for strictly Sofic systems (SSS) [20]. These are illustrated in Figure 1. The effect of extrinsic noise on this type of convergence process was studied some time ago. It was found that extrinsic noise induces an effective Markovian process [25].

Periodic and SSFT processes converge in finite time. SSSs asymptote but do not converge in finite time due to the relaxation of the initial distribution through an infinite number of Cantor sets. Through a structural analysis we see that the difference $\Delta C_\mu(t) = C_\mu - C_\mu(t)$ is a measure of the amount of information in the transient states. For SSSs this quantity only asymptotically vanishes since there are transient cycles that persist for all time, even though their probability decreases asymptotically. This leads to a general definition of (chaotic or periodic) phase locking. We say that the observer has β-locked onto the process when $\Delta C_\mu(t_{lock}) < \beta$; this occurs at the locking time t_{lock}. When the process is periodic, this notion of locking is the standard one of engineering. But it also applies to chaotic processes and corresponds to the observer knowing what state the process is in, even if the next measurement cannot be predicted exactly.

14. Meaning

Enough structure has been developed up to this point to introduce a quantitative definition of the meaning of an observation. The meaning of a message, of course, depends on the context in which its information is made available. If the context is inappropriate the observation will have no basis with which to be understood. It will have no meaning. If appropriate, then the observation will be "understood". And if that which is understood, i.e., the content of the message, is largely unanticipated then the observation will be more significant than a highly likely, "obvious", message.

In the present discussion context is set by the model held by the observer at the time of a measurement. To take an example, assume that the observer is capable of modeling with respect to dynamic states. And, in particular, the observer has estimated a stochastic finite automaton and has been following the process sufficiently long to know the current state with certainty. Then at a particular time the observer measures symbol $s \in \mathbf{A}$. If that measurement forces a disallowed transition, then it has no meaning. Indeed, the response is for the observer to reset the machine to the initial state of its ignorance. If, however, the measurement is allowed, i.e., it is anticipated, then the amount of meaning is $-\log p_{\rightarrow_s v}$. Here $\rightarrow_s v$ denotes the machine state to which the measurement brings the observer's knowledge of the process's state and $p_{\rightarrow_s v}$ is the corresponding morph's probability. The meaning itself, i.e., the content of the measurement, depends on the morphs to which the model's states correspond.

Naturally, similar definitions of meaning can be developed between any two levels in a reconstruction hierarchy. Here we wish to emphasize the main components of meaning, as we have defined it: (i) it is an information-like quantity and (ii) it derives fundamentally from the relationship *across* levels. A given message has different connotations depending on the level. Meaning appears as a change in connotation. This definition answers a question posed previously [30].

15. Infinitary Machines

Recall that it is possible to infer machines with an infinite number of states or morphs from an infinite tree representation. As in the transition from an infinite tree representation to a finite machine, we can ask for a measure that indicates that the finitary machine representation is inadequate. Observing that breakdown we also look for some regularity in the infinite "finitary" machine that we can factor out.

The measure we use for this innovation step plays the same role as entropy in going from trees to finitary machines. We look at the growth rate at the total information contained in the representation [21, 22] $c_\alpha = \lim_{L\rightarrow\infty} L^{-1} 2^{C_\alpha(L)}$. In the topological case, we consider the growth rate of the number of states $c_0 = \lim_{L\rightarrow\infty} L^{-1} \|\mathbf{V}(L)\|$. There is also an information gain version of this:

$$\begin{aligned} c_0 &= \lim_{L\rightarrow\infty} \left[2^{C_\alpha(L)} - 2^{C_\alpha(L-1)}\right] \\ &= \lim_{L\rightarrow\infty} \sum_{v\in\mathbf{V}(L-1),v'\in\mathbf{V}(L)} p_v \log \frac{p'_v}{p_v} \end{aligned}$$

In the last form, $v = \{v'\}/\sim_{L-1}$ is a distinct state at length $(L-1)$ that splits into states $\{v'\}$ at length L reconstruction. When this growth rate is positive then another higher level representation is called for. In this case we have register machines that perform string productions in registers whose lengths grow monotonically. Register machines are inferred from morphs consisting of regular patterns of state transitions in the "infinite" finitary machine [21, 22].

We should also note one of the highest levels, that of the universal Turing machine (UTM). As already noted, the more powerful the representation the more difficult it is to estimate minimal models. At the level of UTMs there is no general procedure for inferring the minimal program from a given data stream. An important difference with our approach is that we are, in effect, considering a class of computations based on deterministic Turing machines with access to a random register: the random oracle. This is the Bernoulli-Turing machine (BTM) [21]. Modeling with respect to BTMs trades off deterministic computation against random guessing. Thus, at every level of the hierarchy, very regular and highly random processes are simply described and so have low complexity.

16. Concluding Remarks

This discussion has explored the idea of hierarchically reconstructing models by focusing of the particular example of moving from a data stream to a tree representation to a finitary machine representation, and finally to a string register machine representation. Each level has an equilibrium statistical mechanics that indicates via a minimization, or variational, principle what the optimal model at each level is. In going between levels the significant measures are growth rates of complexity, i.e., the growth of information contained in a representation. There is a statistical metamechanics in which the innovation from infinite representations at one level lead via a condensation or phase transition to a finite representation at a higher level. We gave a quantitative definition of meaning that is appropriate to the type of incremental learning considered here.

Hierarchical reconstruction hints at how to approach complexity on much larger scales. It indicates some of the necessary structure of evolutionary processes that appear to play off randomness and order. Some portion of natural language undoubtedly manifests such a developmental tension: the need for structural regularity, for example, in order to support communication between individuals and the need for a diverse and rich language to support expressiveness and specificity.

This essay has reviewed the attempt to weave together the structural framework of computation and formal language theories and the combinatorial notions of information theory into an approach to modeling periodic and chaotic nonlinear dynamical systems. The tools ultimately rely on semigroup theory, symbolic dynamics, statistical mechanics, and ergodic theory. Even Bayesian statistics appears in the parameter estimations and in the thermodynamic analogies. The statistical mechanics of inductive inference and learning appear to give a unified framework for these very different disciplines. Similar approaches to learning in neural networks have been developed recently [31].

We close by reframing the question posed at the beginning concerning the naturalness of the drive toward higher complexity: Is it a physical property? Or is it the figment of an organism's need to model its environment? These are captured in the single philosophical question [32], where does intentionality lay? From the present discussion's radical and shameless mechanistic bias it is difficult to see how intentionality and the complexity drive could not be at root a property of physical nature. The difficulty in understanding this remains, of course, in posing the question in a scientific manner that does not strip it of its philosophical content. We continue to wonder ... from what does the anticipation of knowledge spring?

Acknowledgements

The author thanks Karl Young and the participants of the seminar on Foundational Issues in Cognitive Science for useful discussions and Professor Carson Jeffries for his continuing support. This work was funded in part by ONR contract N00014–86-K-0154.

References

1. J.P. Crutchfield and K. Young. Inferring statistical complexity. *Phys. Rev. Lett.* **63**, 105 (1989).
2. C.E. Shannon and W. Weaver. *The Mathematical Theory of Communication.* University of Illinois Press, Champaign-Urbana, 1962.
3. A.N. Kolmogorov and V.M. Tikhomirov. ϵ-entropy and ϵ-capacity of sets in function spaces. *Usp. Math. Nauk.* **14**, 3 (1959) (*Math. Rev.* 22, No.2890).
4. R.J. Solomonoff. A formal theory of inductive control. *Info. Control* **7**, 224 (1964).
5. A.N. Kolmogorov. Three approaches to the concept of the amount of information. *Prob. Info. Trans.* **1**, 1 (1965).
6. G. Chaitin. On the length of programs for computing finite binary sequences. *J. ACM* **13**, 145 (1966).
7. J.G. Kemeny. The use of simplicity in induction. *Phil. Rev.* **62**, 391 (1953).
8. E.T. Jaynes. Where do we stand on maximum entropy? In *Essays on Probability, Statistics, and Statistical Physics.* Ed. E.T. Jaynes. Reidel, London, 1983, p.210.
9. J. Rissanen. Stochastic complexity and modeling. *Ann. Statistics* **14**, 1080 (1986).
10. J.P. Crutchfield and B.S. McNamara. Equations of motion from a data series. *Complex Systems* **1**, 417 (1987).
11. D. Angluin and C.H. Smith. Inductive inference: theory and methods. *Comp. Surveys* **15**, 237 (1983).
12. J.P. Crutchfield and N.H. Packard. Symbolic dynamics of one-dimensional maps: Entropies, finite precision, and noise. *Int. J. Theor. Phys.* **21**, 433 (1982).
13. J.P. Crutchfield. Noisy chaos. PhD Thesis, University of California, Santa Cruz. Published by University Microfilms Intl., Minnesota, 1983.
14. R.E. Blahut. *Principles and Practice of Information Theory.* Addision-Wesley, Reading, 1987.
15. A. Rènyi. On the dimension and entropy of probability distributions. *Acta Math. Hung.* **10**, 193 (1959).
16. D.M. Cvetkovic, M. Doob and H. Sachs. *Spectra of Graphs.* Academic Press, New York, 1980.
17. R. Fischer. Sofic systems and graphs. *Monatsh. Math.* **80**, 179 (1975).
18. J.E. Hopcroft and J.D. Ullman. *Introduction to Automata Theory, Languages, and Computation.* Addison-Wesley, Reading, 1979.
19. N. Chomsky. Three models for the description of language. *IRE Trans. Info. Th.* **2**, 113 (1956).
20. J.P. Crutchfield and K. Young. Finitary ϵ-machines. In preparation, 1989.
21. J.P. Crutchfield and K. Young. Computation at the onset of chaos. In *Entropy, Complexity, and the Physics of Information.* Ed. W. Zurek. Addison-Wesley, Reading, 1990.
22. J.P. Crutchfield and E. Friedman. Language cascades: A universal structure at the onset of chaos. Preprint, 1990.

23. D. Huffman. The synthesis of sequential switching circuits. *J. Franklin Inst.* **257**, 161, 275 (1954).

24. J.P. Crutchfield and N.H. Packard. Noise scaling of symbolic dynamics entropies. In *Evolution of Order and Chaos.* Ed. H. Haken. Springer, Berlin, 1982, p.215.

25. J.P. Crutchfield and N.H. Packard. Symbolic dynamics of noisy chaos. *Physica* **7 D**, 201 (1983).

26. P. Grassberger. Toward a quantitative theory of self-generated complexity. *Int. J. Theor. Phys.* **25**, 907 (1986).

27. J.P. Crutchfield. Inferring the dynamic, quantifying physical complexity. In *Measures of Complexity and Chaos.* Eds. N.B. Abraham, A.M. Albano, A. Passamante, and P.E. Rapp. Plenum, New York, 1989, p.327.

28. S. Wolfram. Computation theory of cellular automata. *Commun. Math. Phys.* **96**, 15 (1984).

29. R. Rammal, G. Toulouse, and M.A. Virasoro. Ultrametricity for physicists. *Rev. Mod. Phys.* **58**, 765 (1986).

30. J.P. Crutchfield. Information and its metric. In *Nonlinear Structures in Physical Systems - Pattern Formation, Chaos and Waves.* Eds. L. Lam and H.C. Morris, Springer, Berlin, 1989.

31. E. Levin, N. Tishby, and S.A. Solla. A statistical approach to learning in layered neural networks. In *Computational Learning Theory.* Eds. R. Rivest, D. Haussler, and M.K. Warmuth. Morgan Kaufmann, San Mateo, California, 1989, p. 245.

32. D.C. Dennett. *The Intentional Stance.* MIT Press, Cambridge, 1987.

A THEORY OF STATE SPACE RECONSTRUCTION IN THE PRESENCE OF NOISE

Martin Casdagli, Stephen Eubank,
J. Doyne Farmer, and John Gibson

Theoretical Division and
Center for Nonlinear Studies
Los Alamos National Laboratory
Los Alamos, NM 87545, USA

and

Santa Fe Institute
1120 Canyon Road
Santa Fe, NM 87501, USA

Abstract

Takens' theorem demonstrates that in the absence of noise a multidimensional state space can be reconstructed from a single time series. This theorem does not treat the effect of noise, however, and so gives no guidance about practical considerations for reconstructing a good state space. We study the problem of reconstructing a state space with observational noise, examining the likelihood for a particular state given a series of noisy observations. We define a quantity called the *distortion*, which is proportional to the covariance of the likelihood function. This is related to the *noise amplification*, which corresponds to the root-mean-square errors for time series prediction with an ideal model. We prove that in the low noise limit minimizing the distortion is equivalent to minimizing the noise amplification.

We derive several asymptotic scaling laws for distortion and noise amplification. They depend on properties of the state space reconstruction, such as the sampling time and the reconstruction dimension, and properties of the dynamical system, such as the dimension and Lyapunov exponents. When the dimension and Lyapunov exponents are sufficiently large these scaling laws show that, no matter how the state space is reconstructed, there is an explosion in the noise amplification – from a practical point of view all determinism is lost, even for short times, so that the time series is effectively a random process.

In the low noise, large data limit we show that the technique of local principal value decomposition (PVD) is an optimal method of state space reconstruction, in the sense that it achieves the minimum distortion in a state space of the lowest possible dimension.

1. Introduction

There are many situations in which we observe a *time series* $\{x(t_i)\}, i = 1, \ldots, N$ which we believe to be at least approximately described by a d-dimensional dynamical system[1] f:

$$s(t) = f^t s(0) \tag{1}$$

The time series is related to the original dynamical system by:

$$x(t) = h(s(t)) \tag{2}$$

We call h the *measurement function.* The observed time series $x(t)$ is D-dimensional, so that $h : \Re^d \to \Re^D$. We are most interested in dimension-reducing measurement functions, where $D < d$, and we will often implicitly assume $D = 1$.

The state space reconstruction problem is that of recreating states when the only information available is contained in a time series. A schematic statement of the problem of reconstructing a state space is given in Figure 1.

State space reconstruction is necessarily the first step that must be taken to analyze a time series in terms of dynamical systems theory. Typically f and h are both unknown, so that we cannot hope to reconstruct states in their original form. However, we may be able to construct a state space that is in some sense equivalent to the original. This state space can be used for qualitative analysis, for example to construct a phase plot or one dimensional map, or for quantitative statistical characterizations, such as fractal dimension, Lyapunov exponents, or the eigenvalues of fixed points. We are particularly interested in state space reconstruction as it relates to the problem of nonlinear time series forecasting, a subject that has received considerable attention in the last few years [1-10].

State space reconstruction was introduced into dynamical systems theory independently by Packard *et al.* [11], Ruelle[2], and Takens [12]. In fact, in time series analysis this idea is quite old, going back at least as far as the work of Yule in 1927 [13]. The important new contribution made in dynamical systems theory was the demonstration that it is possible to preserve geometrical invariants, such as the eigenvalues of a fixed point, the fractal dimension of an attractor, or the Lyapunov exponents of a trajectory. This was demonstrated numerically by Packard *et al.* and was proven by Takens.

The basic idea behind state space reconstruction is that the past and future of a time series contain information about unobserved state variables that can be used to

[1]This is one of several possible ways of representing a dynamical system. f^t is the map that takes an initial state $s(0)$ to a state $s(t)$. The time variable t can be either continuous or discrete. f^t is sometimes called the *time-t map* of the dynamical system.

[2]Private communication.

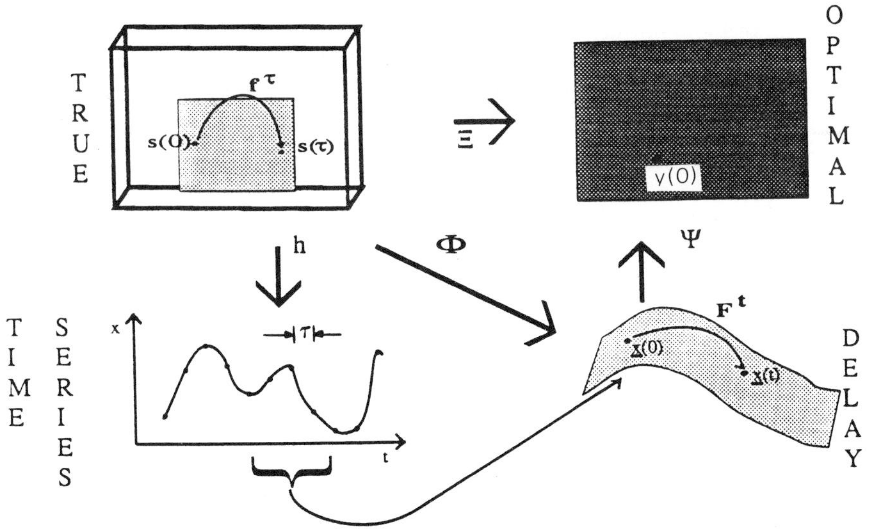

Figure 1: The true dynamical system f, its states s, and the measurement function h are unobservables, locked in a black box. Values of the time series x separated by intervals of the lag time τ form a delay vector $\underline{x}$ of dimension $m = 1 + m_+ + m_-$, where m_+ is the number of values from the future and m_- is the number of values from the past. Φ maps the original d-dimensional state s into the delay vector $\underline{x}$. Ψ further maps the delay vector $\underline{x}$ into a new state y, of dimension $d' \leq m$.

define a state at the present time. The past and future information contained in the time series can be encapsulated in a *delay vector*[3]:

$$\underline{x}(t) = (x(t + \tau m_+), \ldots, x(t), \ldots, x(t - \tau m_-)) \tag{3}$$

The dimension of the delay vector is $m = 1 + m_- + m_+$. The number of samples taken from the past is m_-, and the number taken from the future is m_+. If $m_+ = 0$ then the reconstruction is *predictive*; otherwise it is *mixed.* The time separation between coordinates, τ, is the *lag time.*

Takens studied the reconstruction map Φ, which maps the states of a d-dimensional dynamical system into m-dimensional delay vectors:

$$\underline{x} = \Phi(s) = h(f^{\tau m_+}(s)), \ldots, h(s), \ldots, h(f^{-\tau m_-}(s)) \tag{4}$$

He showed that generically Φ is an embedding when $m \geq 2d + 1$. An *embedding* is a smooth, one-to-one coordinate transformation with a smooth inverse. When Φ is an embedding, $\Phi(\Re^d)$ is diffeomorphic to $\Re^d$.

If Φ is an embedding then a smooth dynamics F is induced on the space of reconstructed vectors:

$$F^t(\underline{x}) = \Phi f^t(\Phi^{-1}(\underline{x})) \tag{5}$$

The reconstructed states can be used to estimate F. F is equivalent to the original dynamics f, and we can use it for any purpose that we could use the original dynamics, such as prediction, computation of dimension, fixed points, etc.

[3]For convenience we assume that the sampling time is uniform.

Takens' proof is important because it gives a rigorous justification for state space reconstruction. However, it gives little guidance on reconstructing state spaces from real-world, noisy data. For example, the measurements $x(t)$ in the proof are arbitrarily precise, resulting in arbitrarily precise states. This makes the specific value of the lag time τ arbitrary[4], and any reconstruction is as good as any other. However in practice, the presence of noise in the data blurs states and makes picking a good lag time critical. Our work "fleshes out" Takens' proof, by examining how states are affected when conditions such as arbitrary precision are relaxed.

There are several such factors which complicate the reconstruction problem for real-world data:

1. *Observational noise.* The measuring instruments are noisy; what we actually observe is $x(t) = \tilde{x}(t) + \xi(t)$, where $\tilde{x}(t)$ is the true value and $\xi(t)$ is noise.
2. *Dynamical noise.* External influences perturb s, so that from the point of view of the system under study the evolution of s is not deterministic. f is thus a *stochastic* dynamical system.
3. *Estimation error.* f and h are both unknown. They can be estimated, but with a finite amount of data some uncertainty remains.

In real problems noise is always present. When we project a d-dimensional state onto a D-dimensional measurement with $d > D$, we throw away information. We can reconstruct some of this missing information from the past and future measurements. However, if the uncertainty of the reconstructed state is much higher than that of the individual measurements, then we have amplified the noise; the system is not as deterministic as it would be if we could observe more information.

State space reconstruction relies on a flow of information from the unobserved variables of the system to the observed variables. This can be qualitatively illustrated with the familiar Lorenz equations:

$$
\begin{aligned}
\frac{dx}{dt} &= 10(y - x) \\
\frac{dy}{dt} &= -xz + 28x - y \\
\frac{dz}{dt} &= xy - \frac{10}{3}z
\end{aligned}
\tag{6}
$$

Assume that we observe x. Since $\frac{dx}{dt}$ does not depend on z directly, information about z depends on the flow of information through y; when z changes it causes $\frac{dy}{dt}$ to change, which causes y and hence $\frac{dx}{dt}$ to change. When $x \approx 0$, since the only coupling is through the xz term a large change in z causes only a small change in x. Equivalently, a small change in x corresponds to a large change in z. Thus noise in the determination of z from measurements of x is acutely amplified when $x \approx 0$.

When noise is present, state space reconstruction becomes a problem in statistical estimation. The formalism that we develop in this paper makes the notion of noise amplification more precise, so that the qualitative analysis of the Lorenz equations in the previous paragraph becomes quantitative. It also provides guidance into the practical problem of reconstructing coordinates so that they minimize noise amplification.

---

[4]Provided it meets the conditions for genericity.

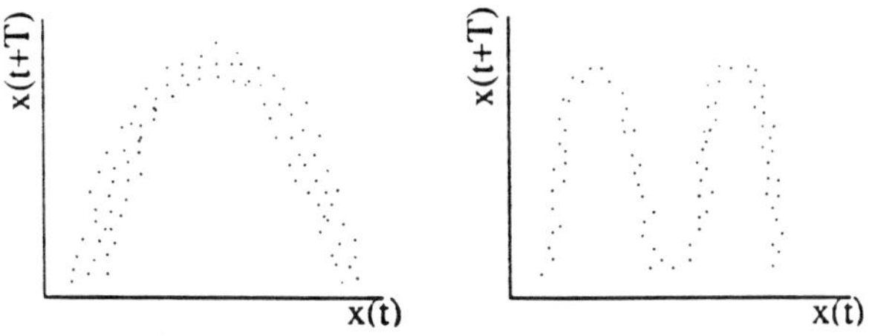

Figure 2: Two hypothetical scenarios for prediction in a one dimensional state space. The horizontal axis is the state at time t, and the vertical axis is the state at time $t+1$ in the future. (a) shows a coordinate system with high noise amplification, while (b) shows a coordinate system with low noise amplification. This is evident from the "thickness" of the distribution of points at any given y. However, since the functional form of (b) is more complicated, with a limited amount of data (b) might result in larger estimation errors than (a).

Noise amplification depends on three factors:

- *The measurement function.*
  One observation may give more information than another.
- *The method of reconstruction.*
  A poor state space reconstruction amplifies noise more than a good state space reconstruction; noise amplification depends on factors such as m and τ.
- *The dynamical system.*
  Noise amplification depends on the flow of information between the individual degrees of freedom, which depends on properties of the dynamical system such as the dimension and Lyapunov exponents.

In assessing predictability it is important to distinguish between estimation error and noise amplification. Figure 2 shows a hypothetical comparison of two prediction problems in the idealized case of a one dimensional state space. The noise amplification is related to the "thickness" of the distribution of points. In Figure 2a the noise amplification is large, and in Figure 2b the noise amplification is small. However, the estimation error in (b) might be larger than that of (a).

Both noise amplification and estimation error cause prediction errors. The estimation error depends on the procedure used to approximate the dynamics. Noise amplification, however, does not. It sets a limit to predictability that is independent of the modeling procedure. In the limit of a large number of data points for most good approximation schemes the estimation error goes to zero. The prediction errors in this limit are given purely by the noise amplification. The noise amplification thus tells us the prediction errors that remain even with a perfect model.

As we shall show, when the dimension and Lyapunov exponents are sufficiently large there can be a complete breakdown of predictability, so that even with a perfect model the time series is unpredictable, even for short times. This is the limit in which a time series becomes a true random process.

Any approach to state space reconstruction uses the information in delay coordinates as a starting point. For some purposes, such as reducing the dimension, it may be desirable to make a further coordinate transformation to a new coordinate system y:

$$y = \Psi(\underline{x}) \tag{7}$$

As described in Section 2, examples of such transformations Ψ are differentiation or the singular value decomposition in PVD. By splitting the reconstruction process into Φ and Ψ, we have conveniently labeled the two parts of the problem. The choice of Φ determines the form of the delay coordinates, which are the raw information we have to work with, while Ψ determines how we use that information. The total reconstruction map $\Xi = \Psi \circ \Phi$ takes the original coordinates s to the reconstructed coordinates y.

We show here that it is impossible to reduce the noise amplification by transforming delay coordinates by Ψ. The minimum possible noise amplification over all Ψ is obtained when $\Psi = \mathbf{1}$ and $y = \underline{x}$. However, it *is* in general possible to compress all the information in $\underline{x}$ into a coordinate y with a lower dimension while keeping the noise amplification the same. This can be desirable for reducing statistical estimation errors, which typically increase with dimension (often exponentially). The local principal value decomposition technique discussed in Section 6 accomplishes this in the minimum possible dimension.

1.1. *Approach and Overview*

The main goal of this paper is to develop a theory which gives insight into practical problems of state space reconstruction in the typical case in which a times series is the only available information. In order to get insight into the problem and develop a theory for its solution we begin by assuming that we know both f and h. In Sections 3 through 6 we develop an understanding of the effect that f and h have on the problem. We are currently investigating the implications of these theoretical results for the case when f and h are unknown, and will report the results in a future publication.

Throughout this paper we assume that the noise is entirely observational. Treating dynamical noise is obviously important, but it is outside the scope of this paper. However, we suspect that many aspects of the framework we have established here can be used to analyze dynamical noise as well.

In Section 2 we review what is currently known about state space reconstruction. We begin by discussing methods currently available for state space reconstruction, such as delay coordinates, derivative coordinates, and principal value decomposition. We then review extensions to Takens' theorem, and present an intuitive discussion of why it is true.

In Section 3 we derive formulas for the likelihood function and compute it for several examples. We use color graphics to develop intuition and to qualitatively illustrate what factors are necessary in order to obtain a good state space reconstruction.

From a practical point of view it is important to have simple criteria for selecting a reconstruction. The likelihood function gives a complete description of a reconstruction, but it is too complicated; we need a number, or a set of a few numbers. In Section 4 we examine several candidates and argue that for this problem criteria based on the variance are more appropriate than other possibilities, such as mutual information. We define two quantities based on the variance, the distortion, which is related to mean-square errors in the state space, and noise amplification, which is related to errors in time series prediction. We discuss the relationship between distortion and noise amplification, showing that minimizing one is equivalent to minimizing the other.

In Section 5 we discuss the problem of deciding how much noise amplification to expect in a given situation. We demonstrate that for a given τ noise amplification is a monotonically non-increasing function of m. We derive the behavior of the noise amplification as a function of m, τ, d, and the Lyapunov exponents, testing our conclusions on a simple example. We show that for predictive coordinates an explosion in the noise amplification occurs when the Lyapunov exponents and dimension are sufficiently large, which causes a transition from behavior that is approximately deterministic for short times to behavior that is effectively random over any time scale. Finally, we use two examples to illustrate several aspects of the behavior of the noise amplification.

In Section 6 we study the effect of making further coordinate transformations to delay coordinates. We demonstrate that in the low noise, large data limit, local principal value decomposition is an optimal state space reconstruction method in the sense that it minimizes the noise amplification with a coordinate system of the smallest possible dimension.

1.2. *Summary of Notation*

The notation we use in this paper is summarized in Table 1.

Table 1: Notation used in this paper

Symbol	Description
$s(t)$	d-dimensional state at time t
f^t	time-t map of dynamical system; $s(t) = f^t(s(0))$
$x(t)$	D-dimensional value of time series at time t. (We often assume $D = 1$.)
h	measurement function $x(t) = h(s(t))$
$S(t)$	$(d - D)$-dimensional measurement surface $S(t) = \{s : x = h(s)\}$
τ	sampling time $t_{i+1} - t_i$
$\underline{x}$	m-dimensional delay vector $(x(t + \tau m_+), \ldots, x(t), \ldots, x(t - \tau m_-))$
y	reconstructed d'-dimensional coordinate based on $\underline{x}$
Φ	delay reconstruction map $\underline{x} = \Phi(s)$
Ψ	map taking delay vector to new coordinate $y = \Psi(\underline{x})$
Ξ	total reconstruction map $\Xi = \Psi \circ \Phi$
$\xi(t)$	noise fluctuation, usually assumed to be Gaussian IID
$\underline{\xi}$	m-dimensional vector of noise fluctuations $(\xi(t + \tau m_+), \ldots, \xi(t), \ldots, \xi(t - \tau m_-))$
$\tilde{\underline{x}}$, $\tilde{s}$	true values of $\underline{x}$, s in absence of noise
$\hat{s}$, $\hat{x}$, $\hat{f}$	best estimate for s, x, f
p	probability density function (identified by its arguments)
$p(x\|y)$	probability density for x given y
Σ	distortion matrix
δ^2	Trace of Σ

2. Review of Previous Work

2.1. *Current Methods of State Space Reconstruction*

The currently used possibilities for state space reconstruction include delay coordinates, derivative coordinates, and global PVD. Each of these is sometimes done in conjunction with filtering. As a matter of experience it is quite clear that the method of reconstruction can make a big difference in the quality of the resulting coordinates, but in general it is not clear which method is the best.

Delay coordinates are currently the most widely used choice. They have the nice property that the statistical properties of each dimension are the same. They have the unpleasant property that in order to use them it is necessary to choose the delay parameter τ. If τ is too small each coordinate is almost the same, and the trajectories of the reconstructed space are squeezed along the identity line; if τ is too large, in the presence of chaos and noise the dynamics at one time become effectively causally disconnected from the dynamics at a later time, so that even simple geometric objects look extremely complicated. Most of the research on the state space reconstruction problem has centered on the problems of choosing τ and m for delay coordinates. The proposals for doing this include information-theoretic quantities [14 – 16], and others [17].

Another method for reconstructing a state space is the *method of derivatives*, originally investigated by Packard *et al.* [11]. The coordinates are derivatives of successively higher order:

$$\begin{aligned} y_1(t) &= x(t) \\ y_2(t) &= \hat{x}'(t) \\ &\vdots \\ y_m(t) &= \hat{x}^{(m-1)}(t) \end{aligned} \tag{8}$$

$\hat{x}^{(j)}(t)$ is a numerical approximation to the jth derivative of $x(t)$. As Takens proved, as long as m is sufficiently large derivatives define an embedding.

There are many different methods for the numerical computation of derivatives, so in this sense the method of derivatives actually defines a family of different methods, depending on the algorithm used. Straightforward methods of numerical differentiation act as a high pass filter, with a response function that is proportional to frequency. The quality of derivative coordinates in the presence of noise can be considerably improved by using a numerical algorithm that uses low pass filtering to balance the response function.

The other method in common use is *principal value decomposition*, also called *principal component analysis*, *factor analysis*, or *Karhunen-Loeve decomposition*. Broomhead and King originally proposed this for reconstructing a state space for chaotic dynamical systems [18]. The simplest way to implement the procedure is to compute the covariance matrix $C_{ij} = \langle x_i x_j \rangle_t$, and then compute its eigenvalues α_i. (x_i represents the ith coordinate of the delay vector $\underline{x}$; $\langle\rangle_t$ denotes a time average.) The eigenvectors of C_{ij} define a new coordinate system, which is a rotation of the original delay coordinate system. The eigenvalues are the average root-mean-square projection of the m-dimensional delay coordinate time series onto the eigenvectors. Ordering them according to size, the first eigenvector has the maximum possible projection, the second has the largest possible projection for any fixed vector orthogonal to the first, and so on.

We have recently shown that PVD coordinates are very closely related to (appropriately low pass filtered) derivative coordinates [19].

At this point there is no clear statement as to which of these methods is superior. Fraser has presented evidence for situations in which delay coordinates are superior to PVD [20]. However, we have observed examples where the opposite is true. The situation at this point is inconclusive, and it is not clear what causes one coordinate system to be better than another. One of our central motives for defining noise amplification is to compare different methods of state space reconstruction. This gives guidance for optimizing the parameters of a particular method, or for comparing two different methods.

Principal value, derivative, and delay coordinates are related to each other by linear transformations. However, the transformation from delay coordinates to the original coordinates is typically *non*linear. As Fraser has demonstrated [20], nonlinear coordinate transformations can be greatly superior. The method of local PVD, discussed in Section 6 implements a nonlinear coordinate transformation, which gives it the potential for better performance.

2.2. *Takens' Theorem Revisited*

In order to understand when delay vectors form an embedding, Takens investigated the equation $\underline{x} = \Phi(s)$. For a univariate time series ($D = 1$) this can be regarded as a set of m simultaneous nonlinear equations in d variables. The transformation Φ maps a d-dimensional surface into an m-dimensional space. If the surface $\Phi(s)$ contains no self intersections then there is a unique solution for s given any x, and Φ is an embedding. If Φ is sufficiently close to a linear transformation then this may be possible with $m = d$. In general, however, for a unique solution we must have $m > d$. Generically, when $2d \geq m$ a d-dimensional surface in m dimensions has self intersections on sets of dimension $2d - m$; when $2d < m$ generically it has no self intersections at all. The case when $d = 2$ and $m = 3$, for example, is shown in Figure 3; in this case there are typically self intersections along one dimensional curves.

When $m = d + 1$ the set of self intersections is typically of dimension $d - 1$, and Φ is an embedding almost everywhere. As m increases the dimension of the set of self intersections decreases, until finally when $m > 2d$ there are no self intersections at all. Thus, $m \geq 2d + 1$ *guarantees* that Φ is an embedding, it is *possible* that it will be an embedding with m as small as $m = d$. See reference [21] for a more complete discussion.

The reconstruction process can also be considered in terms of the constraint that each measurement causes in the original state space. This gives a more dynamical point of view, which turns out to be useful for visualization in higher dimensions, and particularly in the presence of noise.

Let the *measurement surface* $S(t)$ be the set of possible states that are consistent with a given measurement $x(t)$, i.e., $S(t) = \{s(t) : x(t) = h(s(t))\}$. When h is smooth $S(t)$ is a surface of dimension $d - D$. For example, when $d = 2$ and h is projection onto the horizontal axis, the measurement surfaces consist of horizontal lines. The effect of a series of measurements can be understood by transporting them to a common point in time. The state at that time must lie in their intersection $I(t)$:

$$s(t) \in I(t) = f^{-\tau m_+} S(t + \tau m_+) \cap \ldots \cap S(t) \cap \ldots \cap f^{\tau m_-} S(t - \tau m_-) \qquad (9)$$

The intersection $I(t)$ is never empty, since there must be at least one state consistent with all the measurements. If $I(t)$ does not consist of a single point, Φ is not an embedding. An example for the case when $d = 2$ and $m = 3$ is shown in Figure 4.

In most real situations f, h, and consequently Φ are unknown. Nonetheless, as long as there is a smooth one-to-one correspondence between the delay coordinate and the original state we know that there is an embedding, so that the delay coordinate $\underline{x}$ can be used in place of the original coordinate s.

3. Geometry of Reconstruction with Noise

The goal of reconstruction is to assign a state based on a series of measurements. With noise this task is considerably more difficult because the measurements are uncertain, and there are many states that are consistent with a given series of measurements. The probability that a given state occurred can be characterized by a conditional probability density function[5] $p(s|\underline{x})$. This illustrates how the presence of noise complicates the reconstruction problem: without noise a point is sufficient to characterize what is learned from a measurement, but with noise this requires a function giving the probability of all possible states. For chaotic dynamics the properties of $p(s|\underline{x})$ can be very complicated, as has been demonstrated by Geweke.

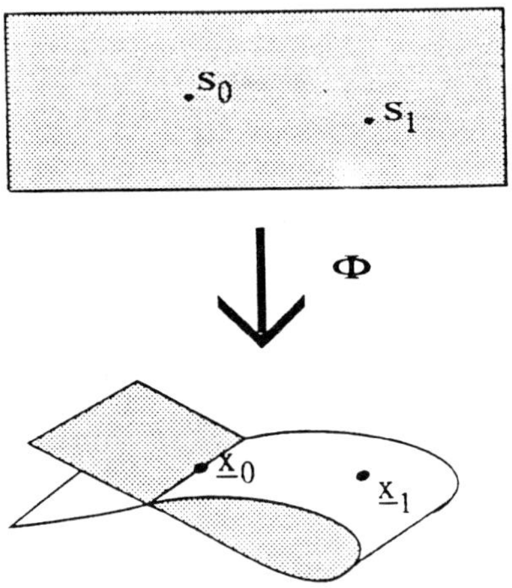

Figure 3: Solutions of the equation $\underline{x} = \Phi(s)$ when $d = 2$ and $m = 3$. There are typically self intersections along curves. For example, the state s_0 is mapped onto a self intersection, while s_1 is not. Except for special values of s like s_0, Φ defines an embedding.

---

[5] We use probability density functions rather than measures only because we want to keep the discussion accessible to the widest audience possible. All of the statements given here can be recast in more rigorous terms using measures.

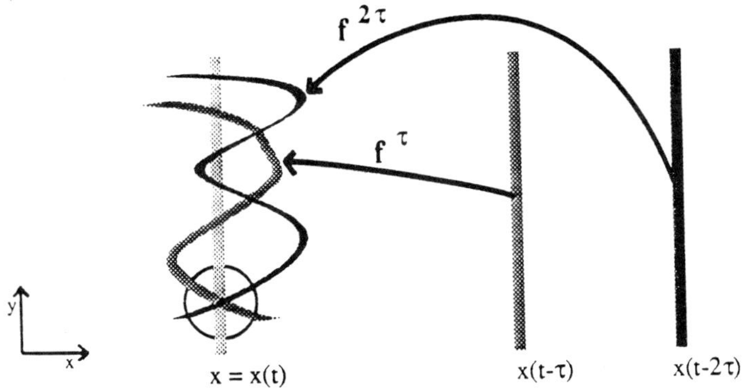

Figure 4: *A dynamical view of reconstruction in terms of the evolution of measurement surfaces.* Suppose that the measurement function h corresponds to projection onto the horizontal axis, so that $h(s) = x$. A measurement at time t implies that s lies somewhere along the light gray vertical line defined by $x = x(t)$. Similarly, a measurement at time $t - \tau$ implies that it was on the darker line $x = x(t - \tau)$, and a measurement at time $t - 2\tau$ implies that it was on the darkest line $x = x(t - 2\tau)$. To see what this implies when they are taken together, each curve can be mapped forward by f to the same time t. If their intersection is a not single point, then the reconstruction is not an embedding.

In this section we derive several formulas for $p(s|\underline{x})$ when h and f are known. We compute $p(s|\underline{x})$ for several examples, to illustrate qualitatively how it depends on $\underline{x}$ and on the properties of the reconstruction problem.

3.1. *The Likelihood Function*

We can derive $p(s|\underline{x})$ from Bayes' theorem, making use of the fact that $p(\underline{x}|s)$ is relatively simpler. According to the laws relating conditional and joint probability:

$$p(s|\underline{x})p(\underline{x}) = p(\underline{x}|s)p(s) \tag{10}$$

This can be rearranged as:

$$p(s|\underline{x}) \propto p(s)p(\underline{x}|s) \tag{11}$$

The factor $p(\underline{x}|s)$ on the right is often called the *likelihood function*, since it represents the likelihood of a given series of observations. The *prior* $p(s)$ encapsulates any information that we had before these observations occurred. If we are studying a chaotic attractor, for example, and we know its natural measure, then we can take this as our prior. If we have no prior knowledge, however, then this term can be taken to be constant. The term on the left represents what we know about s after taking the observations $\underline{x}$ into account, and is called the *posterior*.

When f and h are known we can write down a formula for the likelihood function. Assume the noise ξ is zero mean:

$$p(\underline{x}|s) = p(\underline{x} - \underline{\tilde{x}}) \tag{12}$$

where $\underline{\tilde{x}}$ is the "true" value of $\underline{x}$, in the absence of noise. It is related to the state s by $\underline{\tilde{x}} = \Phi(s)$. If we furthermore assume that the noise is IID (each fluctuation is statistically Independent, and Identically Distributed with probability p):

$$p(\underline{x}|s) = \prod_{i=-m_-}^{i=m_+} p(x(t+i\tau) - h(f^{\tau i}(s))) \tag{13}$$

3.1.1. *Gaussian Noise*

If we assume that $p(\xi)$ is a Gaussian of variance ϵ^2, this becomes:

$$p(\underline{x}|s) = \prod_{i=-m_-}^{i=m_+} \frac{1}{\sqrt{2\pi}\epsilon} \exp\left[-\frac{(x(t+i\tau) - \tilde{x}(t+i\tau))^2}{2\epsilon^2}\right] \tag{14}$$

If we assume the Euclidean norm, using the definition of Φ this can be rewritten as

$$p(\underline{x}|s) = A \exp\left[-\frac{1}{2\epsilon^2}\|\underline{x} - \Phi(s)\|^2\right] \tag{15}$$

where A is a normalization constant.

Thus, the probability for $\underline{x}$ given the true value of s is quite simple: it is an isotropic Gaussian centered on the true delay vector $\underline{\tilde{x}} = \Phi(s)$. The probability for s given $\underline{x}$, in contrast, is much more complicated; using Bayes' theorem (Eq. (11)) gives

$$p(s|\underline{x}) = A' p(s) \exp\left[-\frac{1}{2\epsilon^2}\|\underline{x} - \Phi(s)\|^2\right] \tag{16}$$

where A' is another normalization constant. Although this looks quite similar to Eq. (15), it is actually quite different, as it is interpreted as a function of s rather than $\underline{x}$. Because of the nonlinear function Φ, it is not a Gaussian.

3.1.2. *Uniform Bounded Noise*

Another case that is easily treated is that of uniform bounded noise:

$$p(\xi) = \begin{cases} \frac{1}{2\epsilon} & \text{if } |\xi| \leq \epsilon \\ 0 & \text{if } |\xi| > \epsilon \end{cases} \tag{17}$$

The effect of a given measurement can be visualized geometrically in terms of the *measurement strip* $S_\epsilon(t) = \{s : |x(t) - h(s)| < \epsilon\}$. The measurement strip is the support of p, and is similar to the measurement surface discussed earlier, except that it is "thickened" by ϵ. Following Eq. (13) the likelihood function can be computed in a manner analogous to Eq. (9). The state must lie inside the intersection of the measurement strips:

$$s(t) \in I_\epsilon(t) = f^{-\tau m_+} S_\epsilon(t + \tau m_+) \cap \ldots \cap S_\epsilon(t) \cap \ldots \cap f^{\tau m_-} S_\epsilon(t - \tau m_-) \tag{18}$$

The likelihood function is uniform over the domain defined by $I_\epsilon(t)$, and zero outside this domain. For an invertible dynamical system a simple tool for determining whether or not a given point $s \in I_\epsilon(t)$ is to test whether it satisfies the condition:

$$f^{\tau m_+}(s) \in S_\epsilon(t + \tau m_+) \wedge \ldots \wedge s \in S_\epsilon(t) \wedge \ldots \wedge f^{-\tau m_-}(s) \in \wedge S_\epsilon(t - \tau m_-) \tag{19}$$

3.1.3. *Chaotic Geometry*

We have performed several numerical experiments using the above formulae for likelihood functions, in particular for two dimensional dynamical systems such as the Ikeda map. This has allowed us to investigate the effects of varying the number of measurements, varying the noise level, and the effect of homoclinic tangencies. We have found the use of color graphics particularly helpful in exploring how these various effects interact with each other. We intend to illustrate these findings in a future publication.

4. Criteria for Optimality of Coordinates

To choose between different reconstructed coordinate systems we must first decide what we mean by "best". Since our primary interest is in prediction, in principle we could just use each coordinate system to estimate a dynamical system $\hat{f}$, and test to see which coordinate system gives the best predictions. In practice, however, this is cumbersome, since it is necessary to test many different statistical estimation procedures as well as many different coordinate systems. The errors would contain the effects of both estimation error and observational noise, and the result of such an experiment might give very little theoretical insight. Instead, we are going to begin by developing a formalism for understanding the effects of observational error alone, and we defer the problem of estimation error to a later publication.

The presence of noise forces us to think about the time series in probabilistic terms. The predictive value of a given reconstructed coordinate y is given by the conditional probability density of a future value of the time series given y:

$$p(x(T)|y(0)) \quad = \quad \text{probability} \quad x(T) \quad \text{given} \quad y(0) \tag{20}$$

$p(x(T)|y(0))$ tells us what we can predict about $x(T)$ given $y(0)$, assuming a perfect estimation procedure. This applies to either iterative or direct estimation procedures [1, 5]. Any criterion for evaluating the effect of observational noise must be based on $p(x(T)|y(0))$.

4.1. *Error Measures*

To choose between different coordinate systems, we must decide which property of $p(x(T)|y(0))$ to optimize. This amounts to deciding what we mean by the "best" predictions. There are several natural possibilities:

- *Maximum expectation.*
  We can view this as a gaming problem in which we must bet on the value of $x(T)$. We naturally want to maximize our expectation, which we can do by choosing the embedding which gives $p(x(T)|y(0))$ with the largest possible maximum value. However, since $x(T)$ is a continuous variable, we can never predict it exactly, and this is not well defined unless a coarse graining is specified. Also, this only optimizes the rate of return with an infinite bank.
- *Mutual information.*
  Let H represent the entropy function $H(x) = \int p(x) \log p(x) dx$. We can choose the coordinate system that maximizes the mutual information $I(x, y) = H(x) - H(x|y)$, where $H(x|y)$ is the entropy associated with the conditional probability

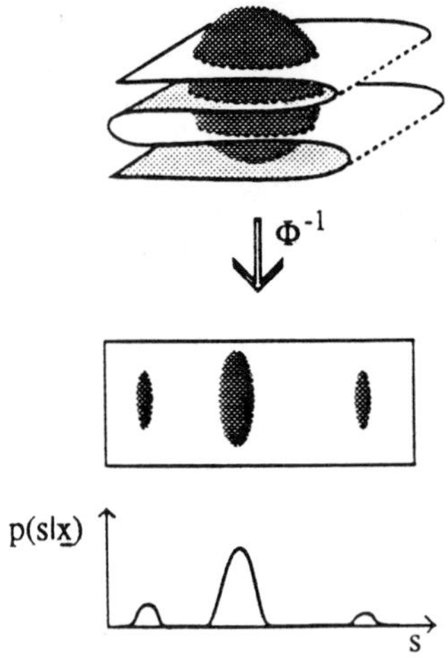

Figure 5: *The conditional probability of $s(t)$ given $\underline{x}(t)$.* Observational noise induces an uncertainty in the delay vector, here symbolized as a "noise ball". The noise ball, in turn, induces a conditional probability density for the true state given the delay vector. The variance of this density quantifies the quality of the embedding. Note that a sharp, multimodal density function can have a low entropy but a large variance.

density $p(x|y)$. Note that since $H(x)$ is fixed, this is equivalent to minimizing the conditional entropy $H(x|y)$. This criterion has the disadvantage that for continuous variables it may have little to do with prediction error. For example, a $p(x|y)$ with sharp peaks at different values of x may have a very low entropy but a high variance[6]. This makes the mutual information very insensitive to whether or not a reconstructed coordinate system forms an embedding. Several authors have looked at various criteria relating to mutual information [23, 14, 15, 20].

- *Variance.*
  The variance

$$\mathrm{Var}(x|y) = \int x^2 p(x|y)dx \tag{21}$$

is a lower bound on the mean-square prediction error $E(x - \hat{x})^2$ as follows (see [24] for details). It can be shown that the mean square prediction error is minimized by taking the predictor $\hat{x} = E(x|y)$. The prediction error of this ideal predictor is then $\mathrm{Var}(x|y)$, which is a lower bound for any other

[6]At any finite level of resolution, $x(T)$ may be thought of as a "message", with a corresponding number of bits, as originally proposed by Shaw [22, 23]. The entropy tells us our uncertainty in predicting this message. However, it weighs the low order bits equally with the high order bits. In predicting a continuous variable, however, an error in the highest order bit is usually much worse than one in the lowest order bit. The inability of the entropy to make this distinction makes it a poor measure of the quality of predictions.

predictor. For continuous variables this is a very natural way to evaluate the quality of predictions. Furthermore, it is easy to calculate analytically and estimate numerically. We shall use this measure.

- *Other measures*,
  such as mean-absolute error or the geometric mean error have the advantage that, compared to mean-square error, they do not emphasize outliers. Our choice of variance as compared to these other measures is primarily one of convenience.

4.2. *Noise Amplification*

The observational errors have a variance of ϵ^2. We define the *noise amplification at a given noise level* ϵ, σ_ϵ, as the ratio of the variance in the future value $x(T)$ given the present reconstructed coordinate $y(0)$, to the variance of the observational errors:

$$\sigma_\epsilon(T) = \frac{1}{\epsilon}\sqrt{\mathrm{Var}(x(T)|y(0))} \tag{22}$$

The quantity σ_ϵ has the advantage that it can be estimated directly from a time series[7], and so can be used as an operational test for the quality of a given set of coordinates.

The quantity σ_ϵ depends on both ϵ and $p(\xi)$. We can remove this dependence by assuming the noise is Gaussian and taking the limit as $\epsilon \to 0$. We will call this simply the *noise amplification* σ:

$$\sigma(T) = \lim_{\epsilon \to 0} \sigma_\epsilon(T) \tag{23}$$

This limit may not always exist; in particular, when the state space reconstruction is not an embedding it will tend to infinity. However we will show in Section 4.4, that for state space reconstructions that are embeddings the limit exists, and in the case of Gaussian noise, it depends solely on geometric factors, specifically, the dynamical system f, the measurement function h, the embedding dimensions m_+ and m_-, the state $y(0)$, and the prediction time T. In Section 4.6 we will illustrate how in some situations the limit depends on properties of an underlying attractor, and the realization of the noise perturbations. At small noise levels, σ can be used to provide an estimate of the "true" noise amplification σ_ϵ. Geometrically, σ_ϵ measures the "thickness" of Figure 2 in the vertical direction at the state $x(t) = y(0)$.

Taking the limit as the noise goes to zero is quite different from simply *setting* the noise to zero, as was effectively done by Takens [12]. When the noise is set to zero, all reconstructions that are embeddings are equivalent. In the *limit* as the noise goes to zero, however, two embeddings may have quite different noise amplifications.

If we are interested in a geometric object such as a chaotic attractor that has an ergodic measure, we can also eliminate the dependence on the state $y(0)$ by taking an averaging over the values of $y(0)$ with respect to this measure. We will call this the *average noise amplification*:

$$\langle\sigma\rangle = \langle\sigma\rangle_{y(0)} \tag{24}$$

[7]Disregarding the factor of $1/\epsilon$.

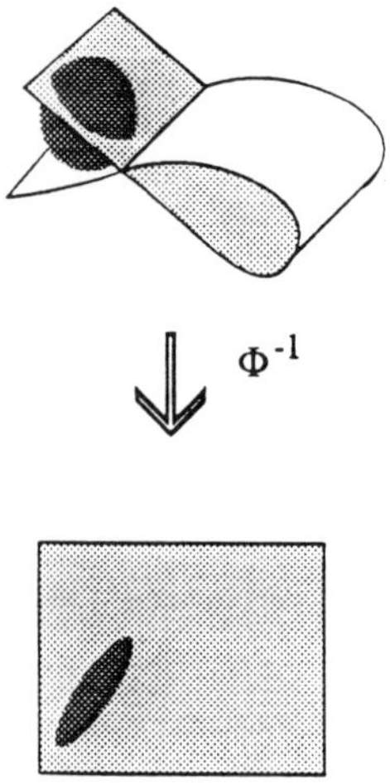

Figure 6: Since the errors in the time series are identically distributed, the probability density function for noisy delay coordinates is isotropic (top). This induces a non-isotropic distribution in the original space S (bottom).

4.3. *Distortion*

The noise amplification has the disadvantage that it depends on the time T. This problem can be overcome by defining a quantity Σ_ϵ that is related to noise amplification, called the ***distortion matrix at a given noise level*** ϵ as follows:

$$\Sigma_\epsilon = \frac{1}{\epsilon^2}\mathrm{Var}(s|y) \tag{25}$$

We define a related scalar quantity δ_ϵ called the ***distortion at a given noise level*** ϵ[8] by Eq. (26):

$$\delta_\epsilon = \frac{1}{\epsilon}\sqrt{\mathrm{Var}\ (\ \|s\|\ |\ y\ )} = \sqrt{\mathrm{trace}\ \Sigma_\epsilon} \tag{26}$$

Finally, by analogy with Eq. (23), we define the ϵ-independent quantities $\Sigma = \lim_{\epsilon\to 0} \Sigma_\epsilon$ and $\delta = \lim_{\epsilon\to 0}\ \delta_\epsilon$. The motivation for defining the distortion in this way comes from considering the geometrical effects of reconstruction. In delay coordinates the probability distribution corresponding to the noise is isotropic. For example, for Gaussian noise a surface on which the probability density function $p(x)$ is a constant is an m-dimensional sphere, as shown in Figure 6. Assuming that Φ is an embedding in a neighborhood of this sphere, in the low noise limit this sphere will map into a d-dimensional ellipsoid in the original state space S. The distortion Σ is a $d \times d$ symmetric real matrix, whose eigenvalues are proportional to the squares of the principal axes of this ellipsoid.

[8]The term "distortion" was originally used for another closely related quantity defined by Fraser [20].

The distortion has two disadvantages when compared with the noise amplification. First, it depends on the coordinates used to describe the dynamical system; for example, rescaling s changes the distortion. Second, it is not observable, and cannot be computed from a time series alone. Nonetheless, the distortion is a valuable tool for understanding the behavior of different coordinate reconstructions. This is because of some theoretical relationships between distortion and noise amplification which we will describe in Section 4.5. These results tell us that, in order to understand the effect of noise on the embedding process, it should be helpful to examine how the uncertainty in measurements of the time series translates into uncertainty in the original state space.

From a practical point of view the distortion may seem less relevant than the noise amplification. However there are some cases where we may know parametric forms for f and h, for example from scientific laws, and want to estimate the "hidden variables" s, or the unknown parameters of f and h from a noisy time series. Note that unknown parameters are just like dummy hidden variables. This is the problem in extended Kalman filtering, and has also been considered by Breiden and Hübler [25]. Then the distortion matrix is of direct interest in quantifying the uncertainties in the estimates for the hidden variables. For example, in the Lorenz system, we considered how accurately it was possible to infer the value of z given values for x.

4.4 *Low Noise Limit*

In the low noise limit, when Φ is an embedding the probability density $p(s|\underline{x})$ becomes much simpler. For example, take Eq. (15), which assumes Gaussian noise and a uniform prior. We can rewrite it as

$$p(\underline{x}|s) = Ae^{-\frac{Q(s)}{2\epsilon^2}} \tag{27}$$

where A is a normalization constant and $Q(s) = \|\underline{x} - \Phi(s)\|^2$. The most likely value of s (the maximum likelihood solution $\hat{s}$) occurs at the maximum value of $p(s|\underline{x})$, where $DQ(\hat{s}) = 0$. If Φ is an embedding, then there exists a unique maximum for sufficiently small noise levels. When ϵ is small, $p(s|\underline{x})$ is concentrated near its maximum, and it is possible to get a good approximation for $p(s|\underline{x})$ by expanding Q in a Taylor series about $\hat{s}$. To leading order in $s - \hat{s}$ this is[9]:

$$p(s|\underline{x}) \approx C \exp\left[-\frac{1}{2\epsilon^2}(s-\hat{s})^\dagger D\Phi^\dagger D\Phi(s-\hat{s})\right] \tag{28}$$

where the derivatives $D\Phi^\dagger$ and $D\Phi$ are evaluated at $s = \hat{s}$, and C is a normalization constant. By inspection, from the definition of the distortion matrix we see that:

$$\Sigma = (D\Phi^\dagger D\Phi)^{-1} \tag{29}$$

Note that if Φ is an embedding then $D\Phi$ is of full rank, and Σ is well defined. The uncertainty in the estimate of s is thus an anisotropic Gaussian centered on the maximum likelihood estimate $\hat{s}$. The distortion of this Gaussian is given by the

[9]To differentiate Q, we take advantage of the fact that it is of the form $Q = v^\dagger v$, where $v = \underline{x} - \Phi(s)$. Differentiating gives $DQ = Dv^\dagger v + v^\dagger Dv = 2Dv^\dagger v$, and $D^2Q = 2[D^2 v^\dagger v + Dv^\dagger Dv]$. But v is of order ϵ, so the dominant term is $D^2Q(\hat{s}) = 2D\Phi^\dagger D\Phi$.

eigenvalues of Σ. The smaller the eigenvalues of Σ, the more well defined the initial state.

Since Φ is the vector function whose components are $\Phi_i = h(f^{i\tau})$, according to the chain rule the components of the derivative are $D\Phi_i = DhDf^{i\tau}$. When the measurement function h is one dimensional, $D\Phi$ is the $m \times d$ matrix:

$$D\Phi = \begin{pmatrix} DhDf^{\tau m_+} \\ \vdots \\ Dh \\ \vdots \\ DhDf^{-\tau m_-} \end{pmatrix} \tag{30}$$

Since s is d-dimensional, as long as Φ is an embedding $D\Phi$ has d nonzero singular values. The inverse squares of these singular values are equal to the eigenvalues of Σ.

In control theory $D\Phi$ is called the *observability matrix*. A system is observable if the observability matrix has full rank, which is one of the conditions for Φ to be an embedding. Whether $D\Phi$ has full rank evidently depends on detailed properties of the coupling between variables in f, and on the measurement function h. For example if the dynamical system f has a representation such that it splits into two non-interacting subsystems, and the measurement function is a constant (for example zero) on one of the subsystems, then intuitively one would expect that this subsystem is unobservable. Indeed in such an example all the columns of the observability matrix corresponding to this subsystem are zero, and it is not of full rank. On the other hand if the measurement function depends on both subsystems, then, by Takens' theorem, generically full rank will be attained. We will consider such an example in Section 5.

Finally, we can use Eqs. (28) and (29) above to derive Eq. (31) for the noise amplification by transforming variables to $\tilde{x}(T) = h(f^T(s(0)))$. In the low noise limit we can take h and f to be approximately linear, so that a small variation of $\tilde{x}(T)$ about its mean value $\hat{x}(T)$ is $x(T) - \hat{x}(T) \approx DhDf^T(s(0) - \hat{s}(0))$. Then, since $s(0)$ has a Gaussian density with covariance matrix Σ, and covariance matrices transform under linear transformations L according to $\Sigma \rightarrow L\Sigma L^\dagger$, it follows that $x(T)$ (the *noisy* future observation) has a Gaussian density with variance:

$$\sigma(T) = 1 + DhDf^T\Sigma(Df^T)^\dagger Dh^\dagger \tag{31}$$

Intuitively this makes sense; the uncertainty in the initial state is first altered by the derivative of the dynamics, and then projected down onto the time series and finally convolved with noise. Also it is straightforward to verify that Eq. (31) is invariant with respect to the representation of the underlying state space dynamics and measurement function.

4.5. *Relation Between Noise Amplification and Distortion*

In general, when we observe a time series we cannot observe the original coordinates, and so it is impossible to compute the distortion from the time series. Fraser originally posed the question of whether or not it is possible to minimize the distortion of a reconstruction by using only the information available in a time series [20]. We demonstrate that this is indeed possible in the low noise limit, by demonstrating a relation between the distortion matrix Σ and the noise amplification $\sigma(T)$, which *is* computable from a time series.

Define an ordering on distortion matrices by $\Sigma_1 \leq \Sigma_2$ if $\Sigma_2 - \Sigma_1$ is positive semi-definite[10]. Consider the set of all reconstructions $y = \Psi(\underline{x})$, where $\Psi : \Re^m \rightarrow \Re^{d'}$ and m and d' are fixed. Then our result firstly states that if there exists a y^* such that $\Sigma(y^*) \leq \Sigma(y)$ for all y, then y^* will also satisfy $\sigma_T(y^*) \leq \sigma_T(y)$ for all y and T[11]. The converse is also true generically: any reconstruction y' that minimizes $\sigma(T)$ over Ψ for all T will also minimize Σ. Thus, since $\sigma(T)$ is an observable, in principle it can be minimized by finding a transformation that gives a simultaneous minimum for several different times. As we will see in Section 6, this is not as difficult as it might seem.

Derivation. We can use Eq. (31) to demonstrate that any reconstruction y^* that minimizes the distortion Σ will also minimize the noise amplification $\sigma(T)$ for any time T as follows. Let $y^* = \Psi^*(\underline{x})$. Then $v(\Sigma(y) - \Sigma(y^*))v^\dagger \geq 0$ for any d-dimensional vector v and any reconstruction y. By taking $v^\dagger = DhDf^T$, we have $\sigma(T, y) - \sigma(T, y^*) \geq 0$ for all T.

To demonstrate the converse, we proceed as follows. Let $y' = \Psi'\underline{x}$. As will be shown in Section 6, there exists a transformation y^* such that $\Sigma(y^*) \leq \Sigma(y)$ for all y. It suffices to show that $\Sigma(y') = \Sigma(y^*)$. But by definition of y' we know $\sigma(y', T) \leq \sigma(y^*, T)$ for all T, and by the the first part of the demonstration we know that $\sigma(y', T) \geq \sigma(y^*, T)$ for all T. It follows that $v_T^\dagger M v_T = 0$ for all T, where $v_T^\dagger = DhDf^T$, and $M = \Sigma(y') - \Sigma(y^*)$ is necessarily a positive semi-definite matrix. To complete the demonstration we must show that $M = 0$. Now transform to coordinates so that $M = diag(m_1, ..., m_d)$. We obtain a contradiction if one or more of the m_i are non-zero, because suppose (without loss of generality) that $m_1 > 0$. Then $v_T M v_T^\dagger \geq m_1 \|v_T^{(1)}\|^2 > 0$, where $v_T^{(1)}$ denotes the first component of v_T in the new coordinates. Note that there must generically exist a T such that $\|v_T^{(1)}\|^2 > 0$, because a finite subset of the vectors v_T make up an observability matrix of the form $D\Phi$, which by Takens' theorem is generically of full rank, so that the v_T span $\Re^d$.

4.6. *Numerical Example: The Lorenz Equations*

In this subsection we illustrate the above ideas using the Lorenz equations as an example.

4.6.1. *Low Noise Limit Distortion*

In Section 1 we gave an intuitive sketch of the flow of information between variables in the Lorenz system. We argued that when $x \approx 0$, the observations of x tell us little about z. The distortion makes this notion precise. To illustrate how the flow becomes restricted as x nears zero we numerically compute the distortion along a typical trajectory of the Lorenz attractor, using five dimensional delay coordinates with $m_+ = 0$ and $m_- = 4$, and $\tau = 0.01$ (by keeping the τ small, we guarantee that

[10]By definition a $d \times d$ matrix M is positive semi-definite if $v^\dagger M v \geq 0$ for all d-dimensional vectors v.

[11]In Section 6 we will show that such a y^* exists, is generically unique up to invertible coordinate transformations, and show how to compute it straightforwardly from f and h.

all the coordinates in the delay vector may be near zero simultaneously). Since the measurement function is projection onto the x axis, Dh is the row vector $(1,0,0)$. The derivative matrix $Df^{-i\tau}$ of the map associated with the Lorenz equations can be found by integrating the equations for the differentials, i.e., as is done in computing Lyapunov exponents for an ODE. For numerical stability, we are often forced to integrate forwards along an orbit segment, and we then use singular value decomposition to invert the resulting matrices. In order to visualize the distortion's x-dependence, we plot δ against the x coordinate, as shown in Figure 7. The graph is multi-valued, since δ depends on y and z as well as x.

To illustrate the dependence of the distortion on the time lag τ used, we arbitrarily fix a state $s = (-1.8867, -5.1366, 24.7979)$, and plot δ against τ. See Figure 8. We choose three different embedding dimensions as follows. The upper curve is for $m_+ = 0, m_- = 2$, and because of the low embedding dimension, there are singularities. The middle curve is for $m_+ = 0, m_- = 4$, and the singularities have vanished. Note also that as τ increases, there is very little advantage in using a higher embedding dimension. Intuitively, this is because the motion on the Lorenz attractor is chaotic, and measurements in the far past fail to give new information in the unstable direction. We will return to this topic in Section 5. Finally, the lower curve is for a non-predictive embedding with $m_+ = 5, m_- = 4$. Significant noise reduction has been achieved since future coordinates do give information in the unstable direction.

Note that in all three cases, the distortion blows up at $\tau = 0$. This is to be expected, since in this limit, measurements become redundant. In fact a general result of Section 5 implies that for this example $\delta \to \tau^{-2}$ as $\tau \to 0$. On the other hand, for this chaotic example, intuitively we should expect to see the distortion increasing as τ increases, due to irrelevancy. However, this is clearly not reflected in the numerics. In fact, the low noise limit approximation must ultimately break down as τ increases, even for small noise levels. This situation may be visualized in Figure 6. As τ increases, there will be more and more folds, and the induced distribution for $p(s|\underline{x})$ will become multimodal, causing an explosion in the distortion

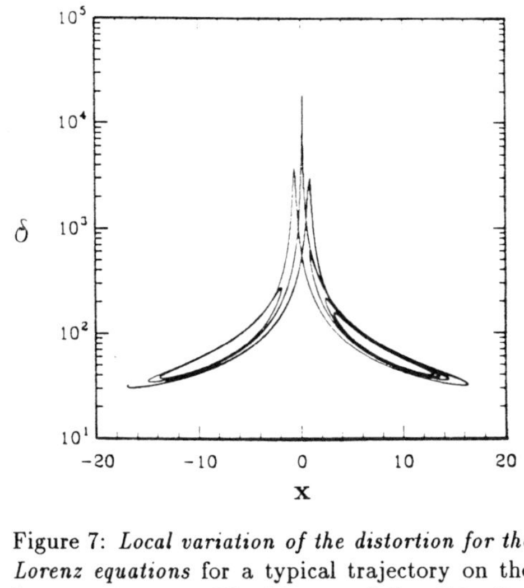

Figure 7: *Local variation of the distortion for the Lorenz equations* for a typical trajectory on the Lorenz attractor. The blowup of the distortion along $x = 0$ is a result of the poor information flow from z to x when $x = 0$.

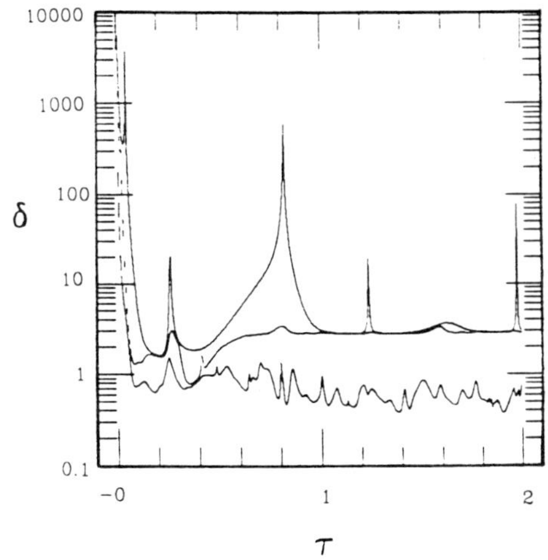

Figure 8: *The distortion as a function of* τ for three different embedding dimensions.

at finite resolution. This effect cannot be obtained by the purely local analysis of Section 4.4. We return to this problem in Section 4.7. However, the above results show that it is a very complicated problem to settle on an optimal value for the time delay τ.

In Section 5 we will consider the dependence of the distortion on the embedding dimension m in more detail.

4.6.2. *Finite Noise Distortion*

In this subsection, we investigate numerically the accuracy of the low noise limit formulae above for approximating the distortion at finite resolution δ_ϵ. A similar investigation could be done for the noise amplification. Recall that the noise amplification at finite resolution measures the "thickness" of Figure 2. One could attempt to measure this thickness directly by a lengthy numerical simulation, and compare the result to the formula for noise amplification. We will now describe an algorithm for this idea in the case of distortion which we refer to as a Monte Carlo simulation.

We use the exact likelihood function $p(\underline{x}|s)$ of Eq. (15), and take the prior $p(s)$ to be the natural measure $p_\nu(s)$ on the attractor to obtain $p(s|\underline{x})$. We assume that the dynamics has sufficiently nice mixing properties so that Eq. (32) holds for almost all initial conditions s_0 in the basin of attraction, where ϕ is any smooth function and Δt is held fixed at some small value:

$$\int \phi(s)p_\nu(s)ds = \lim_{N\to\infty} \frac{1}{N} \sum_{i=1}^{N} \phi(f^{i\Delta t}(s_0)) \tag{32}$$

Then taking $\phi_1(s) = \|s\|^2 p(\underline{x}|s)$ and $\phi_2(s) = s\ p(\underline{x}|s)$, we obtain Eq. (33), where $w_i \equiv \exp(-\|\underline{x} - \Phi(f^{i\Delta t}(s_0))\|^2/2\epsilon^2)$:

$$\epsilon^2\delta_\epsilon^2 = (\sum_{i=1}^{\infty} w_i \|f^{i\Delta t}(s_0)\|^2 / \sum_{i=1}^{\infty} w_i) - \|\sum_{i=1}^{\infty} w_i f^{i\Delta t}(s_0)/\sum_{i=1}^{\infty} w_i\|^2 \tag{33}$$

This is turned into a numerical approximation by truncating after N terms, where N is varied until satisfactory convergence has been achieved. Note that the smaller ϵ is taken, the larger N must be taken for convergence. This approximation is clearly much more CPU intensive than the analytical formula of Section 4.4 for the low noise limit δ.

Figure 9 illustrates the results of such a computation for the case of the Lorenz equations. The two solid lines are plots of δ_ϵ against τ for noise levels $\epsilon = 0.5$ and $\epsilon = 0.25$, which represents a signal to noise ratio of about 20 and 40. A predictive embedding was chosen with $m_+ = 0$ and $m_- = 4$. The noisy delay vector $\underline{x}$ was generated from a state corresponding to Figure 8. The dotted line is the corresponding plot of δ against τ taken from Figure 8.

We make the following observations about this figure.

1. The distortion at finite resolution appears to have converged well for a value of ϵ as high as 0.5, for a wide range of values of τ.
2. For the range of τ over which convergence at finite resolution has been achieved, δ provides an upper bound for δ_ϵ. This upper bound is also very sharp over a wide range of values of τ. We believe that the failure of δ to bound δ_ϵ for large values of τ is due to the phenomenon of irrelevancy and bimodality mentioned in Section 4.6. We also believe that the failure of δ to bound δ_ϵ sharply for all small values of τ is due to the fact that the Monte Carlo simulation for δ_ϵ is carried out on a trajectory, effectively including the prior information of being on an attractor. In our calculation of Σ, we use a uniform prior, so that we should only expect $\delta = \sqrt{\text{trace } \Sigma}$ to provide an upper bound on δ_ϵ.

We have performed other Monte Carlo simulations and the results indicate that the situation can be more complicated in other examples. Firstly, we performed

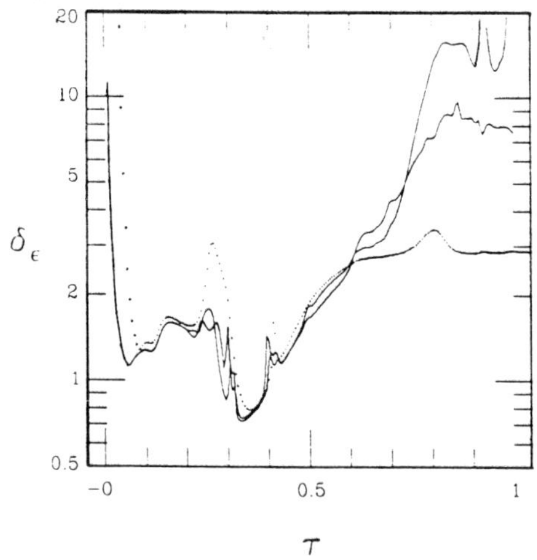

Figure 9: *The distortion δ_ϵ at finite resolution ϵ as a function of τ* for $\epsilon = 0.5$ and $\epsilon = 0.25$ are plotted in solid lines from a Monte Carlo simulation. The dotted line is the corresponding plot for the low noise limit of the distortion using the analytical formula of Section 4.4.

simulations with the Ikeda map, and Gaussian noise. We observed that the limit of δ_ϵ as $\epsilon \to 0$ sometimes fails to exist. We believe this is due to the highly fractal structure of the underlying attractor. This is not a problem at realistic noise levels for the Lorenz example, because in that case, the fractal structure is only apparent at an extremely small resolution. Secondly, we performed simulations for the Lorenz example, but using uniform noise. We observed that the limit of δ_ϵ exists, but is dependent on the realization of the noise used to produce the delay vector $\underline{x}$. To obtain a well defined limit requires taking an ensemble average over many realizations of the noise. Fortunately, as demonstrated in Section 4.4, this problem does not arise for Gaussian noise.

5. Limits to Predictability

For a given noise level the noise amplification tells us how much loss of predictability occurs purely because of the reconstruction process; it therefore sets a limit to prediction that is independent of the number of data points or the modeling technique. The distortion[12] (or equivalently, the noise amplification) depends on the state space reconstruction, for example, on the parameters m_+, m_-, and τ. It also depends on the properties of the underlying dynamical system such as the dimension and Lyapunov exponents, and on the measurement function. In this section we show that there are some general scaling laws that make it possible to estimate the way the distortion will change as these parameters are varied. These scaling laws set upper bounds to predictability.

To study the dependence on the reconstruction it is sufficient to consider delay coordinates. As we prove in Section 6, this is because delay coordinates provide a lower bound on distortion, in the sense that a coordinate transformation of delay coordinates cannot reduce the distortion. One fact that is immediately apparent is that *gathering more information can only decrease the distortion.* This follows from an elementary property of conditional probabilities. Suppose we are given two delay vectors $\underline{x}^{(1)}$ and $\underline{x}^{(2)}$ for which $\underline{x}^{(1)} \subset \underline{x}^{(2)}$, i.e., $\underline{x}^{(2)}$ is of higher dimension than $\underline{x}^{(1)}$, and contains $\underline{x}^{(1)}$ as a subset. Then

$$\Sigma(s|\underline{x}^{(2)}) \leq \Sigma(s|\underline{x}^{(1)}) \tag{34}$$

in the sense of Section 4. Thus, to reduce the distortion the dimension of the reconstructed space should be as high as possible.

As a practical matter, however, finite data resources usually impose a limit on the state space dimension. It is therefore important to know *which* information is most useful. For uniform lag times this translates into choosing the best values for τ, m_+, and m_-. The scaling laws derived in Sections 5.1 and 5.2 provide insight into this question.

Another fact that is intuitively obvious is that when τ is sufficiently small successive measurements become almost *redundant*, in the sense that in the absence of noise they approach the same value; the difference in their value is mainly due to

[12]In this section we study the distortion rather than the noise amplification because distortion does not depend on the prediction time. However, from the results of the previous section, the results will apply to either quantity.

measurement noise. In this case images of the measurement surfaces are roughly parallel in the neighborhood of the true state. Let t_r denote the *redundancy time*, above which measurement surfaces intersect at a significant angle. Then we expect that if the *window width* $w = m\tau$ is much less than t_r, then the distortion will be very large. To avoid this we should choose $w > t_r$. On the other hand, for a chaotic system, let t_i denote the *irrelevancy time*, of order $\log \epsilon/\lambda$, where λ is the largest Lyapunov exponent. Intuitively, it should be expected that measurements made outside a window width w much greater than t_i will be irrelevant, in the sense that the images of the measurement surfaces will line up along the unstable direction, and so give no information in that direction. To avoid this we should choose $w < t_i$. In the case that $t_i < t_r$, one of the above conditions on w must be violated, and one would expect a very large distortion. In this section we will investigate the extent to which the above intuition is born out quantitatively, by deriving general scaling laws for the distortion, and working out some examples.

5.1. *Scaling Laws*

When m is sufficiently large or τ is sufficiently small the distortion behaves according to well defined scaling laws. There are two regimes. One of these occurs when the window width $w = m\tau$ is small, and the other occurs when the window width is large.

5.1.1. *Small Window Width Limit*

The scaling is the same whether or not the dynamics is chaotic. The scaling law is

$$m\tau \to 0 \quad : \quad \delta = O(m^{-1/2}(m\tau)^{(1-d)}) \tag{35}$$

where "$O()$" denotes "the order of". Note that for $d > 1$ the distortion blows up in the limit as $\tau \to 0$, with an exponent that increases with dimension.

Example: The Lorenz equations. In Figure 10 we plot the distortion δ as a function of the embedding dimension m, with τ fixed at 0.005, and s fixed at the same value as for Figure 8. A predictive and a non predictive embedding are shown. Observe that for small m, in both cases the scaling goes as $m^{-3/2}$, as predicted by Eq. (35). At larger m, a different behavior is apparent, as will be discussed in Section 5.1.2. For another example, see Figure 9 of Section 5.2.

Derivation. Expand $D\Phi$ in a Taylor series in time around $t = 0$. For convenience assume a predictive embedding, with the first row simply Dh. Then the rows of $D\Phi$ are of the form

$$D\Phi_{i+1} = a^{(0)} + a^{(1)}(i\tau) + a^{(2)}(i\tau)^2 + \ldots \tag{36}$$

where $i = 0, \ldots, m-1$ labels the row, and the $a^{(j)}$ are fixe d d-dimensional row vectors. For sufficiently small values of τ the embedding surfaces are approximately linear, and there is a unique crossing when $m \geq d$. If we truncate the Taylor series at order $d-2$ the matrix cannot be of full rank, since there are only $d-1$ independent vectors $a^{(j)}$. Consequently the dth singular value is zero to order τ^{d-2}. But if we truncate the Taylor series at order $d-1$ the matrix will generically be of full rank at almost all states s because the d d-dimensional vectors $a^{(j)}$ involved in the expansion are typically independent. Therefore the dth singular value is typically of order $(m\tau)^{(d-1)}$.

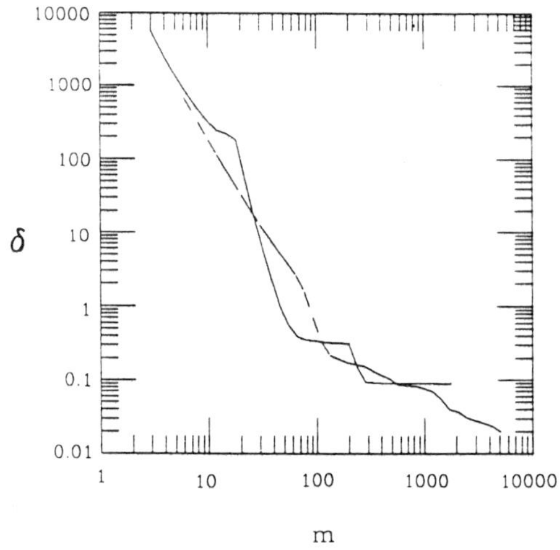

Figure 10: *The distortion as a function of m.* The solid curve is for a predictive embedding with $m_+ = 0$, and the dashed curve is for a nonpredictive embedding with $m_+ = m/2$ and $m_- = m/2 - 1$.

The dominant eigenvalue of Σ is the square of the inverse of the d largest singular values of $D\Phi$, which implies the τ scaling in Eq. (35).

The m scaling comes from the law of large numbers. If we fix the window width (at a small value) and increase m, then the variance decreases as m^{-1} because of the assumed independence of the measurement errors. These two arguments taken together give the scaling law of Eq. (35).

Remark. In the small window limit, it can also be demonstrated that the singular vectors of $D\Phi$ converge onto Legendre polynomials for almost all states s. It was previously shown by us [19] that in the limit of low noise and small window width, global Broomhead and King coordinates converge onto Legendre polynomials, with the singular spectrum satisfying similar scaling laws. The ideas behind the proofs are very similar, with the observability matrix replacing the covariance matrix C_{ij} defined in Section 2.1.

5.1.2. *Large Window Width Limit*

In the limit as $m \to \infty$ when the window width is large, there are different scaling behaviors, depending on whether or not the dynamics is chaotic.

Nonchaotic systems. When τ is large the measurement surfaces are no longer nearly parallel. Each measurement can be treated as independent, and according to the law of large numbers the scaling is:

$$m \to \infty \quad : \quad \delta = O(m^{-1/2}) \tag{37}$$

In general it is intuitively clear that this also increases with d, since when d is large the information in the time series is spread over more coordinates. In the example in Section 5.2, it is shown that $\delta = \sqrt{2}\, d^{3/2} m^{-1/2}$.

Chaotic systems with predictive coordinates. As described at the beginning of this section, for a chaotic system measurements in the distant past provide no information about the position along the unstable direction. While information is provided in the stable direction, this information is at a fine scale of resolution that is typically below instrumental error. Since the uncertainty of the position along the unstable manifold is the limiting factor, this information is irrelevant for prediction.

In the limit of sufficiently large τ and small λ it is possible to derive a scaling law for the distortion matrix. This is possible because in this case the eigenvectors of the distortion matrix line up with the stable and unstable manifolds. The eigenvalues of the distortion matrix have three different behaviors:

$$\textbf{Unstable manifold} \quad (\lambda_i > 0)$$
$$m \to \infty \quad : \quad \Sigma_{ii} = O(1 - e^{-2\lambda_i d' \tau}) \tag{38}$$
$$\textbf{Neutral manifold} \quad (\lambda_i = 0)$$
$$m \to \infty \quad : \quad \Sigma_{ii} = O(m^{-1}) \tag{39}$$
$$\textbf{Stable manifold} \quad (\lambda_i < 0)$$
$$m \to \infty \quad : \quad \Sigma_{ii} = O(e^{2m\lambda_i \tau}) \tag{40}$$

In the above equations we have transformed the distortion matrix Σ to the appropriate coordinates. The distortion in the unstable manifold approaches a constant, while in the neutral manifold it goes to zero as a polynomial, and in the stable manifold it goes to zero exponentially with m. The mean-square error in a prediction is related to the trace of the distortion, which is dominated by the largest eigenvalues. As we demonstrate in the derivation below, this approaches a constant.

Chaotic systems with mixed coordinates. With mixed coordinates the situation is quite different from that of predictive coordinates, since future information makes it possible to pinpoint the position along the unstable manifold precisely, and all the eigenvalues go to zero as $m \to \infty$. The calculation of the distortion follows closely that for predictive coordinates, except that all the sums and products must be taken from $\frac{-m_+}{d'}$ to $\frac{m_-}{d'}$. The results for the stable and neutral manifolds remain essentially the same, but the unstable manifold is now dominated by the contributions from the future terms. We get instead:

$$\textbf{Unstable manifold} \quad (\lambda_i > 0)$$
$$m \to \infty \quad : \quad \Sigma_{ii} = O(e^{-2m_+ \lambda_i \tau}) \tag{41}$$
$$\textbf{Neutral manifold} \quad (\lambda_i = 0)$$
$$m \to \infty \quad : \quad \Sigma_{ii} = O(m^{-1}) \tag{42}$$
$$\textbf{Stable manifold} \quad (\lambda_i < 0)$$
$$m \to \infty \quad : \quad \Sigma_{ii} = O(e^{2m_- \lambda_i \tau}) \tag{43}$$

The main difference is that all the eigenvalues $\Sigma_{ii} \to 0$ as $m \to \infty$; the eigenvalues for the stable and unstable manifold go to zero exponentially. The neutral manifold thus provides the leading order contribution to the distortion.

The above relationships are apparent in Figure 10. The relationships are also not valid when $d'\tau \gg \frac{\log \epsilon}{\lambda}$; when this assumption is violated the behavior is entirely different, as we discuss in Section 5.2.

The following derivation of Eqs. (38)-(40) is admittedly rather loose; to turn these into a more rigorous statement may involve placing restrictions on quantities such as the measurement function and the nature of the dynamical system. It should probably be omitted at a first reading.

Derivation of Eqs. (38-40). The m-dimensional delay vector $\underline{x}^{(m)}$ can be broken into a series of lower dimensional delay vectors rooted at different times. To derive the scaling we transport all of them to the same time and examine their joint likelihood function. Let d' be the minimum dimension for which delay vectors define a global embedding, and for convenience pick m so that it is an integer multiple of d'. Let $\underline{x}_j^{(d')}$ be the d'-dimensional delay vector rooted at time $-jd'\tau$, $\underline{x}_j^{(d')} = (x(-jd'\tau), \ldots, x(-(jd' + d' - 1)\tau))$. Assume the measurement errors are Gaussian with variance ϵ^2, and let $\underline{\xi}_j^{(d')} = (\xi(-jd'\tau), \ldots, \xi(-(jd' + d' - 1)\tau))$ be the vector of d'-dimensional measurement errors rooted at time $-jd'\tau$. Let F be the induced d'-dimensional dynamics in delay space. In the limit as $\epsilon \to 0$ the vector of measurement errors rooted at time $-jd'\tau$ transported to time 0 is $\underline{\acute{\xi}}_j^{(d')} = DF^{jd'\tau}(\underline{\xi}_j^{(d')})$. The noise p.d.f. $p(\underline{\xi}_j^{(d')})$ is an isotropic Gaussian of variance ϵ^2; following a calculation similar to that of Section 4.4, to leading order in ϵ

$$p(\underline{\acute{\xi}}_j^{(d')}) = A \exp\left[\frac{1}{2\epsilon^2}(\underline{\acute{\xi}}_j^{(d')})^{\dagger} \Theta_j^{-1}(\underline{\acute{\xi}}_j^{(d')})\right] \tag{44}$$

where A is a normalization constant, and Θ_j is a $d' \times d'$ dimensional matrix $\Theta_j = (DF^{jd'\tau}(\underline{x}_j^{(d')}))^{\dagger} DF^{jd'\tau}(\underline{x}_j^{(d')})$.

Let $\underline{\acute{x}}_j^{(d')}$ be the noise free delay vector such that $\underline{\acute{x}}_j^{(d')} - \underline{x}^{(d')}(0) = \underline{\acute{\xi}}_j^{(d')}$. The set $\{\underline{\acute{x}}_j^{(d')}\}, j = 0, 1, \ldots, \frac{m}{d'}$ contains the same information as the m-dimensional delay vector $\underline{x}^{(m)}$. Furthermore, $\{\underline{\acute{\xi}}_j^{(d')}\}$, $j = 0, 1, \ldots, \frac{m}{d'}$ is a collection of independent random variables. Following similar reasoning to that of Section 3.1, the above statements plus Bayes' theorem (with a uniform prior) imply

$$\begin{aligned} p(\underline{x}^{(d')}(0)|\underline{x}^{(m)}) &= Ap(\{\underline{\acute{x}}_j^{(d')}\}|\underline{x}^{(d')}(0)) \\ &= p(\{\underline{\acute{\xi}}_j^{(d')}\}) \\ &= A' \prod_{j=0}^{\frac{m}{d'}-1} \exp\left[\frac{1}{2\epsilon^2}(\underline{\acute{\xi}}_j^{(d')})^{\dagger} \Theta_j^{-1}(\underline{\acute{\xi}}_j^{(d')})\right] \\ &= A' \exp\left[\frac{1}{2\epsilon^2} \sum_{j=0}^{\frac{m}{d'}-1} (\underline{\acute{x}}_j^{(d')} - \underline{x}^{(d')}(0))^{\dagger} \Theta_j^{-1}(\underline{\acute{x}}_j^{(d')} - \underline{x}^{(d')}(0))\right] \end{aligned} \tag{45}$$

where A and A' are normalization constants. The distortion can be obtained by expanding in a Taylor series, as in Section 4.4. Hence we obtain:

$$\Sigma = \left[\sum_{j=0}^{\frac{m}{d'}-1} \theta_j^{-1}\right]^{-1} \tag{46}$$

It follows from the definition of the Lyapunov exponents that when $d'\tau$ is sufficiently large Θ_j approaches a matrix whose eigenvalues are $e^{2j\lambda_1 d'\tau}, \ldots, e^{2j\lambda_{d'} d'\tau}$. Furthermore, for large $d'\tau$ the eigenvectors approach limiting values, independent of j. In this case we can evaluate Eq. (46) in the basis of eigenvectors:

$$\left[\sum_{j=0}^{\frac{m}{d'}-1} \Theta_j^{-1}\right]^{-1}_{ii} = \sum_{j=0}^{\frac{m}{d'}-1} e^{-2j\lambda_i d'\tau} = \frac{1 - e^{-2m\lambda_i\tau}}{1 - e^{-2\lambda_i d'\tau}} \tag{47}$$

When $\lambda_i > 0$ the numerator approaches 1 as $m \to \infty$, and by inverting we obtain Eq. (38). When $\lambda_i < 0$ the second term in the numerator dominates and we obtain Eq. (40). When $\lambda_i = 0$ the summation in the previous equation is no longer valid; however, the sum is clearly of order m, and we obtain Eq. (39).

Note that while $\underline{x}^{(d')}(0)$ is related to s by a coordinate transformation, because Σ is not invariant under coordinate transformations the distortion is not in general the same. Nonetheless, we expect their scaling properties to be the same.

Note also that in this derivation, by taking delay vectors of dimension d' we are assuming that the predictability changes very little over times $d'\tau$, i.e., $d'\tau \ll \frac{\log \epsilon}{\lambda}$. When this assumption breaks down the scaling is radically different, as we demonstrate in the example of the next section.

5.2. *A Solvable Example*

In this section we investigate the distortion for an example that is sufficiently simple that the observability matrix can be calculated explicitly. Consider a system of $d/2$ negatively damped harmonic oscillators:

$$\frac{d}{dt}\begin{pmatrix} u_i \\ v_i \end{pmatrix} = \begin{pmatrix} \lambda_i & -\omega_i \\ \omega_i & \lambda_i \end{pmatrix}\begin{pmatrix} u_i \\ v_i \end{pmatrix} \qquad i = 1, .., \frac{d}{2} \tag{48}$$

The state space dimension d is even. u_i and v_i are both taken *modulo* 1, corresponding to (piecewise smooth) motion on a torus. $\lambda_i > 0$ are the Lyapunov exponents; for convenience we will sometimes take $\lambda_i = \lambda =$ constant. We take the measurement function to be:

$$h = \frac{2}{d}\sum_{i=1}^{\frac{d}{2}} u_i \tag{49}$$

We will consider a predictive reconstruction with $m_+ = 0$.

This example is admittedly rather contrived. The oscillators are independent, so measurements only give information about the whole system because the measurement function involves a combination of all of the degrees of freedom. In a more typical example the flow of information depends on the coupling of the unobserved degrees of freedom to the observed degrees of freedom. Nonetheless, as we shall see, even this very simple example exhibits nontrivial behavior.

This system has the following analytic solution:

$$\begin{aligned} u_j(t) &= u_j(0)\; e^{-\lambda_j t}\; \cos\omega_j t \\ v_j(t) &= v_j(0)\; e^{-\lambda_j t}\; \sin\omega_j t \end{aligned} \tag{50}$$

Applying the definition of Φ and differentiating, the observability matrix can be calculated explicitly

$$\begin{aligned} D\Phi_{i,2j-1} &= \frac{2}{d}\; e^{-(i-1)\lambda_j\tau}\; \cos(i-1)\omega_j\tau \\ D\Phi_{i,2j} &= -\frac{2}{d}\; e^{-(i-1)\lambda_j\tau}\; \sin(i-1)\omega_j\tau \end{aligned} \tag{51}$$

where i ranges from 1 to m and j ranges from 1 to $d/2$. Note that $D\Phi$ is constant throughout the state space.

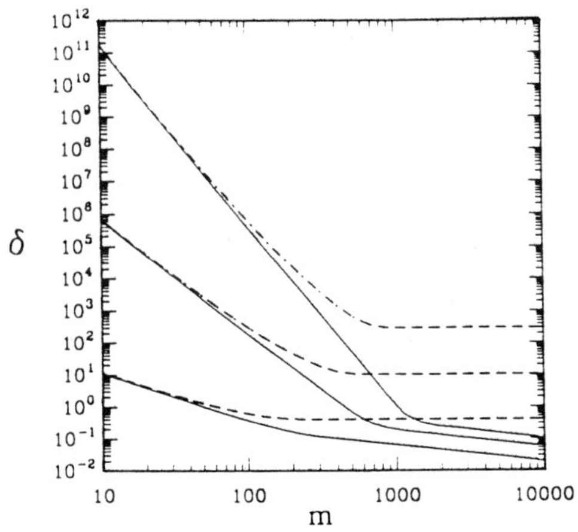

Figure 11: *The distortion δ plotted as a function of the delay coordinate dimension m* with a fixed delay time $\tau = 0.01$, for the system defined by Eqs. (48) and (49). For the solid curves $\lambda = 1$ and the system is chaotic, while for the dashed curves $\lambda = 0$ and the system is not. Three different dimensions are shown, $d = 2, 4$, and 6.

To compute the distortion we must first evaluate $D\Phi^\dagger D\Phi$. The odd rows and columns are:

$$D\Phi^\dagger D\Phi|_{2i-1\ 2j-1} = \frac{4}{d^2}\sum_{k=0}^{m} e^{-k(\lambda_i+\lambda_j)\tau}\cos k\omega_i\tau\cos k\omega_j\tau \tag{52}$$

There are similar expressions for the other terms, with sin cos and sin sin instead of cos cos. The distortion can be obtained from the singular value decomposition of $D\Phi^\dagger D\Phi$.

In Figure 11 we plot δ as a function of m for several different values of the dimension and Lyapunov exponents.

This illustrates several of the features derived in Section 5.1.

- **Small w:**
  For small values of m the window width w is also small. The chaotic and nonchaotic cases behave approximately the same. As m decreases the distortion increases as a power law with the predicted exponent $1/2 - d$.
- **Large w:**
  For the chaotic case the distortion approaches a constant while for the non-chaotic case the distortion decreases according to $m^{-1/2}$, independent of the dimension.

Note that, as it must, the distortion decreases monotonically with m.

The behavior of the plateau at $m = \infty$ for chaotic systems can be investigated by taking the limit $\tau \to 0$ in Eq. (52) and approximating by an integral. This gives:

$$D\Phi^\dagger D\Phi|_{2i-1\ 2j-1} \approx \frac{2(\lambda_i+\lambda_j)}{\tau d^2}\{[(\lambda_i+\lambda_j)^2+(\omega_i+\omega_j)^2]^{-1}+[(\lambda_i+\lambda_j)^2+(\omega_i-\omega_j)^2]^{-1}\} \tag{53}$$

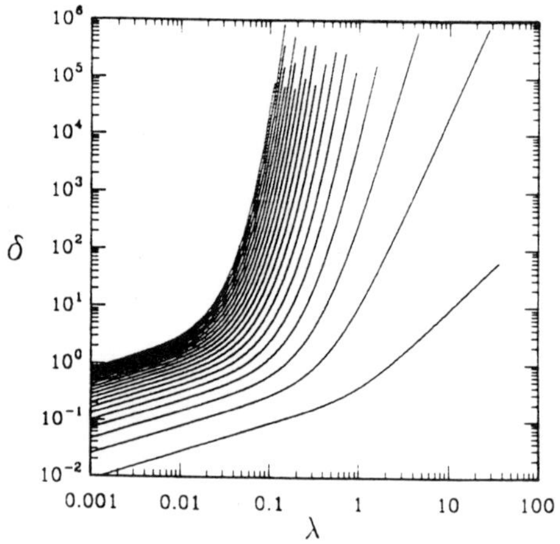

Figure 12: The distortion δ plotted as a function of the Lyapunov exponent λ for dimensions $d = 2, 4, \ldots, 20$. The curves with the lowest distortion have the lowest dimension. Notice that there are two scaling regimes, one for low λ and another for higher λ. In the high λ regime the enormous noise amplification means that even a small noise level ϵ makes the system behave effectively as a random process. There is no predictability, *even for short times.*

The behavior of the plateau under changes in parameter values is investigated by using Eq. (53) with $\lambda_i = \lambda = const$, and the frequencies uniformly spaced so that $\omega_i = \frac{2}{d}, \frac{4}{d}, \ldots, \frac{d}{d}$. The result is shown in Figure 12.

There are two scaling regimes, one for low λ, and one for high λ. In the low λ regime the motion is effectively predictable and $\delta = O(\lambda^{1/2})$. In the high λ limit, however, the distortion appears to diverge at a rate that increases with dimension, $\delta = O(\lambda^{d-1/2})$. Note that the crossover between the two scaling limits occurs at a lower value of λ when the dimension increases. The scaling law for the high (respectively low) λ limit can be obtained by substituting $m\tau = O(1/\lambda)$ in the small (respectively large) window width scaling law of Eq. (35) (respectively (37)). However, there are some problems with this argument, as the prefactor of Eq. (35) might depend on λ and alter the λ scaling behavior. Indeed we have investigated other high dimensional examples, and the behavior can be more involved than that described above, though we often see a crossover effect.

5.3. *When a Time Series Becomes a Random Process*

The divergence of the distortion in the high dimension limit seen in the previous example is particularly significant, because it demonstrates how noise amplification causes a system that is sufficiently chaotic and sufficiently high dimensional to become a random process. With a noise amplification of 10^6, unless the noise level $\epsilon < 10^{-6}$ (a very rare occurrence), the dynamics is fundamentally unpredictable, to first order in ϵ, even for short times. Note that in the previous example the distortion exceeds 10^6 when $d > 20$ and $\lambda > 0.1$. In this case there is simply not enough information in the time series to make the motion deterministic, on any time scale. We add the *caveat* that when the distortion is extremely large, there may be important effects of second order in ϵ which are beyond the above local analysis.

6. Coordinate Transformations

Up until now we have assumed that the reconstructed coordinates are simple delay coordinates, so that the reconstruction map $\Xi = \Phi$. Delay coordinates have the advantage of being simple and direct. However, the question arises of whether we can get better results by transforming to new coordinates. In general we may want to consider other coordinates $y = \Psi(\underline{x})$ where we further transform the delay coordinates so that the total reconstruction map $\Xi = \Psi \circ \Phi$.

6.1. *Effect on Noise Amplification*

There are two senses in which we might hope to make the coordinates "better": The first is that we might attempt to reduce the noise amplification by reducing noise, thereby locating the state more precisely. The second is that we might hope to reduce the dimension of the coordinate system, which reduces estimation error.

We will first address the question of changing the noise amplification. Two basic facts are apparent:

- *Invertible coordinate transformations cannot change the noise amplification.* This is evident from that the conditional probability density $p(x(T)|\Psi(\underline{x}(0)))$ is a function of $x(T)$ alone; $\Psi(\underline{x}(0))$ is not an argument of p, but rather a label that identifies this as a particular member of a family of different functions. As long as the function Ψ is one-to-one, it leaves the corresponding function p unchanged.
- *Non-invertible coordinate transformations cannot decrease the noise amplification.* If more than one state $\underline{x}$ is mapped into the same state $\Psi(\underline{x})$, this generally has the effect of broadening p. This is evident since:

$$p(x(T)|y) = \sum_{\{\underline{x}:y=\Psi(\underline{x})\}} p(x(T)|\underline{x}) \tag{54}$$

  Summing probability densities either increases the variance or leaves it unchanged. Hence, the noise amplification either increases or remains the same.

Thus, we see that we cannot decrease the noise amplification by a change of coordinates. In order to decrease the noise amplification we must alter the original information set, by changing Φ. For example, we can increase the dimension of the original delay space. However, from the point of view of noise amplification a coordinate transformation on the original delay coordinates is at best neutral.

Changing coordinates can be quite useful, however, for improving the estimation problem. This is particularly true for reducing the dimensionality. The estimation problem generally becomes exponentially worse as the dimension increases. Thus, we wish to find coordinates that make the dimension as small as possible while leaving the noise amplification unchanged.

6.2. *Local Analysis*

In the low noise limit, to first order in ϵ the transformation Ψ can be approximated locally by its derivative $D\Psi$ (the constant term plays no role in the following). An expression for $p(s|D\Psi(\underline{x}))$ can be derived using a generalization of the argument of Section 3.3 as follows. Assuming a uniform prior, we have $p(s|D\Psi(\underline{x})) \propto p(D\Psi(\underline{x})|s)$. But $p(D\Psi(\underline{x})|s) = p(D\Psi\vec{\xi})$, where $\vec{\xi} = \underline{x} - \Phi(s)$. Hence we obtain Eq. (55) by transforming the isotropic Gaussian distribution of the noise $\vec{\xi}$ through the linear map $D\Psi$.

$$p(s|D\Psi(\underline{x})) \approx A \exp\left[-\frac{1}{2\epsilon^2}(D\Psi\underline{x} - D\Psi\Phi(s))^\dagger (D\Psi D\Psi^\dagger)^{-1}(D\Psi\underline{x} - D\Psi\Phi(s))\right] \quad (55)$$

As before, in the limit that ϵ is small we can expand this in a Taylor series. The arguments parallel those leading to Eq. (28), except that Φ is replaced by $\Psi \circ \Phi$. The result is that

$$p(s|y) \approx C \exp\left[-\frac{1}{2\epsilon^2}(s - \hat{s})^\dagger \Sigma^{-1}(s - \hat{s})\right] \quad (56)$$

where:

$$\Sigma = \left(D\Phi^\dagger D\Psi^\dagger (D\Psi D\Psi^\dagger)^{-1} D\Psi D\Phi\right)^{-1} \quad (57)$$

Note that, as expected, a locally invertible coordinate transformation will not alter the distortion, since $(D\Psi D\Psi^\dagger)^{-1} = (D\Psi^\dagger)^{-1} D\Psi^{-1}$. In the next section we show how to minimize Σ with respect to Ψ.

6.3. *Optimal Reconstruction*

We will now show that it is possible to compress the information contained in a delay vector $\underline{x}$ into a smaller number of dimensions, while retaining all the available relevant information. Local principal value decomposition provides a way of achieving this.

$D\Phi$ is an $m \times d$ matrix which maps variations in the d-dimensional state, δs, into variations in the delay vector, $\delta\underline{x}$. For $m > d$, singular value decomposition expresses $D\Phi$ as the product of three linear transformations, U, W, and $V^\dagger$:

$$D\Phi = UWV^\dagger \quad (58)$$

The first of these, $V^\dagger$, is represented by an orthogonal $d \times d$ matrix that performs a rotation onto the principal axes. The second transformation W is represented by a diagonal $d \times d$ matrix that stretches or contracts the principle axes; its diagonal elements w_i are called the *singular values* of $D\Phi$. The third transformation U is represented by a column orthogonal $m \times d$ matrix that maps onto the m-dimensional delay space, so that $U^\dagger U = \mathbf{1}$, the identity in d dimensions.

Inserting this into Eq. (57) we get:

$$\Sigma = \left(D\Phi^\dagger D\Phi\right)^{-1} = \left(VWU^\dagger UWV^\dagger\right)^{-1} = VW^{-2}V^\dagger \quad (59)$$

$$\delta \equiv \sqrt{\text{trace } \Sigma} = \sqrt{\text{trace } (VW^{-2}V^\dagger)} = \sqrt{\text{trace } W^{-2}} = \left\{\sum_i w_i^{-2}\right\}^{1/2} \quad (60)$$

The eigenvalues of the distortion matrix are the inverse squares of the singular values, since V can be viewed as a similarity transformation which diagonalizes the distortion matrix:

$$V^{\dagger}\Sigma V = W^{-2} \tag{61}$$

The singular values w_i describe how well the observations determine the original state s along each of the principal axes of Σ. If w_i is small, then the observations are highly uncertain along the corresponding axis. The best coordinates are obviously those that make w_i as large as possible for all i.

Note that the singular values depend only on the way in which we construct the original delay coordinates (and of course on the dynamical system and the measurement function). In order to reduce the uncertainty of our coordinates we must gather more information, for example, by increasing the dimension of the delay space or by choosing a better value of τ. The search for an optimal τ is nontrivial, and increasing the dimension of the delay space worsens estimation error. Ideally, we would like a reconstruction algorithm which incorporates all the information in a given window in the lowest possible dimension.

Local principal value decomposition coordinates are the best possible coordinates which can be derived from the information set represented by a given delay embedding, in the sense that they compress all the available information into the smallest possible dimension, d. We define local PVD coordinates by introducing a transformation $\Psi = U^{\dagger}$ from the delay space of dimension m to a space of dimension $d < m$ which projects onto the d principal axes determined by singular value decomposition. They are local in the sense that a principal value decomposition at each point in the delay space produces a different U. Geometrically, the transformation $U^{\dagger}$ maps noisy delay vectors back onto the tangent space to the embedded state space $\Phi(\Re^d)$, and is thus a natural candidate for an optimal reconstruction.

Although the optimality of PVD coordinates is almost intuitively clear, to make sure it is understood we give a proof: First, we show that any reconstruction in fewer than d dimensions has infinite distortion, then we show that the PVD coordinates have the same distortion as the delay embedding. First, suppose that the total reconstruction map Ξ maps points from d dimensions to $d' < d$. Performing a singular value decomposition on its transpose,

$$D\Xi^{\dagger} = UWV^{\dagger} \tag{62}$$

yields for the distortion matrix:

$$\Sigma = (UW^{-2}U^{\dagger})^{-1} \tag{63}$$

But this is a $d \times d$ dimensional matrix with at most d' nonzero singular values, since W is $d' \times d'$. The distortion, as above, is the square root of the sum of squares of the inverses of all d singular values, hence it must diverge. Obviously, we cannot embed the data in fewer than d dimensions with a finite distortion. Consider instead the map from d to d dimensions defined by $\Xi = \Psi \circ \Phi$, where $\Psi = U^{\dagger}$. The distortion matrix in this case is just

$$\begin{aligned}\Sigma &= (VWU^{\dagger}\Psi^{\dagger}\Psi UWV^{\dagger})^{-1} = (VWU^{\dagger}UU^{\dagger}UWV^{\dagger})^{-1} \\ &= (VW^{2}V^{\dagger})^{-1}\end{aligned} \tag{64}$$

the same as for the delay coordinates themselves. Local PVD coordinates give us a reconstruction which takes advantage of all the information available in a high dimensional embedding yet minimizes estimation error by minimizing the embedding dimension.[13]

7. Conclusion

7.1. *Results*

Takens' theorem establishes that delay coordinates from a dynamical system are diffeomorphic to the system's original coordinates if their dimension is sufficiently large. The particular values of τ and m are not important when the data is infinitely precise, as long as m is large enough. However, in the presence of noise, the quality of an embedding is highly dependent on these parameters. The distortion and noise amplification quantify the quality of an embedding, and we give analytical formulae for them in terms of the dynamics and the measurement function in the low noise limit. The scaling laws for these quantities give fundamental bounds on the predictability of a dynamical system and show how a deterministic system becomes a random process.

Reconstruction techniques such as PVD always start with delay vectors. But since the distortion of a delay reconstruction provides a lower bound on the distortion of any further transformation, the only advantage of such transformations is dimension reduction. Local PVD is an optimal method of state space reconstruction, since it retains the distortion of its delay vectors while projecting them into the lowest possible dimension.

7.2. *Open Questions*

Although we have developed a theory of state space reconstruction in the presence of noise, in this paper we have not addressed the practical issues which arise when constructing nonlinear predictive models from time series. We are currently conducting numerical experiments in order to find out if our local principal value reconstruction technique has advantages over existing techniques for modeling. However, these experiments are subject to a number of complications which prevent their inclusion at this date. For example, forecasting error includes both noise amplification and estimation errors due to finite data sets. We have found that estimation error often dominates forecast error for quite reasonable noise levels and data set sizes. Further, estimation error varies with the reconstruction, making it difficult to distinguish the contribution of either effect to the overall error. We intend to publish the results of these experiments and an analysis of the estimation problem in **Physica D**.

We have also left several theoretical questions unanswered as follows. We expect that our theory will need to be modified in the case of dynamical or correlated noise. Also, the example we use to illustrate the breakdown of predictability as dimension and Lyapunov exponents increase is very special. We intend to explore this phenomenon for more realistic examples.

---

[13]Of course, any linear transformation which can be obtained from $U^\dagger$ by a rotation will have the same property.

Acknowledgements

We are grateful for support from the Department of Energy and the Air Force Office of Scientific Research under grant AFORS–91–0015 and the National Institute of Mental Health under grant 1–R01–MH47184–01.

References

1. M. Casdagli. Nonlinear prediction of chaotic time series. *Physica D* **35**, 335–356 (1989).
2. J. Cremers and A. Hübler. Construction of differential equations from experimental data. *Z. Naturforsch.* **42a**, 797–802 (1987).
3. J.P. Crutchfield and B.S. McNamara. Equations of motion from a data series. *Complex Systems* **1**, 417–452 (1987).
4. J.D. Farmer and J.J. Sidorowich. Predicting chaotic time series . *Phys. Rev. Lett.* **59**, 845–848 (1987).
5. J.D. Farmer and J.J. Sidorowich. Exploiting chaos to predict th e future and reduce noise. In *Evolution, Learning and Cognition.* Ed. Y.C. Lee. World Scientific, Singapore, 1988.
6. P. Grassberger. Information content and predictability of lumped and distributed dynamical systems. Tech. Rep. WU-B-87-8, University of Wuppertal, 1987.
7. A. Lapedes and R. Farber. Tech. Rep. LA-UR-87-2662, Los Alamos Nat. Lab., 1987.
8. A.I. Mees. Modelling complex systems. Math. Dep. University of Western Australia, 1989.
9. H. Tong and K.S. Lim. Threshold autoregression, limit cycles, and cyclical data. *J. of the Roy. Stat. Soc. B* **42**, 245–292 (1980).
10. M.B. Priestley. State dependent models: A general approach to nonlinear time series analysis. *J. of Time Series Analysis* **1**, 47–71 (1980).
11. N.H. Packard, J.P. Crutchfield, J.D. Farmer, and R.S. Shaw. Geometry from a time series. *Phys. Rev. Lett.* **45**, 712–716 (1980).
12. F. Takens. Detecting strange attractors in fluid turbulence. *Dynamical Systems and Turbulence.* Eds. D. Rand and L.-S. Young. Springer, Berlin, 1981.
13. G.U. Yule. *Phil. Trans. Roy. Soc. Lond. A* **226**, 267 (1927).
14. A.M. Fraser and H.L. Swinney. Independent coordinates for strange attractors from mutual information. *Phys. Rev. A* **33**, 1134–1140 (1986).
15. A.M. Fraser. Information and entropy in strange attractors. *IEEE Trans. on Inf. Theory* **IT-35**, 245–262 (1989).
16. H.D.I. Abarbanel, R. Brown, and J.B. Kadtke. Prediction and system identification in chaotic nonlinear systems: Time series with broadband spectra. U.C. San Diego, 1989.
17. A. Cenys and K. Pyragas. Estimation of the number of degrees of freedom from chaotic time series. Tech. Rep. Inst. of Semicond. Phys., Acad. of Sci. LSSR, Vilnius, 1987.

18. D.S. Broomhead and G.P. King. Extracting qualitative dynamics from experimental data. *Physica D* **20**, 217–236 (1987).

19. J.F. Gibson, M. Casdagli, S. Eubank, and J.D. Farmer. Principal component analysis and derivatives of time series. Tech. Rep. Los Alamos Nat. Lab., 1990.

20. A.M. Fraser. Reconstructing attractors from scalar time series: A comparison of singular system and redundancy criteria. *Physica D* **34**, 391–404 (1989).

21. T. Sauer, J. Yorke, M. Casdagli, and E. Kostelich. Embedology (1990).

22. R.S. Shaw. Strange attractors, chaotic behavior, and information flow. *Z. Naturforsch.* **36a**, 80–112 (1981).

23. R.S. Shaw. *The Dripping Faucet as a Model Dynamical System.* Aerial Press, Santa Cruz, 1984.

24. M.B. Priestley. *Spectral Analysis of Time Series.* Academic Press, New York, 1981.

25. J. Breiden and A. Hübler. Reconstructing equations of motion from experimental data with hidden variables. Tech. Rep. Univ. of Illinois (1990).

APPLIED CHAOS: QUANTIFYING COMPLEX SYSTEMS

LEONARD A. SMITH[1]

Department of Applied Mathematics
and Theoretical Physics
University of Cambridge
Cambridge CB3 9EW, U.K.

Abstract

This contribution discusses dynamic reconstructions and their application to the identification of, quantification of and discrimination between complex systems. Phase space reconstructions which reproduce the flow of dynamical systems in time are constructed from (multiple probe) time series data. These dynamic reconstructions are used to quantify the unknown system. In the case of chaotic systems, this is accomplished through the isolation of unstable periodic orbits. The method is applied to data from the Ikeda map, where it is shown that the data requirements of this approach are modest relative to those required for other types of analysis. The application of this approach to systems where the underlying dynamics are stochastic is also discussed.

1. Introduction

In recent years there has been a rapid increase in the application of nonlinear dynamical systems ideas to the analysis of complex time series. These ideas have provided a new paradigm which has proven useful in many fields of study. Originally, reconstructions were "static" in that they attempted to reconstruct the geometry from a time series [1]. Unfortunately, methods of analysis developed for the study of low dimensional low noise systems have sometimes been applied to data sets with which they cannot produce meaningful results. This is particularly true of dimension calculations [2] where many results may be disqualified on simple geometrical constraints [3]. While much work has been done on determining the reliability of such algorithms [4], a major draw back of these approaches is that they fail in subtle ways. More recently, there have been propositions to build dynamic reconstructions for predicting chaotic systems [5, 6] (additional references are given in Ref. [7]). One advantage

[1]Present address: Department of Engineering, University of Warwick, Coventry CV4 7AL, UK

Information Dynamics, Edited by H. Atmanspacher and
H. Scheingraber, Plenum Press, New York, 1991

of these new approaches is that they may be used to quantify the dynamics of the system instead of attempting to analyze the geometry. A second property is that they fail robustly; the value of which cannot be overstressed. In this paper, I develop a reconstruction and apply it not for predicting, but rather to quantify the chaotic system underlying the observations. The reconstructed dynamical system will then be used to locate unstable periodic orbits and quantify their stability. This information may then be used, for example, to determine the origin of a time series known to have come from one of several chaotic systems.

It will be shown that this approach yields excellent results with reasonable data requirements. Note however that the particular predictor constructed here is not claimed to be the best; the fundamental point of this paper is to demonstrate that reconstructing the dynamics from a time series yields a testable and, in some cases, reliable model upon which analysis and system identification techniques can be usefully applied.

2. The Ikeda Map

Before describing the reconstruction, I introduce the system which will be used to generate most of the time series described herein, namely the Ikeda map [8]. (Initial studies with the Henon map showed it to be unsuitable as described in Ref. [7].) The Ikeda map is

$$\begin{aligned} x' &= 1 - \mu(x\cos(t) - y\sin(t)) \\ y' &= \mu(x\sin(t) + y\cos(t)) \end{aligned} \tag{1}$$

where $t = 1/(1 + x^2 + y^2)$. This system displays chaotic behavior over a range of values for the parameter μ including the values chosen here.

In order to quantify the similarity of a dynamic reconstruction with the original system, we compute the spectrum of unstable periodic orbits of each. This spectrum provides information on the global properties of the system as well [9, 10]. Two approaches for determining these orbits were employed. First, a trajectory was followed for n iterations; if the nth iteration of the map was within a distance δ from the initial point, a gradient flow/Newton-Raphson scheme was implemented to search for a periodic orbit. Typically, δ was about one tenth the diameter of the attractor; large values of δ resulted in more computational time per iteration of the map, but tended to find different orbits in fewer iterations while small δ lead to the testing of only near returns. Alternatively, initial points may be taken on a fixed grid. This approach has the advantage of "immediately" sampling regions of the attractor which are visited only rarely, but at the cost of searching regions far from the attractor. As it is very difficult to ascertain whether all the periodic orbits have been found, a combination of these approaches is recommended. In addition, searching from small, random displacements from known periodic orbits eases the identification of nearby orbits in the case where one of them dominates the global search.

The results of these searches ($\mu = 0.83$) yield the following numbers of periodic orbits of period 1 through 12 of (1, 1, 2, 3, 2, 3, 4, 6, 14, 13, 22, 40) . We have searched for orbits up to period 23, at this time well over 4000 orbits have been found. This number is still increasing slowly. Obviously, it is quite difficult to determine if all the periodic orbits of given length have been located. The number of periodic points of any given period provides an estimate of the entropy of the map [10]. In this case we have the estimates near 0.5. Equations (1) also have a fixed point near (1.532, -2.071); as this point appears far from the attractor of interest, we have omitted it from the listing above.

3. The Reconstructed Flow

The basic question in reconstructing a flow or map is one of interpolation in a high dimensional space with irregularly spaced data points. As noted below, the data may be combined from several different probes. To each observation, $\mathbf{x_i}$, there will correspond an image or future vector, $\mathbf{y_i}$. Following Casdagli [11], I shall use a radial basis function as the basis for interpolation.

Consider a deterministic system with phase space dimension M_s. Given n distinct base points $(\mathbf{x}_i, i = 1, 2, ..., n)$, consider a single component of each image vector $\mathbf{y_i}$. Let this component be labeled s_i $(i = 1, 2, ..., n)$. The goal is to determine a predictor $f(\mathbf{x_i}) : R^{M_s} \rightarrow R$ such that:

$$f(\mathbf{x_i}) = s_i \tag{2}$$

Following Powell [12], consider $f(x)$ of the form

$$f(\mathbf{x}) = \sum_{i=1}^{n} \lambda_i \phi(||\mathbf{x} - \mathbf{x_i}||) \tag{3}$$

where $\phi(r)$ are radial basis functions and the λ_i are constants which are uniquely determined by Equations (1) provided the matrix

$$A_{ij} = \phi(||\mathbf{x_i} - \mathbf{x_j}||) \tag{4}$$

is nonsingular. This is always the case when the $\mathbf{x_i}$ are distinct and the $\phi(r)$ are radial basis functions (see Refs. [11, 12]). We will restrict attention to the particular cases $\phi(r) = r$ and $\phi(r) = (r^2 + c^2)^{-1/2}$. Reconstructions with the former are more robust in the presence of noise, while those with the latter have lower data requirements (in the low noise case).

In this way, a prediction for one component, s_i, of the image vector $\mathbf{y_i}$ has been constructed. This procedure must then be repeated for each component of the image vector. The resulting set of functions will be referred to collectively as the rbf-map. Note that, with experimental data, the requirement that the base points be distinct may not be satisfied when the data are digitized at low resolution. Even with 64 bit accuracy, it is found that the reconstruction is improved if a minimum distance between the $\mathbf{x_i}$ is maintained.

In the past the $\mathbf{x_i}$ have often been constructed by the method of delays from a single time series. An alternative approach is to use different measurement functions of the system for each element of $\mathbf{x_i}$ (potentially combined with time delays). Here the measurement functions may be distinct probes or a projection of the signal onto a singular value decomposition or Fourier basis. The introduction of measurements in different physical units requires careful thought in determining the scaling of distances in different coordinate directions of the reconstructed phase space. One approach is to choose these scales by optimizing the prediction error profile described in Ref. [7]. Note that the question of rescaling arises in the simple example of a set of coupled o.d.e.'s where the measurement functions may be taken as x, x', x'', and so on, or as nonlinear combinations of these values.

Once an rbf-map has been generated, characteristics of the reconstructed map are accessible. For example, a gradient flow/Newton Raphson style technique may be developed (using finite difference estimates of the local derivatives) to locate the periodic orbits of the rbf map and determine their stability. Note that the stable eigenvalues are best estimated by a reconstruction of the series under time reversal. A comparison of the periodic orbits found in a rbf-map with those found with a knowledge of the Ikeda map is given in the next section.

4. Discussion and Conclusions

Several rbf maps were constructed based on a 2^7 observations taken from 2^8 iteration trajectories from the Ikeda map. In a $\phi(r) = (r^2 + 1)^{-1/2}$ reconstruction, all known periodic orbits on the attractor with period less than 8 were present with their locations accurate to 3 decimal digits and their unstable eigenvalue to within 2%. Estimates of the unstable eigenvalues of orbits up to period 10 were within 5% of their true values. (In some reconstructions erroneous orbits were also found, these however were always far from the attractor.) Note that this accuracy is more than enough to distinguish series originating from the $\mu = 0.83$ case from those from the $\mu = 0.85$ case.

It is desirable to compare the data requirements of this approach to that of other methods for retrieving similar data, for example, by comparing the amount of data required to quantify the unstable periodic orbits by direct observation. This is particularly interesting due to the independent usefulness of this data as noted above.

The most straightforward method for detecting periodic orbits is simply to scan the data stream for near returns; if the $(n+1)^{th}$ iteration of a point is sufficiently near that point, that location is a candidate for a periodic orbit. Clearly, this approach requires a careful definition of "sufficiently near" which should depend on the particular orbit.

Consider the dynamics within a radius ℓ of a period n point. In the limit $\ell \to 0$, the system's dynamics are governed by the linearization of the map about the periodic point; the linear dynamics may be characterized through the stable and unstable eigenvalues, λ_s and λ_u. As a trajectory moves away from a periodic point, each nth iteration will be approximately a factor λ_u farther away from the point. Call this (location dependent) ratio of distances the effective value of λ_u. We estimate the size of the linear region by determining the distance along the unstable manifold at which this effective value differs by more than some threshold from the true value, say 5%. This then defines a visitation radius for each periodic orbit: if the properties of the orbit are to be deduced directly from near visitations, a passage within this radius must be observed. Thus the visitation time provides an estimate of the amount of data required to quantify a periodic orbit to desired accuracy. Note that this is only a lower bound on the length of trajectory required. If the difficulty of finding period n orbits is denoted by the average visitation time of the least visited period n orbit, then the visitation times for orbits of period 1 to 8 ($\mu = 0.83$) are approximately $(2^7, 2^{8.5}, 2^{12}, 2^{11.5}, 2^{12.8}, 2^{15.7}, 2^{16}, 2^{18})$. These results are based on an average over 20 initial conditions, each of which is iterated up to 2^{22} times. This was not long enough for every initial condition to visit some orbits of period 8 and greater; orbits which were accurately located by the 2^7 point rbf-map described above. This demonstrates that the construction of dynamic flows is a vastly more efficient method to determine the spectrum of periodic orbits.

Rbf-maps from experimental data of a nonlinear electronic oscillator [13] have also been constructed. In this case it is not possible to compare the periods and eigenvalues with the "true" values. Examination of the time series near the predicted periodic orbits in addition to the comparison of results from maps constructed from different surfaces of section show that consistent estimates are obtained. A detailed description of these results is presented elsewhere [14].

Finally we note that this approach is applicable to systems where multiple measurement functions are available simultaneously (e.g., where several sensors are in use). Even in the stochastic case, different measurement functions will be related through their dependence on the current state of the system in such a way as to create forbidden regions in the reconstruction space. An application of these ideas to multiple sensor systems has been proposed [15].

In conclusion, we have seen that a system's dynamics, as characterized by the unstable periodic orbits, are quantitatively reproduced in a dynamic reconstruction using radial basis functions. The data requirements of this method are modest in comparison with other approaches and data from systems with similar dynamics may be distinguished using this approach. Applications to experimental data and higher dimensional systems appear to support the belief that these methods shall provide a powerful new tool for the analysis of nonlinear dynamical systems.

Acknowledgments

I am happy to acknowledge discussions with P. Cvitanovic, M. Muldoon, I. McIntosh, M. Powell, and C. Sparrow. This work was supported by the Science and Engineering Research Council and the U.S. Navy under Grant # N00014-88-J1197.

References

1. N.H. Packard, J.P. Cruchfield, J.D. Farmer, and R.S. Shaw. Geometry from a time series. *Phys. Rev. Lett.* **45**, 712 (1980).
2. P. Grassberger and I. Procaccia. Characterization of strange attractors. *Phys. Rev. Lett.* **50**, 346 (1983).
3. L.A. Smith. Intrinsic limits on dimension calculations. *Phys. Lett.* **133 A**, 283 (1988).
4. J. Theiler. Estimating fractal dimension. Preprint 1989.
5. J.D. Farmer and J. Sidorovich. Predicting chaotic time series. *Phys. Rev. Lett.* **59**, 8 (1987).
6. M. Casdagli. Nonlinear prediction of chaotic time series. *Physica* **35 D**, 335 (1989).
7. L.A. Smith. Quantifying chaos with predictive flows and maps: Locating unstable periodic orbits. In *Measures of Complexity and Chaos.* Ed. N.B. Abraham *et al.* Plenum Press, New York, 1990.
8. K. Ikeda. *Opt. Commun.* **30**, 257 (1979).

9. D. Auerbach, P. Cvitanovic, J.-P.Eckmann, G. Gunaratne, and I. Procaccia. Exploring chaotic motion through periodic orbits. *Phys. Rev. Lett.* **58**, 2387 (1987).
10. P. Cvitanovic, P. Gunaratne, and I. Procaccia. Topological and metric properties of Henon type attractors. *Phys. Rev.* **A 38**, 1503 (1989).
11. M.J.D. Powell. Radial basis functions for multivariate interpolation: a review. University of Cambridge Preprint (1985).
12. Michelli. *Constr. Approx.* **2**, 11 (1986).
13. S. Gaito and G. King. Chaos on a catastrophy manifold. In *Measures of Complexity and Chaos.* Ed. N.B. Abraham *et al.* Plenum Press, New York, 1990.
14. L.A. Smith and G. King. Unstable periodic orbits in an experimental chaotic attractor. Preprint 1989.
15. L.A. Smith, K.R. Godfrey, P. Fox, and K. Warwick. A new technique for fault detection in multisensor probes. *Abstract Volume of IEE Conference Control 91*, Edinburgh, 1991.

MODELLING FRACTALS AND TURBULENCE

MOGENS JENSEN

NORDITA
DK-2100 Copenhagen, Denmark

Abstract

In these notes we review some recent work on models of fractals and turbulence. For both cases we look for scaling phenomena of various quantities. The scaling behavior of strange attractors obtained from small scale convection experiments is described within the multifractal formalism. The attractors are modelled by low-dimensional dynamical systems and good agreement between theory and experiment is found. To model turbulent phenomena we use many-dimensional dynamical systems called coupled-map-lattices. In a model for boundary-layer-induced turbulence the scaling behavior of the fluctuations is calculated and qualitative agreement with large-scale convection experiments is found. Finally, we model vortex motion by a discrete Ginzburg-Landau model and measure numerically when such motion becomes turbulent.

1. Modelling Fractals

1.1. *Two–Frequency Systems*

A system with two coupled frequencies shows very rich behavior. When an internal frequency ω_1 (for example an unstable mode) couples non-linearly to a second oscillation of frequency ω_2 (introduced from outside), there is a resonance every time a harmonic of one frequency is sufficiently close to a harmonic of the other frequency:

$$n\omega_1 \simeq m\omega_2 \qquad \text{or} \qquad \frac{\omega_1}{\omega_2} \simeq \frac{m}{n} \tag{1}$$

A resonance remains stable even when some of the parameters (as ω_2 or its amplitude A_2) are slightly changed. A typical phase diagram is shown in Fig.1 [1]. The K-parameter is related to the strength of the coupling between the frequencies and the Ω-parameter is proportional to the external frequency ω_2. Any resonance, with $\frac{\omega_1}{\omega_2}$ rational, fills up a finite area and forms a tongue which widens as K is increased. The figure is an "Arnold-tongue diagram" for systems with two frequencies [2]. The diagram in Fig.1 has been constructed from a very simple dynamical system, a circle map [1-5], of the form:

$$\theta_{n+1} = \theta_n + \Omega - \frac{K}{2\pi}\sin(2\pi\theta_n) \tag{2}$$

Here θ_n is an "angular" variable and K and Ω are constants. This mapping can be found from the dynamics on a cut through the two-torus generated by the two frequencies ω_1, ω_2 [5]. The cut of a torus results in a curve which is equivalent to a circle, therefore the term "circle-map". On this cut it is easy to see that the ratio between the frequencies, $\frac{\omega_1}{\omega_2}$, is equal to the winding number [2, 3]:

$$W = \lim_{n\to\infty} \frac{\theta_n - \theta_0}{n} \tag{3}$$

The lines in Fig.1 are the borders of the Arnold-tongues where W is rational. As K increases beyond 1 the tongues will start to overlap, which causes hysteresis and chaotic behavior. Therefore K=1 defines the critical line for the onset of chaos. The corresponding diagram for the experiment, described in detail in the next chapter, is shown in Fig.2.

Until now we have discussed properties of rational winding numbers. What happens if the winding number is irrational? Then the frequencies do not resonate and the motion is quasiperiodic (incommensurate). The parameter values in the $K-\Omega$ diagram for which W is a given irrational number lie on a smooth line. As K increases, these lines are squeezed more and more between the tongues and finally, at K=1, the quasiperiodic motion is no longer possible and becomes chaotic. This outlines a route to chaos via quasiperiodicity. In this way the K=1 line is the "reverse" to the real axis: On the real axis the irrationals have full measure and the rationals zero measure. On the K=1 line the rationals have broadened to the full measure and the irrationals have zero measure.

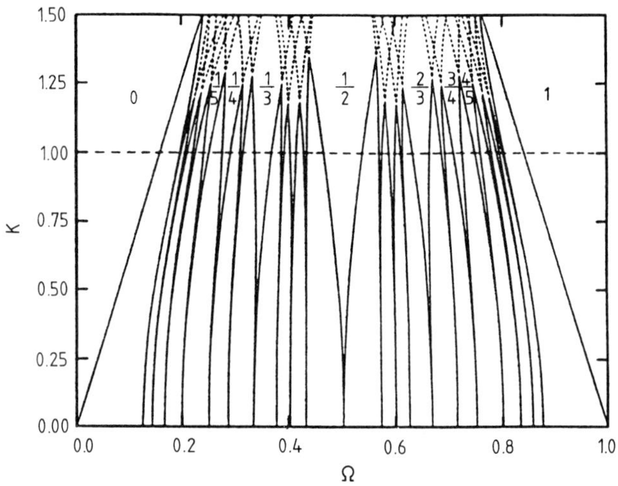

Figure 1: An "Arnold-tongue" diagram for a system with two frequencies [1].

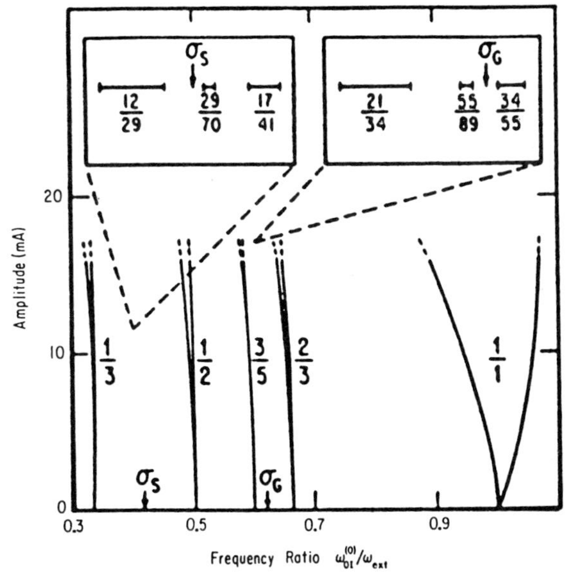

Figure 2: The phase-diagram, A_2 versus ω_2, for the convection experiment on mercury. Note the Arnold tongues and the similarity to Fig.1. The inserts are magnifications around two irrationals, the golden mean and the silver mean [6, 7].

In the following we focus on a particular irrational number, the golden mean $w^\star = \frac{\sqrt{5}-1}{2}$. The reason for choosing this point is that the number has a particular nice continued fraction expansion (consisting of all 1's) [5].

1.2. *The Experimental Setup*

The Rayleigh-Benard experiment to be described has been developed by Stavans, Heslot, and Libchaber [6,7]. The experimental cell is shown in Fig.3a and the general set-up in Fig.3b. The cell has a height of 0.7 cm and a width of 1.4 cm such that precisely two convective rolls are present (Fig.4a). The cell is filled with fluid mercury which has a very low Prandtl number. In the experiment, only the temperature of the upper plate is regulated. The temperature is controlled by the regulation bolometer near the heating coil (Fig.3b). The temperature stability is better than 10^{-5} ^oC. A temperature signal of the flow is measured by the bolometer at the bottom plate.

The convection starts when the temperature difference between the top and the bottom is about 3.0^oC and two convective rolls appear. When the temperature difference is increased further, the convective rolls become unstable against an oscillatory instability. The wavy instability travels along the rolls and creates an oscillatory temperature variation at the bottom bolometer. Thus it is easy to extract the internal frequency $\omega_{OI} = \omega_1$ of this oscillation. The typical period of the oscillation is a few seconds.

The second frequency is introduced into the system from outside. Two thin line electrodes are attached to the top and bottom plates. The external frequency is generated by rectangular ac current pulses whose width is about 1/10 of the period of the oscillatory instability. A typical value of the pulse amplitude is 20 mA. The whole system is embedded in a magnetic field of 200 G parallel to the convective rolls. As mercury is conductive, a Lorentz force acts on the fluid horizontally and transverse to the rolls axes. The action of the Lorentz force is to bend the convective rolls, thereby creating an ac vertical vorticity component in the flow. By varying the amplitude and the frequency of the injected current, Stavans *et al.* [6, 7] mapped out a number of Arnold tongues, as shown in Fig.2. As expected the width of the tongue decreases rapidly as the denominator of the rational number increases. It was possible to observe tongues with denominators up to 200.

To approach the golden mean the experimentalists follow a series of rational tongues, with winding numbers w_n, converging in the limit to $w^\star$. These rationals are written in terms of Fibonacci numbers F_n, $w_n = \frac{F_{n-1}}{F_n}$, where $F_{n+1} = F_n + F_{n-1}$, F_0=0, F_1=1 [4,5]. As the amplitude of the ac signal is increased the experiment approaches the critical point where chaos sets in. For a certain value of the external frequency and amplitude, the experiment is now at criticality with golden mean winding number. From the temperature measured in units of the external period by the bottom bolometer, we obtain a strobed signal T_n. Fig.4 shows the attractor with

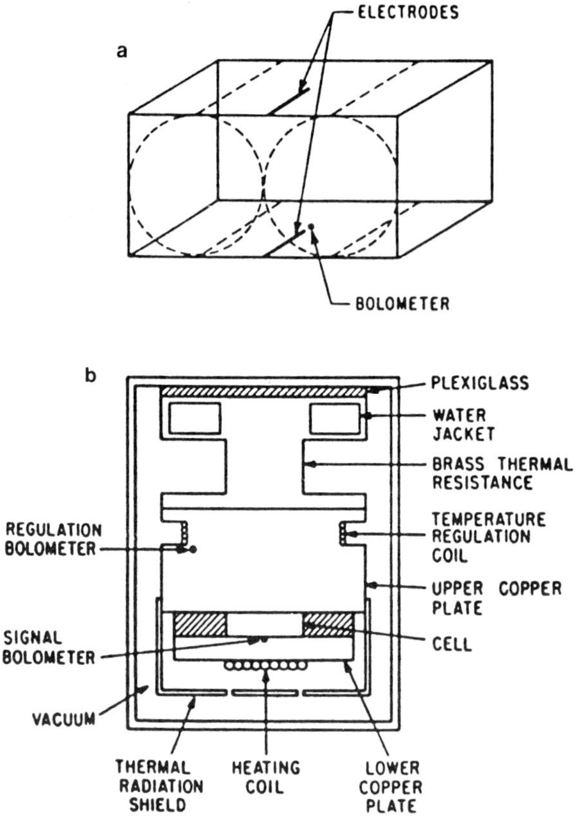

Figure 3: a) A schematic diagram of the Rayleigh-Benard cell. The height of the cell is 0.7 cm. An ac current is passed through the electrodes and the experimental signal is taken from the bottom bolometer. b) The experimental set-up containing the cell in Fig.3a. The set-up is discussed in details in Refs. [6, 7].

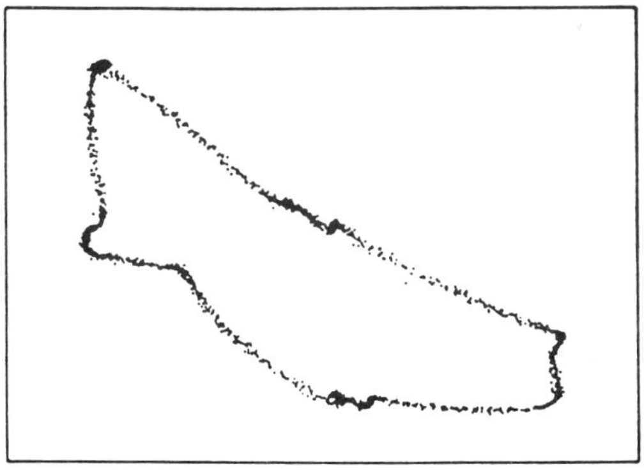

Figure 4: The experimental attractor with golden mean winding number. Plotted is a phase space in the strobed temperature signal, T_{n+1} versus T_n. Note the variation in density along the attractor.

T_{n+1} plotted against T_n. We observe an interesting structure with a large variation in density. Our goal is to study the scaling properties of this attractor in detail, i.e. its multifractal properties which is done in Chap.1.4.

Similar to the experiment, we construct an attractor for the circle map model. The winding number should be $w^\star$ at the point where chaos sets in. Criticality is for K=1 and Ω must be chosen such that W, Eq.(3), is equal to $w^\star$. The golden mean is again approached with Fibonacci tongues [4] giving the value Ω_{gm}=0.6066.... The corresponding attractor with the angular variable, θ_n, plotted on the circle is shown in Fig.5. Again there is a large variation in density, a variation we wish to quantify through the multifractal formalism and compare the results with the results obtained from analyzing the experimental attractor, Fig.4.

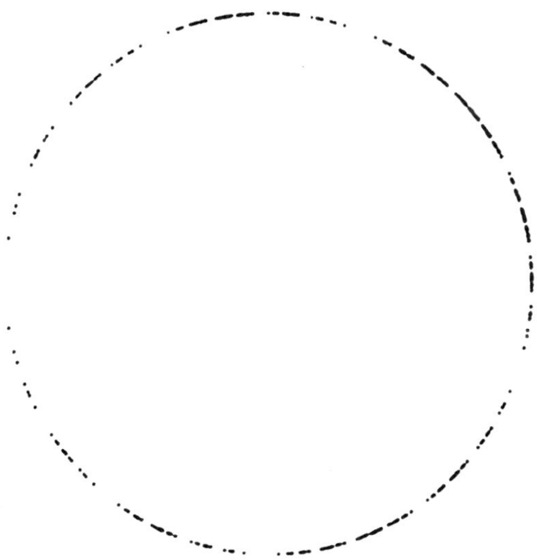

Figure 5: The "theoretical" attractor of the circle map, Eq.(2), with golden mean winding number and K=1. Note also here the variation in density along the attractor.

1.3. *Multifractal Formalism*

A multifractal is composed of infinitely many interwoven "single" fractals. The nature of such behavior has been widely discussed [8-15]. One dimension is insufficient for a characterization of a multifractal. More information is given by a continuum of dimensions, the Rènyi dimensions D_q [16,17]. A similar characterization is obtained by the spectrum of scaling indices, the $f(\alpha)$ spectrum, defined by Halsey *et al.* [10]. The D_q-function and the $f(\alpha)$ spectrum are related by a Legendre transformation.

We start with a fractal measure which in the case of the attractors in Figs.4,5 is given by the density of the points. The set is partitioned into N pieces of sizes l_i and the probability of the ith piece is denoted p_i. Then a partition sum is constructed [10]:

$$\Gamma(q,\tau) = \sum_{i=1}^{N} \frac{p_i^q}{l_i^\tau} \tag{4}$$

As $l = \max_i l_i \to 0$, three possibilities may occur. If $\tau > \tau(q)$ then $\Gamma(q,\tau)$ diverges to infinity; if $\tau < \tau(q)$ then $\Gamma(q,\tau)$ converges to zero. At $\tau = \tau(q)$ the partition sum tends to a constant defining a functional relation between τ and q. $\tau(q)$ is related to the generalized dimensions by $\tau(q) = (q-1)D_q$. Next we assume that the density in the ith piece scales like a power law in the $l \to 0$ limit:

$$p_i \sim l^{\alpha_i} \tag{5}$$

Eq.(5) defines the pointwise dimension, α_i, for any point of the set. Let us say that the pointwise dimension of a specific point on the set is equal to α_1. This value may be found on a subset of the whole set of a dimension $f(\alpha_1)$ [10]. So, the set with dimension $f(\alpha_1)$ is one of the interwoven subfractals. Similarly, another value of the scaling exponent, α_2 say, is found on a subset of dimension $f(\alpha_2)$ and so on. In general, a continuous $f(\alpha)$ spectrum is defined over an interval in α. The partition sum Eq.(4) is transformed into an integral in α. As a typical length scale we use $l = \max_i l_i$. The number of times α assumes a value in the interval $[\alpha'\ ,\ \alpha' + d\alpha']$ is then:

$$d\alpha' \rho(\alpha') l^{-f(\alpha')} \tag{6}$$

Inserting this and the scaling ansatz Eq.(5) into Eq.(4) we obtain:

$$\Gamma(q,\tau) = l^{-\tau} \int d\alpha' \rho(\alpha') l^{q\alpha' - f(\alpha')} \tag{7}$$

In the limit $l \to 0$, the minimal value of the exponent to l under the integral will dominate. This value is found via a saddle-point approximation:

$$\frac{d}{d\alpha'} \left[q\alpha' - f(\alpha') \right]_{\alpha' = \alpha(q)} = 0 \tag{8}$$

This leads to the following Legendre transformation which is used to calculate the $f(\alpha)$ spectrum [10]:

$$\begin{aligned} \frac{d\tau}{dq} &= \alpha & \qquad \tau(q) &= \alpha q - f \\ \frac{df}{d\alpha} &= q & \qquad \frac{d^2 f}{d\alpha^2} &< 0 \end{aligned} \tag{9}$$

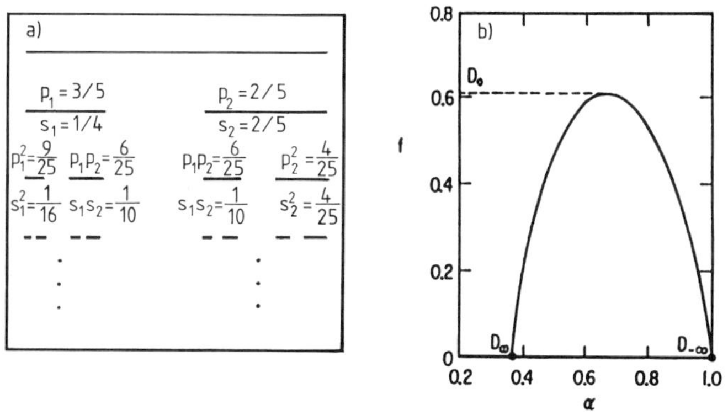

Figure 6: a) A two-scale Cantor set constructed from two length scales s_1, s_2 and two values of the measure, p_1, p_2 (note that in the formalism, $s_i \sim l_i$). b) The corresponding $f(\alpha)$ spectrum [10].

We observe that the $f(\alpha)$ spectrum is convex and that the slope at each point is q. As $q \to \infty$ the largest p_i (i.e., the most concentrated part of the multifractal) dominate the partition sum Eq.(4). This corresponds to the place where the $f(\alpha)$ curve vanishes with infinite slope (if there is no phase transition, see Chap.1.5) which is at its left-most part for the minimum α value. As $q \to -\infty$ the smallest p_i dominate (i.e., the least concentrated part) and at the corresponding α-value the right-most part of the $f(\alpha)$ curve vanishes with (negative) infinite slope.

The simplest multifractal one can imagine is a two-scale Cantor set, shown in Fig.6a. On the second level the partition function for this set is very simple

$$\Gamma = \left(\frac{p_1^q}{l_1^\tau} + \frac{p_2^q}{l_2^\tau}\right) \tag{10}$$

and will on higher levels follow a binomial expansion [10]. We calculate $\tau(q)$ and obtain the $f(\alpha)$ spectrum shown in Fig.6b. The spectrum has the expected convex form, is smooth and continuous and its maximum point is the dimension D_0. The minimum value of α is $\alpha_{min} = \frac{\log p_1}{\log l_1}$ and the maximum value is $\alpha_{max} = \frac{\log p_2}{\log l_2}$.

1.4. *Universality of $f(\alpha)$ Spectra: Theory and Experiment*

We now analyze the attractor of the circle map with golden mean winding number, Fig.5. The attractor is partitioned into small boxes of sizes l_i and the measures p_i in every box must be calculated. In practice we assign the same probability to each box, $p_i = \frac{1}{N}$ where N is the number of points on the attractor. The length scales are then the distances from a point to its nearest neighbor. The partition sum, Eq.(4), gives the function $\tau(q)$ and via the Legendre transform, Eq.(9), the $f(\alpha)$ spectrum. This spectrum is shown in Fig.7 and it quantifies the global scaling structure of the attractor. The interesting point is that the spectrum is universal for the class of critical circle maps with golden mean winding number at the onset of chaos [18]. Therefore it is compared to experiments at the same point of criticality, i.e., at the onset of chaos and golden mean winding number. We already discussed a set with these properties, the attractor from the Rayleigh-Benard experiment shown in Fig.4. The set is partitioned into small boxes and the density of points in each box is calculated.

In practice one embeds the experimental signal in 3-d (to avoid accumulations caused by 2-d projections) [19]. The corresponding $f(\alpha)$ values is shown as the dots in Fig.7. The error bars are due to finite amount of data points (~5000) and a long time drift in the experiment. We see that within the error bars the theory and the experiment agree [19].

What are the conclusions that can be drawn from Fig.7? The results indicate that the forced Rayleigh-Benard experiment and the critical circle map are in the same universality class. Although the theoretical attractor, Fig.5, and the experimental attractor, Fig.4, look quite different, they belong from a scaling point of view to the same class of multifractals. Thus the $f(\alpha)$ formalism is a convenient tool to quantitatively characterize fractal sets and compare them to each other. The universality of the $f(\alpha)$ spectrum with golden mean winding number has been confirmed in a completely different experiment, a driven photo conductor [20]. The system is driven so hard that it is in a regime of negative differential resistance which gives rise to a spatial modulation of charge in the photo conductor giving a very stable internal oscillation (ω_1). The conductor is driven by an external ac current of frequency ω_2. The two frequencies couple and we have the situation described earlier. The system is driven to chaos at the ratio of the frequencies equal to the golden mean. A strobed attractor (of ~ 100000 points!) is obtained in the current signal and the corresponding $f(\alpha)$ spectra show striking agreement between theory and experiment. Therefore this experiment also belongs to the universality class described by the critical circle map.

1.5. *Tips and Fjords on Fractal Aggregates*

We now turn to study the multifractal properties of fractal aggregates, like diffusion limited aggregates or viscous fingers in a porous medium. We want to quantify the "branchiness" of such structures and develop a quantitative description of their growth. Several calculations of $f(\alpha)$ spectra for aggregates have been performed, see Fig.8. These calculations usually rely on various box-counting methods using the

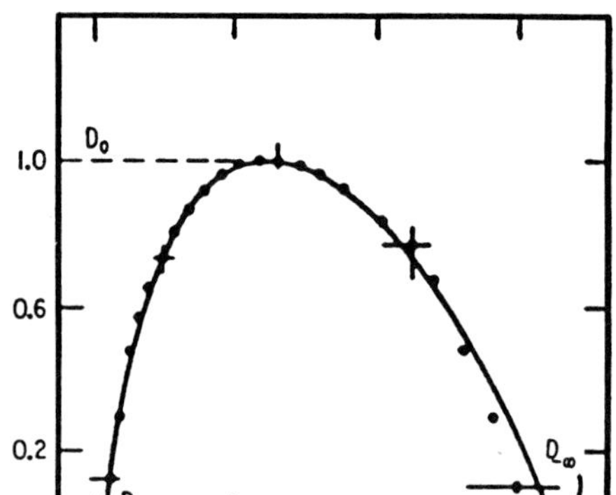

Figure 7: The $f(\alpha)$ spectrum for the "theoretical" attractor, Fig.5, is shown as the solid line. The $f(\alpha)$-values of the experimental attractor, Fig.4, are shown as the dots [19].

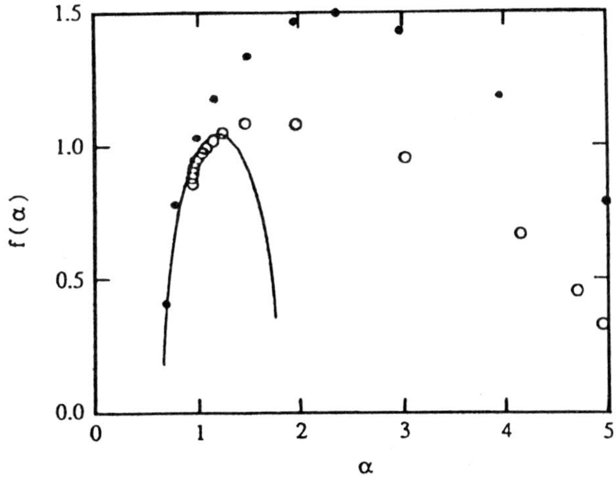

Figure 8: Multifractal $f(\alpha)$ spectra calculated from an ensemble of DLA clusters (black dots) [23], an experiment on non-Newtonian fluids (circles) [24], and an experiment on viscous fingers (curve) [25].

partition sum, Eq.(4), over the boxes. It is typically found that the convergence for the tips (i.e., the $\alpha' s < 1$) is very good but the convergence for the fjords (i.e., the $\alpha' s > 1$) is extremely poor. In Fig.8 we see that some calculations obtain very large α values in the fjord regime and altogether there are big discrepancies between the various calculations.

We argue it is not at all surprising when these calculations disagree in the fjord regime. This is due to the fact that in probabilistic diffusion or growth there can be fjords which are highly screened from the growth field. The typical behavior is that the more screened a fjord is, the larger the corresponding α value. Actually, it is even possible that some fjords are completely screened, causing $\alpha_{max} \to \infty$: this appears as a phase transition in the corresponding $f(\alpha)$ spectrum [26,21]. On the other hand, the tip regime is very stable and insensitive to changes of the model and we expect to find a more or less universal growth-spectrum for typical aggregates. This is consistent with previous results shown in Fig.8: their $f(\alpha)$ spectra agree up to α=1, i.e. in the tip regime. We base our intuition on calculations with fractal sets found in the theory of iterations of conformal mappings, using them as a probe for fractal aggregates [21,22]. There we can calculate the $f(\alpha)$ spectra with accuracy, not attainable by a box counting method.

The use of conformal mappings and their Julia sets as models for diffusion aggregates is based on an analogy between a diffusion field and electrostatics. A DLA cluster, for instance, is generated by diffusing particles and the diffusion is described by the Laplace equation. Thus we have a field which is a solution to the Laplace equation with the boundary condition that the field is zero on the aggregate. The probability of growth is then the gradient of the field at the boundary. This is equivalent to finding the electrostatic potential around the aggregate with the boundary condition that it is zero at the aggregate. The gradient of the potential is the electric field, thus the electric field is the growth measure, or the harmonic measure [27].

As a simple prototype conformal map we consider

$$f(z) = z^2 + c \tag{11}$$

with complex parameter c [28,29]. For a specific value of c, the corresponding Julia set is the boundary of the set:

$$\{z| \lim_{n\to\infty} |f^n(z)| < \infty\} \tag{12}$$

We are interested in cases where the Julia set has resemblance to fractal aggregates. Thus, first of all the set should have branches on top of branches and be "skinny" (i.e. have small or zero measure). Examples of such sets are obtained for c values corresponding to the Misiurewicz points [30]. Figure 9 shows an example of such a case, it is chosen to have five "arms" in order to stress the resemblance with probabilistic diffusion aggregates (by varying c one can obtain any number of arms). Our goal is to solve the Laplace equation around the set, i.e., to find the potential lines. The potential at the point z is given by [31]:

$$U(z) = \lim_{n\to\infty} \frac{1}{2^n} \log |f^n(z)| \tag{13}$$

Notice that $U = 0$ on the Julia set, i.e., it is grounded. The pointwise dimension α_i at any point z_i on the Julia set, is:

$$U(z_i + \epsilon) \sim \epsilon^{\alpha_i} \quad , \quad \epsilon \to 0 \tag{14}$$

The minimum value α_{min} is found at one of the fixed points, $z_0 = \frac{1+\sqrt{1-4c}}{2}$, of Eq.(11) [21]:

$$\alpha_{min} = \frac{\log 2}{\log f'(z_0)} \tag{15}$$

α_{min} <1 and this point is thus located at the tip with the strongest divergence of the electric field ($\vec{E} = \nabla U$).

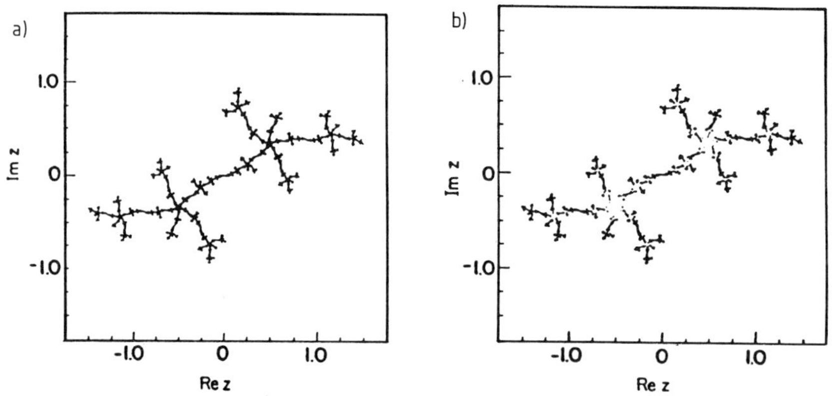

Figure 9: a) A Julia set with five arms for the map, Eq.(11), with $c = -.636754 - .685031i$. Note its visual resemblance with a fractal aggregate. b) The same Julia set found from inverse iterates of the map, Eq.(13). The density of the inverse iterates is proportional to the electric field and is thus equivalent to the harmonic measure on a "real" aggregate [21].

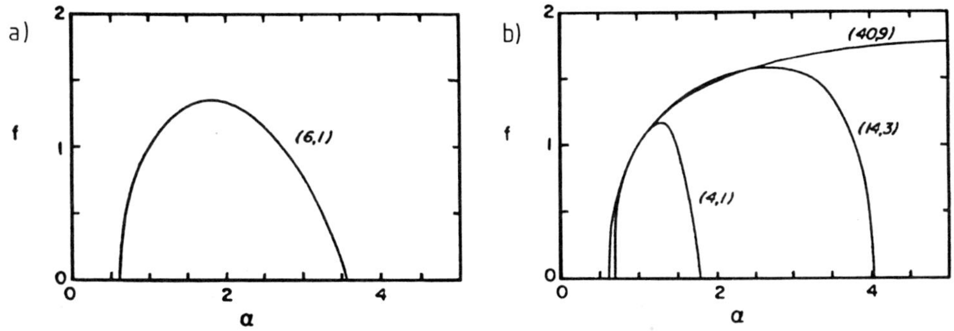

Figure 10: a) The $f(\alpha)$ spectrum for the Julia set in Fig.9. b) The $f(\alpha)$ spectra for a series of Julia sets with increasing density of tiny branches. Note that α_{max} becomes increasingly large. For the parameter values, see Ref. [21].

In practice, we do not have to do the cumbersome job and calculate the α value in each point on the set from Eq.(14). Starting at the tip z_0 and iterating the map Eq.(11) backward we obtain all other tips on the set, and the corresponding α's are found from the derivative of the map in a given point [21]. Similarly, we obtain all other fjords from backward iterates of one fjord. Figure 9b shows a plot of all these backward iterates. Calculating the $f(\alpha)$ spectrum for these type of sets is thus very easy; we use backward iterates and obtain the spectrum shown in Fig.10a. The analogy to the diffusion aggregates is even stronger: the density of inverse iterates on Fig.9b is equal to the electric field on the surface and is thus exactly equivalent to the harmonic measure of a growing aggregate!

To obtain a comparison with a real aggregate we should add many more tiny branches to the Julia set in Fig.9. The corresponding $f(\alpha)$ curves are shown in Fig.10b [21]. The results are interesting! We see that the tip regime is basically unchanged in the series. The fjord regime however varies wildly: α_{max} eventually diverges to infinity. This shows that some fjords are completely screened. By analogy, in real aggregates some fjords are expected to be completely inactive and not grow at all. $\alpha_{max} \to \infty$ can be interpreted as a phase transition in the $f(\alpha)$ spectrum at the Hausdorff dimension (i.e., at q=0). This observation may explain the experimental findings in Fig.8. There is no reason that the calculations should agree for $\alpha > 1$ as some fjords are screened and the calculation here is meaningless. Note, however, that the tip regimes computed from the Julia sets are in agreement with the tip regimes of the experiments in Fig.8. The tip structure is responsible for the growth and does not care about the fjords that are left behind, more or less screened. The possibility even exists that some fjords will have exponentially small field giving $\alpha_{max} = \infty$ (for instance, a rectangular channel has exponentially small field from solving Laplace equation) [32].

2. Modelling Turbulence

2.1. *Chaos, Turbulence, and Coupled Maps*

In the previous paragraphs we described fractal structures generated at the onset of chaos in low-dimensional dynamical systems. From the beginning of the "chaos-revolution" many believed that the understanding of chaos in low-dimensional systems would lead to a different attack on the problem of turbulence. This has not yet appeared to be the case, and the reason is probably that chaos and turbulence are two quite different things. A low-dimensional chaotic system is in general coherent in space, i.e., a time signal from the system behaves chaotically in time but each point in space behaves completely coherently. A good example of such system is the "small" convection box of Libchaber and co-workers discussed in Sect. 1.2. This box is intentionally kept very small to maintain the coherence and allows for studies of low-dimensional chaos. Contrary to this, a turbulent state shows incoherent structures in space. We can also say that many spatial modes will be excited. Therefore experimentalists usually study turbulence in "big" systems to encounter the spatial incoherence. As a turbulent system (for instance a convection box) becomes larger, more and more spatial modes will be excited, eventually providing infinitely many active modes. So, whereas a low-dimensional chaotic state may exhibit a strange attractor of low fractal dimension, a fully developed turbulent state may have an infinite dimensional attractor.

Even though we have argued above that there is a fundamental difference between chaos and turbulence many researchers have tried to connect the two phenomena in various ways. One approach is to couple many low-dimensional dynamical systems (i.e., maps) on large spatial structure and in this way study how the spatial coherence disappears. These so–called coupled-map-lattices (CML) have mainly been advocated by Kaneko [33] and they have proven quite useful to develop intuition about the temporal and spatial behavior in larger systems. We shall in a few examples use a similar approach, but before defining the models let us say a few words about the limitations. CML's do not have conservation laws which are very fundamental for say Navier-Stokes turbulence (at least down to some small "dissipation" scale). Usually, also only one field is introduced in CML's thus omitting the interplay between, for instance, velocity and temperature. With these remarks in mind we shall look at two models, one for boundary layers and one for vortex motion.

2.2. *A Model for Boundary - Layer - Induced Turbulence*

When a fluid or a gas in a container is heated from below, a convective state will commence as the temperature gradient exceeds a critical value. In the convective state, the hot fluid (or gas) moves upwards and the cold fluid downward, but close to the bottom and top plates there can be a small region, a thermal boundary layer, where heat is not transported by convection but only by diffusion [34]. As the temperature gradient is increased further up to high values of the Rayleigh number ($\sim 10^8$), this boundary layer becomes unstable and begins to emit patches ("hot plumes") into the laminar convective regime [35]. Figure 11 shows a picture of such thermal plumes [36]. The motion of these plumes may cause large temperature fluctuations which are associated with a "hard" turbulent state [35].

Here we discuss a very simple phenomenological model which shows features qualitatively similar to the experiment. The model is a coupled-map lattice system introduced in Ref. [37]. We make use of maps that exhibit both turbulent and laminar behavior and are of the form [38]:

$$f_r(x) = \begin{cases} rx & x \leq 0.5 \\ r(1-x) & 0.5 < x \leq 1 \\ x & x > 1 \end{cases} \tag{16}$$

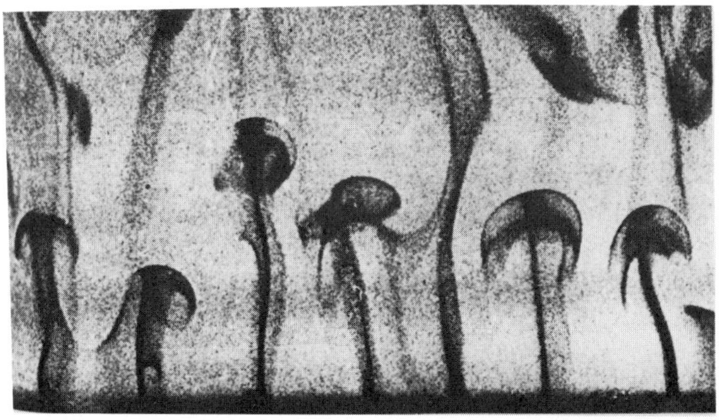

Figure 11: A picture of thermal plumes rising from an unstable thermal boundary layer [36].

The motion is chaotic when $x \leq 1$, which plays the role of a turbulent (or "hot") state, and the motion is laminar for $x > 1$. When coupled diffusively in 1-d and 2-d, Chaté and Manneville found that, as the coupling strength ϵ exceeds a critical value ϵ_c, a turbulent site will percolate through the system [37]. When ϵ is below ϵ_c any initial state will turn into a laminar state.

Our model also uses a diffusive ϵ-coupling but in addition we introduce a convective term of strength v that displaces "hot" fluid (i.e., sites with $x \leq 1$) upwards:

$$\begin{aligned} x_{n+1}^{(i,j)} &= \frac{\epsilon}{4}\left(f_r(x_n^{(i-1,j)}) + f_r(x_n^{(i+1,j)}) + f_r(x_n^{(i,j-1)}) + f_r(x_n^{(i,j+1)}) - 4f_r(x_n^{(i,j)})\right) \\ &\quad + v\left(f_r(x_n^{(i,j-1)}) - f_r(x_n^{(i,j)})\right) + f_r(x_n^{(i,j)}) \end{aligned} \tag{17}$$

(i,j) is a point on a $N \times N$ lattice and n is the time step.

Next a boundary condition at the bottom is introduced in order to play the role of a constant "hot" temperature at the bottom plate. Since the sites with $x \leq 1$ are the "hot" sites we enforce this condition by the constraint:

$$x^{(i,1)} = x_B \quad , \quad x_B < 1 \tag{18}$$

This means the x-value in the first row is kept fixed at x_B (in the following we set x_B=0). The dynamical behavior of the model is simulated on a computer. It is initiated in a laminar state $x_0^{(i,j)} = 1.1 + \eta$ (where η is a small amplitude noise term). The time evolution can be visualized by marking the hot sites, i.e., sites where $x^{(i,j)} \leq 1$. In that way it is easy to observe the hot plumes that travel through the laminar regime. For small values of the convection term (i.e., where v is less than a critical value v_c), the simulation shows that the "hot" boundary condition at the bottom will introduce a "hot" boundary layer of depth of a few lattice lengths. Above the boundary layer the system is in its laminar state. This thermal boundary layer is identified by calculating the average number of hot sites in each layer above the bottom. This number shows a sharp gradient over the width of the boundary layer and then goes to zero in the laminar state, just like the sharp temperature gradient in experimentally observed thermal boundary layers.

As the strength of the gradient term exceeds a critical value v_c, the boundary layer becomes unstable and starts to emit patches (hot plumes) into the laminar regime. The shapes and sizes of the plumes can vary a lot. Figure 12 shows a snapshot of the simulation. For ϵ=0.12 used in Fig.12 we find that $v_c \simeq 0.018$ by monitoring the value of v for which the boundary layer begins to emit plumes. These observations are qualitatively similar to the experimental findings. In the visualization of the experiment one observes that the boundary layer becomes unstable against small amplitude traveling waves and as the Rayleigh number is increased, the waves may detach as convective plumes [35].

In order to understand the boundary layer instability in more detail we consider the variable $x_n^{(i,j)}$ within a simple mean-field approximation. A mean-field value m_j of $x_n^{(i,j)}$ is defined at each layer in the lattice (for specific values of r, ϵ and v):

$$m_j = \frac{1}{T}\frac{1}{N}\sum_{i=1}^{N}\sum_{n=1}^{T} x_n^{(i,j)} \tag{19}$$

In other words, we average $x_n^{(i,j)}$ both over space (N sites) and time (T time steps) to obtain a mean-field value m_j for layer j. By definition, $0 \leq m_j \leq \frac{r}{2}$, and of course $m_1 = x_B$. Within the mean field we compare the diffusive effects against the convective effects. At a specific layer in the lattice, $\bar{j}$, chosen in the top of the boundary layer, the following two terms are calculated:

$$Di(r,\epsilon,v,\bar{j}) = \frac{\epsilon}{4}(m_{\bar{j}-1} + m_{\bar{j}+1} - 2m_{\bar{j}}) \tag{20a}$$
$$Co(r,\epsilon,v,\bar{j}) = v(m_{\bar{j}-1} - m_{\bar{j}}) \tag{20b}$$

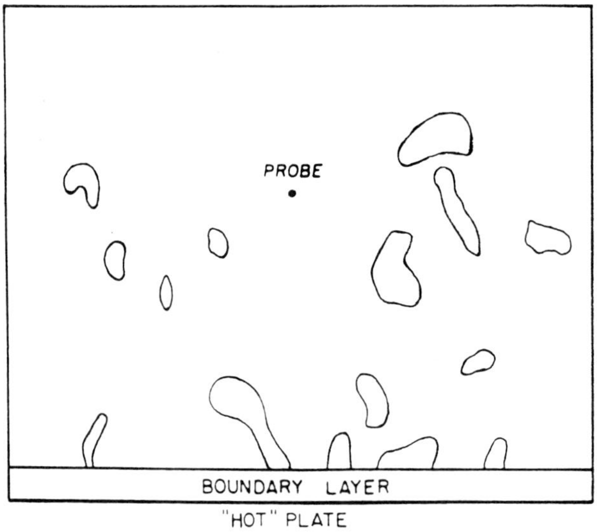

Figure 12: A snapshot of the simulation at ϵ=0.12, v = 0.04 > $v_c \sim 0.018$, r=3.0, and x_B=0 on a 50 × 50 lattice. The patches are the "hot plumes" for which $x^{(i,j)} < 1$. The plumes are released from the boundary layer and drift upwards by convection.

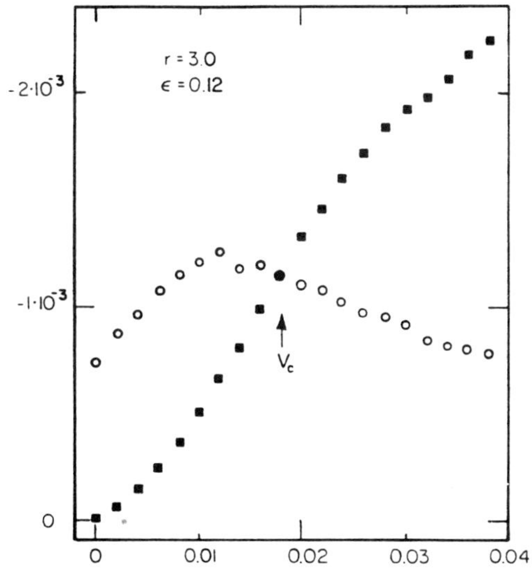

Figure 13: Plot of the diffusive term $Di(3, 0.12, v, 3)$ (Eq.(20a)), shown by circles, and the convective term $Co(3, 0.12, v, 3)$ (Eq.(20b)), shown by squares, *versus* the gradient strength v. The curves cross at the point v_c, where the boundary layer becomes unstable. The size is N=100.

These two terms are the diffusive and convective terms in the local mean field. For ϵ=0.12, the top of the boundary layer is around $\bar{j}$=3 which we use in Eq.(20). Figure 13 shows $Di(3, 0.12, v, 3)$ and $Co(3, 0.12, v, 3)$ plotted *versus* v. The two curves cross at $v \simeq 0.018$. This is the point where, for increasing values of v, the convective term becomes larger than the diffusive term. Above this point the boundary layer is only a diffusive layer but will be unstable against convection and patches from the top of the boundary layer will be released into the laminar regime. Therefore this crossing point is the critical point for the boundary layer instability, v_c.

One of the most striking observations in the experiment [35] was that the statistics of the temperature fluctuations changes from a Gaussian in the "soft" turbulence regime to an exponential in the "hard" turbulence regime. As discussed in Ref. [35] the exponential distribution is caused by the plumes motion around the probe in the center of the cell. Recently, this distribution has been obtained analytically by Sinai and Yakhot [39]. The main purpose of our study is to check whether the fluctuations behave similarly in this simple model. We monitor the fluctuations by placing a probe at a specific point in the center of the cell (here at (i, j)=(25,18) with N=50, see Fig.12). Next the number of time steps, t_p, for each plume to pass the probe, is measured. This time plays the role of a temperature fluctuation, i.e., a long time (large plume) will likely give rise to a large fluctuation in the temperature. As the system evolves many plumes sweep intermittently across the probe. The corresponding distribution of times, $D(t_p)$, is plotted in Fig.14. The straight line indicates an exponential distribution:

$$D(t_p) \sim \exp(-at_p) \tag{21}$$

To check whether this law is robust to changes in the parameters, the distribution is calculated for four different values of v and ϵ. When normalized as in Fig.14 there

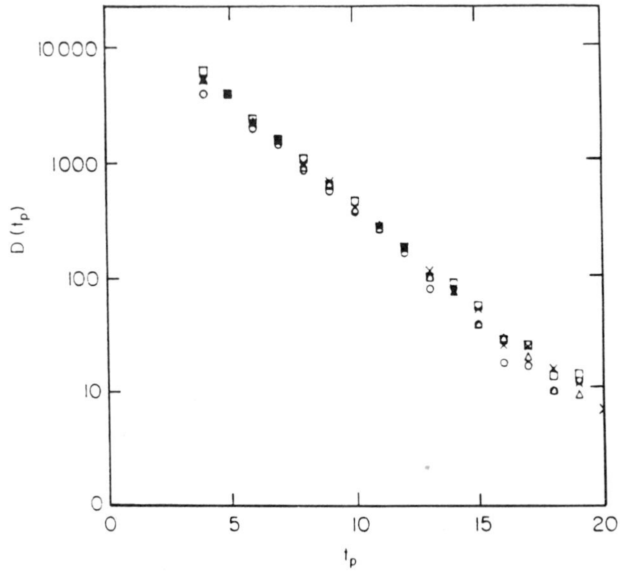

Figure 14: The distribution of passage times, log $D(t_p)$, *versus* t_p. Circles: ϵ=0.12, v=0.035; Triangles: ϵ=0.12, v=0.05; Squares: ϵ=0.12, v=0.08; Crosses: ϵ=0.14, v=0.05. The different curves are normalized to the same value of $D(t_p)$ at t_p=5. Measurements for $t_p \leq 3$ are disregarded. Each calculation is performed over $\sim 10^6$ time steps.

does not seem to be any significant dependence in the constant a on the parameters v and ϵ.

2.3. *Vortex Turbulence in a Ginzburg - Landau Model*

The previous paragraph described a system where both temporal and spatial degrees of freedom are excited and we call the evolution turbulent. An important question is of course: What do we mean by turbulent and how can we quantify it? To answer this we again use the experience obtained from low-dimensional dynamical systems. Here, chaotic behavior is characterized by means of Lyapunov exponents and we say the system is chaotic if one of them is positive. In dynamical systems there are just a few Lyapunov exponents whereas a turbulent system with infinite degrees of freedom will have an infinite number of Lyapunov exponents. We can however still use the same criteria: we say the system is turbulent if one of the "many" Lyapunov exponents becomes positive. In this section we consider a system of turbulent vortex motion where we specifically try to calculate when the turbulence sets in [40,41].

Some chemical reactions, like the Belousov-Zhabotinsky, provide beautiful examples of complex spatial patterns. In a stirred reactor this system can undergo a Hopf bifurcation and go into a temporally periodic spatially uniform state. In the absence of stirring, inhomogeneities appear which affect the local periods and thus tend to dephase the different parts of the system. In two dimensions one observes target patterns, spiral waves, or vortices and the dynamics can be quite complicated [42,43]. Such phenomena have been modelled by the complex Ginzburg-Landau equation [41-48]. This equation is derived by assuming that the "order parameter", A, which is a complex field giving the amplitude and phase of the lowest temporal Fourier mode, is small and slowly varying in space and time:

$$\dot{A} = \mu A - (1+i\alpha)|A|^2 A + (1+i\beta)\nabla^2 A \tag{22}$$

μ, α and β are real numbers. The parameter μ is the usual Landau coefficient, positive μ gives nonzero values of the order parameter. Eq.(22) has a spatially homogeneous solution $A = \sqrt{\mu}e^{-i\alpha\mu t}$ with "rotating" frequency $\omega = \alpha\,\mu$. The periodicity of the phase variable leads to the possibility of topological defects in the form of vortices. At the center of the vortex the phase is singular, but, by letting the modulus vanish, A itself remains well defined. When vortices are present the phase field is not single-valued: The total variation of phase on traversing a loop enclosing one vortex is 2π.

Here we shall determine the transition to turbulence motion of those vortices in a two-dimensional coupled map lattice which closely resembles the complex Ginzburg-Landau PDE [40,41]. We use a map lattice to facilitate the computations on the Ginzburg-Landau equation. The model consists of two parts: a local map $A' = F(A)$ representing the two first terms of (1) and a nonlocal part represented by the diffusion term. The properties of the local map F is quite different from the local map of the model studied in the previous section. There the local map is chaotic (i.e., has a positive Lyapunov exponent), here the local map is completely non-chaotic. Without the diffusion term, Eq.(22) can be written as $\dot{r} = \mu r - r^3$ and $\dot{\phi} = -\alpha r^2$. The time evolution can now easily be reproduced by maps $r_{n+1} = f(r_n)$ and $\phi_{n+1} = \phi_n - \tau\alpha r_n^2$. Here f has an unstable fixed point in 0 and a stable one in $r = \sqrt{\mu}$. f can actually be found exactly from integrating the radial equation: $r^2(t+\tau) = \mu r^2(t)/(\lambda\mu + (1-\lambda)r(t)^2)$ with $\lambda = e^{-2\mu\tau}$.

The diffusion equation has the solution $A(t+\tau_0) = \exp[\tau_0(1+i\beta)\nabla^2]\,A(t)$. As nonlocal map we thus use $\tilde{A} = (1 + \frac{\tau_0}{M}(1+i\beta)\Delta)^M A$, where Δ is a discrete Laplacian and M is an integer that determines the range of the effective interaction. The complete map lattice is now $A_{n+1}(\vec{r}) = F(\tilde{A}_n(\vec{r}))$. We have used periodic boundary conditions on $L \times L$ lattices (L between 25 and 200) and vary α keeping the other parameters fixed. The standard configuration has been $\mu = 0.2$, $\tau = 1$, $\tau_0 = 0.2$ and $\beta = -1$. For these parameters the uniform state (i.e., the state where the field is the same at all sites) is linearly stable for $\alpha < \alpha_2 = 0.82$. We start with random initial conditions, which will have a large number of vortices, and ask when such states become turbulent. Figure 15 shows a typical configuration of a turbulent state of many vortices and anti-vortices.

We determine whether states are turbulent in two ways. The first method computes the largest Lyapunov exponent following the growth of the tangent vector iterated by the Jacobian matrix of the map. In Fig.16 the logarithm of the length of the iterated vector *versus* the discrete time, n, is plotted for a 50×50 lattice. The Lyapunov exponent is the average slope of the curve, which, in Fig.16, is clearly positive – around 0.02 averaged over 7000 iterates. The value of α is here 0.74 which is below the linear stability threshold, α_2. At lower values of α there is a positive slope only up to a finite time T (around 2000 in the inset of Fig.16) and we therefore assign a positive "finite time Lyapunov exponent", λ_f, up to the time T.

The second method of extracting the Lyapunov exponent uses two initial (almost equilibrated) configurations which only differ slightly at one site. The two states are then iterated forward independently and the distance between the states (measured in some appropriate metric) is monitored as a function of time. A plot of the log of

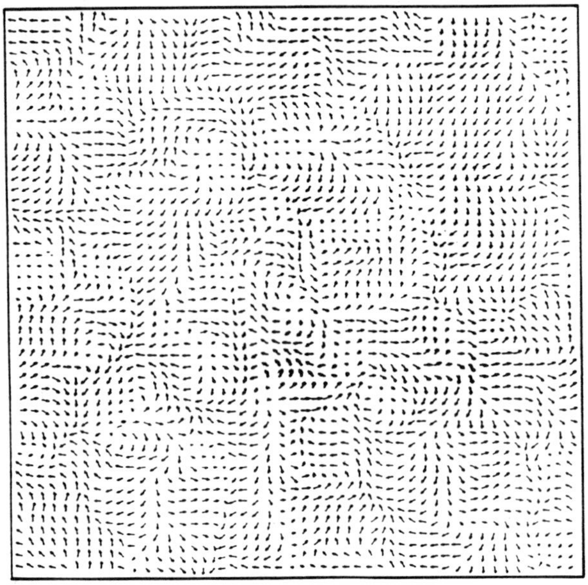

Figure 15: A typical turbulent configuration with many vortices and anti-vortices moving quite a lot between each other. Sometimes a vortex and an anti-vortex collide and annihilate. Sometimes there is spontaneous generation of a vortex-antivortex pair [40].

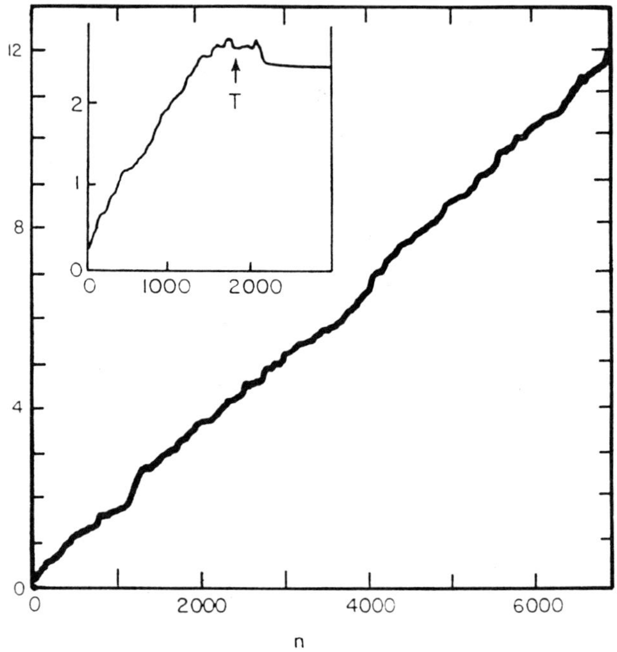

Figure 16: The logarithm of the length of an iterated unit tangent vector versus discrete time n for α=0.74. The inset has α=0.71 and shows that the turbulent state is transient. It only has a positive Lyapunov exponent up to $n = T$ [41].

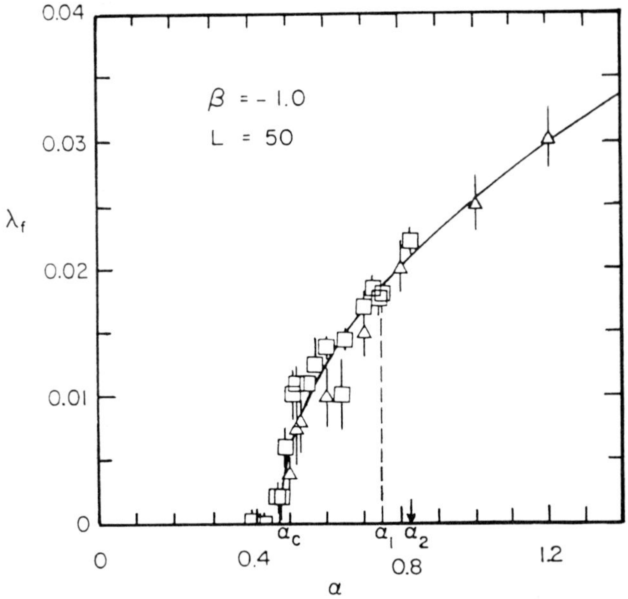

Figure 17: The Lyapunov exponent as a function of the parameter α. The squares were obtained by iterations of a unit tangent vector (as in Fig.16). The triangles were obtained by measuring the distance between two nearby states. For α less than $\alpha_1 \simeq 0.75$ the turbulent state is transient and the Lyapunov exponent is only defined for a finite time T. The true Lyapunov exponent is zero below α_1 as shown by the dotted line. The curve is a power law fit of Eq.(23) with ν=0.50 and α_c=0.48. The threshold of linear instability for the uniform state is at α_2=0.82 [41].

this distance *versus* the time reveals the largest Lyapunov exponent as the slope of the curve. The two methods gave always compatible results.

Figure 17 shows the results of such calculations. The squares represent values of λ_f on a hexagonal lattice obtained by iterating a tangent vector whereas the triangles are obtained on a square lattice using the distance between two almost identical configurations. There seems to be a well defined point, $\alpha = \alpha_c \simeq 0.5$, where λ_f becomes zero and the variation can be fitted fairly well as

$$\lambda_f = (\alpha - \alpha_c)^\nu \tag{23}$$

with $\nu \simeq 0.5$. Clearly, the closer we get to α_c the harder it is to find a good estimate of λ_f, thus the considerable error-bars. The Lyapunov exponents λ_f of the transients show no systematic variation with the system size L.

In the turbulent state, whether transient or not, the number of vortices fluctuates strongly due to creation and annihilation of vortex pairs. The statistical properties, such as the mean vortex density or λ_f, are reasonably independent of the initial condition as opposed to T itself. The final state obtained at $n \gg T$ in the transient turbulent regime $\alpha_c < \alpha < \alpha_1$ contains typically a few entangled vortices. In the entangled state, as described in [40,41], the centers of the vortices are stuck and although they continue to send out waves, the Lyapunov exponent is zero. Opposite vortices attract but the spiral arms can screen each other strongly giving rise to entanglement. The time scale T seems to be determined by the event that one vortex (or maybe two) start outgrowing the others leaving a final state of a few entangled vortices. This event is affected by small details in the state and is thus strongly dependent on the initial conditions. In a larger system more vortices are competing and it is maybe not unreasonable that the turbulent state can persist for a longer time. A recent analytical calculation [49] of the equations of motion for an interacting vortex-antivortex pair leads to a sequence of bound states which might describe entangled configurations. If the transient time $T(L) \to \infty$ as $L \to \infty$ in the whole parameter interval $[\alpha_c, \alpha_1]$, then the transient turbulence will become genuinely turbulent. In that case the infinite system will have a continuous transition probably at a value close to the point α_c, that we already determined. We would then expect the true Lyapunov exponent to scale as (3) close to the transition.

Acknowledgements

I am grateful to Tomas Bohr, Anders W. Pedersen and David Rand for collaborations on the work described in Chap.2.3.

References

1. M.H. Jensen, P. Bak, and T. Bohr. *Phys. Rev. Lett.* **50**, 1637 (1983).
   M.H. Jensen, P. Bak, and T. Bohr. *Phys. Rev.* **A 30**, 1960 (1984).
2. V.I. Arnold. *Am. Math. Soc. Trans., Series 2* **46**, 213 (1965).
3. M.R. Herman. In *Geometry and Topology.* Ed. J. Palis. *Lecture Notes in Mathematics* **597**, 271. Springer, Berlin, 1977.
4. Scott J. Shenker. *Physica* **5 D**, 405 (1982).
5. M.J. Feigenbaum, L.P. Kadanoff, and Scott J. Shenker. *Physica* **5 D**, 370 (1982).
   S. Ostlund, D. Rand, J.P. Sethna, and E.D. Siggia. *Physica* **8 D**, 303 (1983).
6. J. Stavans, F. Heslot, and A. Libchaber. *Phys. Rev. Lett.* **55**, 596 (1985).
7. J. Stavans. *Phys. Rev.* **A 35**, 4314 (1987).
8. B.B. Mandelbrot. *J. Fluid Mech.* **62**, 331 (1974).
9. U. Frisch and G. Parisi. In *Varanna School LXXXXVIII.* Eds. M. Ghil, R. Benzi, and G. Parisi. North-Holland, Amsterdam, 1985, p.84.
   R. Benzi, G. Paladin, G. Parisi, and A. Vulpiani. *J. Phys. A* **17**, 352 (1984).
10. T.C. Halsey, M.H. Jensen, L.P. Kadanoff, I. Procaccia, and B.I. Shraiman. *Phys. Rev.* **A 33**, 1141 (1986).
11. P. Grassberger. *Phys. Lett.* **107 A**, 101 (1985).
12. R. Badii and A. Politi. *Phys. Rev. Lett.* **52**, 1661 (1984).
13. T. Bohr and D. Rand. *Physica* **25 D**, 387 (1987).
14. L. de Arcangelis, S. Redner, and A. Coniglio. *Phys. Rev.* **B 31**, 4725 (1985).
15. R. Blumenfeld, Y. Meir, A.B. Harris, and A. Aharony. *J. Phys. A* **19**, L791 (1986).
16. A. Rènyi. *Probability Theory.* North-Holland, Amsterdam, 1970.
17. H.G.E. Hentschel and I. Procaccia. *Physica* **8 D**, 435 (1983).

18. L.P. Kadanoff. *J. Stat. Phys.* **43**, 395 (1986).
D. Bensimon, M.H. Jensen, and L.P. Kadanoff. *Phys. Rev.* **A 33**, 3622 (1986).
19. M.H. Jensen, L.P. Kadanoff, A. Libchaber, I. Procaccia, and J. Stavans. *Phys. Rev. Lett.* **55**, 2798 (1985).
20. E.G. Gwinn and R.M. Westervelt. *Phys. Rev. Lett.* **59**, 157 (1987).
21. T. Bohr, P. Cvitanović, and M.H. Jensen. *Europhys. Lett.* **6**, 445 (1988).
22. I. Procaccia and R. Zeitak. *Phys. Rev. Lett.* **60**, 2511 (1988).
23. C. Amitrano, A. Coniglio, and F. di Liberto. *Phys. Rev. Lett.* **57**, 1098 (1986).
24. J. Nittmann, H.E. Stanley, E. Touboul, and G. Daccord. *Phys. Rev. Lett.* **58**, 619 (1987).
25. K.J. Måløy, F. Boger, J. Feder, and T. Jøssang. In *Time-Dependent Effects in Disordered Materials*. Eds. R. Pynn and T. Riste. Plenum, New York, 1987, p.111.
26. P. Cvitanović. In *Proceedings of the XIV Colloquium on Group Theoretical Methods in Physics*. Ed. R. Gilmore. World Scientific, Singapore, 1987.
P. Cvitanović. In *Nonlinear Evolution and Chaotic Phenomena*. Eds. P. Zweifel, G. Gallavotti, and M. Anile. Plenum, New York, 1988.
27. N.G. Makarov. *Proc. London Math. Soc.* **51**, 369 (1985).
T.C. Halsey, P. Meakin, and I. Procaccia. *Phys. Rev. Lett.* **56**, 854 (1986).
28. B.B. Mandelbrot. *The Fractal Geometry of Nature*. Freeman, San Francisco, 1982.
29. H.-O. Peitgen and P.H. Richter. *The Beauty of Fractals*. Springer, Berlin, 1986.
30. J. Myrheim and P. Cvitanović. *Phys. Lett.* **94 A**, 329 (1983).
J. Myrheim and P. Cvitanović. *Comm. Math. Phys.* To appear.
31. A. Douady and J.H. Hubbard. CRAS (Paris) **294**, 123 (1982).
32. For further discussions on phase transitions on fractal aggregates, see:
J. Lee and H.E. Stanley. *Phys. Rev. Lett.* **61**, 2945 (1988).
R. Blumenfeld and A. Aharony. *Phys. Rev. Lett.* **62**, 2977 (1989).
B. Fourcade and A.M. Tremblay. *Phys. Rev. Lett.* **64**, 1842 (1990).
33. K. Kaneko. *Prog. Theor. Phys.* **74**, 1033 (1985).
J.D. Keeler and J.D. Farmer. *Physica* **23 D**, 415 (1988).
J. Crutchfield and K. Kaneko. In *Directions in Chaos*, Vol.1. Ed. H. Bao-lin. World Scientific, Singapore, 1988.
34. L.D. Landau and E.M. Lifshitz. *Fluid Dynamics*. Pergamon, Oxford, 1985.
35. B. Castaing, G.H. Gunaratne, F. Heslot, L. Kadanoff, A. Libchaber, S. Thomae, X.-Z. Wu, S. Zaleski, and G. Zanetti. *J. Fluid Mech.* **204**, 1 (1989).
F. Heslot, B. Castaing, and A. Libchaber. *Phys. Rev.* **A 36**, 5870 (1987).
36. E.M. Sparrow, R.B. Husar, and R.J. Goldstein. *J. Fluid Mech.* **41**, 793 (1970).
37. M.H. Jensen. *Phys. Rev. Lett.* **62**, 1361 (1989).
M.H. Jensen. *Physica* **38 D**, 203 (1989).
38. H. Chaté and P. Manneville. *Phys. Rev.* **A 38**, 4351 (1988).
H. Chaté and P. Manneville. *Physica* **32 D**, 409 (1989).
39. Ya.G. Sinai and V. Yakhot. *Phys. Rev. Lett.* **63**, 1962 (1989).
V. Yakhot, *Phys. Rev. Lett.* **63**, 1965 (1989).

40. T. Bohr, M.H. Jensen, A.W. Pedersen, and D. Rand. In *New Trends in Nonlinear Dynamics and Pattern Forming Phenomena.* Eds. P. Coullet and P. Huerre. Plenum, New York, 1989.

41. T. Bohr, A.W. Pedersen and M.H. Jensen. Preprint (1990).

42. A.T. Winfree. *The Geometry of Biological Time.* Springer, Berlin, 1980.

43. Y. Kuramoto. *Chemical Oscillations, Waves and Turbulence.* Springer, Berlin, 1980.

44. A.C. Newell and J.A. Whitehead. *J. Fluid Mech.* **38**, 279 (1969).
A.C. Newell. In *Lectures in Applied Mathematics, Vol.15.* Am. Math. Society, Providence, 1974.

45. H.R. Brand, P.S. Lomdahl, and A.C. Newell. *Physica* **23 D**, 345 (1986).

46. P. Coullet, L. Gil, and J. Lega. *Phys. Rev. Lett.* **62**, 1619 (1989).

47. A.V. Gaponov-Grekhov and M.I. Rabinovich. *Sov. Phys. Usp.* **30**, 433 (1987).

48. L. Kramer, E. Bodenschatz, W. Pesch, A. Weber, and W. Zimmermann. To appear in *New Trends in Nonlinear Dynamics and Pattern Forming Phenomena,* Eds. P. Coullet and P. Huerre. Plenum, New York, 1989.

49. C. Elphick and E. Meron. Universidad Tecnica Federico Santa Maria Preprint (1989).

CHAOTIC DATA AND MODEL BUILDING[1]

Andrew M. Fraser

Systems Science Program
Portland State University
Portland, Oregon 97207, USA

Abstract

The criterion hidden Markov models are trained to minimize is essentially an entropy. Here, such models are trained to fit numerically generated data from the Rössler system and estimates of the KS entropy h_μ are derived from Lyapunov exponent calculations. The values attained for the training objective function are found to be inferior to the values that the h_μ estimates suggest are possible. The nature of the models produced by extensive training is found to be sensitive to the choices of initial model parameters.

1. Introduction

Since the most fundamental measures of model performance are essentially entropies calculated on the basis of the model, model building techniques should be tested on complex data of known entropy. Chaotic data (data from strange attractors of dynamical systems) fit this description exactly. Given a dynamical system and the associated system of differential equations one can estimate the KS entropy h_μ of its attractors. The estimates can be made quite accurate by employing the method of Benettin *et al.* [1]. In this paper I present results of a study in which I applied the hidden Markov modeling techniques used in speech recognition research to data from the Rössler system. There were three significant results of the study:

1. Even after training, the HMMs failed to capture much of the structure in the data. I had expected that fairly simple HMMs could incorporate most of the structure in the data, as measured by the KS entropy. While the log likelihood of an ideal model would approach h_μ, the log likelihood of the best model I found corresponded to an entropy about twice as large.
2. Every different initial condition I tried led to a different model, after training, and I found wide variations in the success of the final models. Thus I suspect

[1]Some of this work will also be reported in the proceedings of the July 1990 meeting of the ISSS.

that the training criterion has many local extrema as a function of its parameters (the transition probabilities).

3. The worst model, after training, was one whose initial parameters were set on the basis of knowledge of which symbol sequences of length six appear in the data. Random initial parameter values provided better models, after training, than these informed guesses. Such results have lead Jelinik's group (Bahl *et al.* [2]) to eschew "expert" input in their models.

2. A Chaotic Time Series

The test data came from the Rössler system [3], i.e., the system of nonlinear equations

$$\begin{aligned} \dot{v}_x &= -v_z - v_y \\ \dot{v}_y &= v_x + a v_y \\ \dot{v}_z &= b + v_z(v_x - c) \end{aligned} \tag{1}$$

with the parameter values $a = 0.15$, $b = 0.20$, $c = 10.0$ (numerical evidence suggests that there is a strange attractor at these parameter values [4]).

By numerically integrating these equations, any initial condition $v(0) \in V \equiv \Re^3$ can be mapped to its image $v(1 \cdot T) = \phi(v(0))$ at a time T later. By choosing a time $T = 1.5$ in conjunction with Eq. 1, I defined a map $\phi : V \to V$ where $V \equiv \Re^3$. Given any $v(0) \in V$ this ϕ defines a unique trajectory $\{v(0), v(1), v(2), \ldots\}$. All of the work in this paper used various projections or subsequences of a single time series that I generated as follows:

1. Starting from an initial condition of $v = (10, 0, 0)$ I applied ϕ 30,000 times to allow the system to relax to the attractor.
2. I recorded 50,000 triples $v(t) \equiv (v(t)_x, v(t)_y, v(t)_z)$.

I used a coarse quantization of the x direction as the partition ξ of V required to reduce time series of vectors to symbol sequences $\{y(0), y(1), y(2), \ldots\}$. The assignment of symbols was made on the following basis:

$$\begin{aligned} y(t) &= 0 \quad \text{if} \quad v_x(t) < -10 \\ y(t) &= 1 \quad \text{if} \quad -10 \le v_x(t) < 0 \\ y(t) &= 2 \quad \text{if} \quad 0 \le v_x(t) < 10 \\ y(t) &= 3 \quad \text{if} \quad 10 \le v_x(t) \end{aligned} \tag{2}$$

I used the forward-backward algorithm[2] to train several HMM's on an observed sequence $\mathbf{y}_O$ of length 8,000. The initial models had 12 or 20 states, all transitions between states were allowed, and any symbol y could be produced on any transition. The parameters of the models were set using a random number generator.

---

[2]K.F. Lee [5] explains the forward-backward algorithm and the Viterbi algorithm in enough detail to make implementation easy. The discussion by Bahl *et al.* [2] is more condensed, but perhaps easier to obtain. The original sources are Baum and Eagon [6] and Viterbi [7].

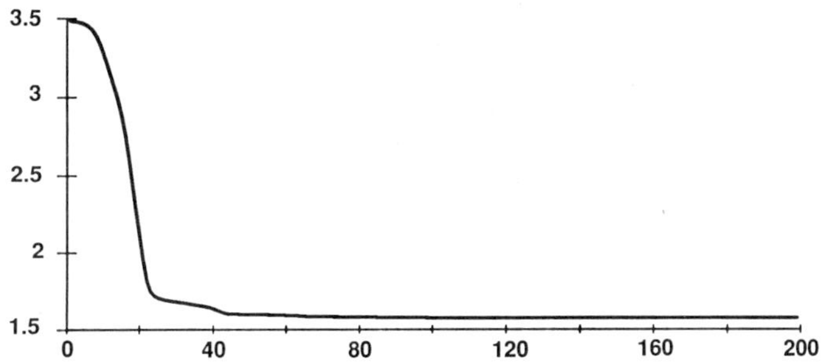

Figure 1: A training characteristic of the forward-backward algorithm. The perplexity is plotted against the number of iterations.

Each iteration of the forward-backward algorithm produces a model for which the likelihood of the data is higher than that given by the previous model. I find it convenient to characterize models in terms of perplexity estimates $PP_i \approx ((\Pr_M(\mathbf{y}_0^{l-1} = \mathbf{y}_O))^{-1/l}$, where $\Pr_M(\mathbf{y}_0^{l-1} = \mathbf{y}_O)$ denotes the probability, calculated on the basis of the model M, of a particular sequence of l observations. By iterating the algorithm, a sequence of models $M_0, M_1, M_2, \ldots$ and a sequence of perplexity estimates $PP_0, PP_1, PP_2, \ldots$ are obtained. The convergence of such perplexity estimates is plotted in Fig.1.

In order to illustrate the structure that the training had found in the symbol sequence, I implemented the Viterbi algorithm to decode most probable state sequences from a given observation sequence. A typical result is displayed in Fig.2 which is based on the three sequences $\mathbf{v}_0^{l-1}$, $\mathbf{y}_0^{l-1}$, and $\mathbf{s}_0^{l-1}$.

Applying the partition ξ (Eq. 2) to $v(t)$ yielded $y(t)$, and $\mathbf{s}_0^{l-1}$ was decoded from $\mathbf{y}_0^{l-1}$ using the Viterbi algorithm. Thus a pair $v(t), s(t)$ was generated for every $t : 0 \leq t < l$. The plot of the "state" s_i consists of the projection on the v_x, v_y plane of all $v(t)$ such that $s(t) = s_i$. The striking feature of these plots is that each state s of the HMM roughly corresponds to an element of a partition of the state space V of the dynamical system. The partition elements are well localized, almost disjoint, and are remarkably unrelated to the quantization partion ξ used to derive $y(t)$ from $v(t)$. It seems that a model which predicts an observation sequence well, will also provide good localization in the unobserved state space of a dynamical system. I think this is essential to the theory of *reconstructing* dynamical systems from scalar time series [8].

3. Model Performance

Table 1 summarizes the effectiveness of different HMM's that I have developed using the dynamical system data. The comparisons are made on the basis of estimates of two types of perplexity. Perplexity $PP_M(Y)$ measures the ability of a model to predict the next y value on the basis of the preceding y values. It is the quantity minimized by the forward-backward algorithm. By changing the seed of the random number generator used to initialize the model M_0, I investigated the uniqueness of the point towards which the forward-backward algorithm converges. As indicated in Table 1 I found that there is no such unique point. The fact that the perplexity of Y is smaller than the perplexity of Y and S, i.e., $PP_M(Y) < PP_M(Y, S)$, indicates that the "Random" models listed in Table 1 are not *unifilar*, that is, a long observation sequence $\mathbf{y}_O$ can be caused by more than one state sequence $\mathbf{s}_d$ [2].

Perplexity is defined by

$$PP(Y) \equiv 2^{h(Y)}$$

where $h(Y)$, the *source entropy* of Y, is in turn defined by:

$$h(Y) \equiv \lim_{l \to \infty} \frac{1}{l} H(\mathbf{Y}_0^{l-1}) \tag{3}$$

The McMillan theorem says that if Y is ergodic, then the estimate

$$h(Y) \approx -\frac{1}{l} \log_2(\Pr(\mathbf{y}_0^{l-1} = \mathbf{y}_O))$$

becomes arbitrarily accurate with arbitrary confidence as the length l of the observed sequence $\mathbf{y}_O$ is made larger. This justifies approximations of the form

$$PP(Y) \approx \Pr(\mathbf{y}_0^{l-1} = \mathbf{y}_O)^{-1/l}$$

for large l. If some redundancy in a source is not captured in a model, then it is of no use. Thus, given a particular model M the perplexity of the source Y with respect to the model

$$PP_M(Y) \approx (\Pr_M(\mathbf{y}_0^{l-1} = \mathbf{y}_O))^{-1/l}$$

is the relevant quantity. Here $\Pr_M$ should be read as "probability calculated on the basis of the model M". Similarly one can write:

$$PP_M(S, Y) \approx (\Pr_M(\mathbf{y}_0^{l-1} = \mathbf{y}_O, \mathbf{s}_0^{l-1} = \mathbf{s}_d))^{-1/l}$$

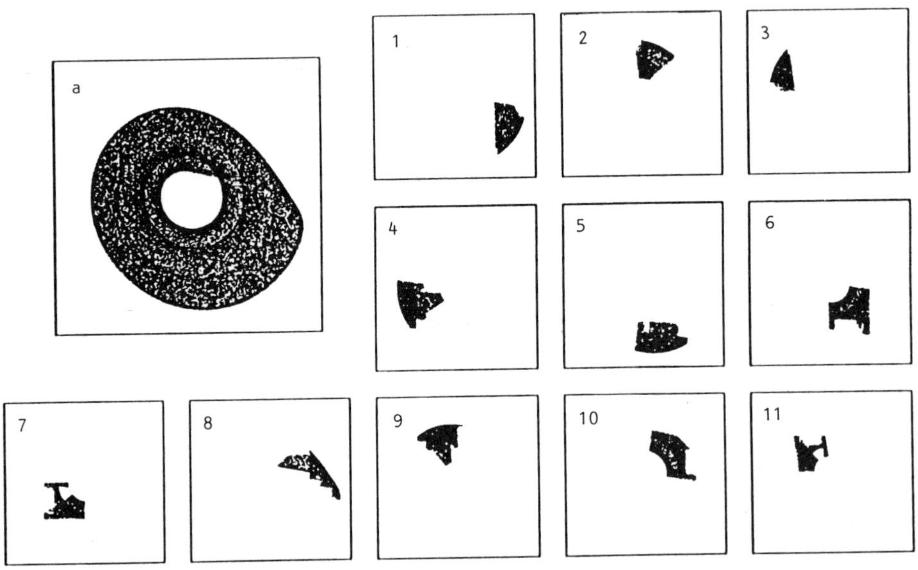

Figure 2: An identification of the hidden states of a HMM $s(t)$ with regions of the vector state space $v(t)$. In plot a, every point $v(t)$ in a trajectory of length 40,000 is projected on the v_x, v_y plane. The ith plot consists of those points $v(t)$ that correspond to symbols $y(t)$ which were decoded to the state $s(t) = s_i$.

Table 1: The perplexity of several HMM's. Each model was trained by 200 iterations of the forward-backward algorithm. $PP_M(Y)$ was obtained by estimating the probability, given the particular model, of sequences of 8,000 observations, $PP_M(Y,S)$ was obtained from the probability of the most likely state and observation sequence as found by the Viterbi algorithm, and the entropy estimates $2^{h_M(Y,S)}$ were calculated using Eq. 3. I set the initial parameter values of the "Expert" model on the basis of knowledge about what sequences of ys are possible. All other models were initialized by a random number generator. Note: The best possible model would have $PP_M(Y) \leq 1.140 = 2^{h_\mu(\phi)}$.

	Training Data		*Fresh Data*	
	$PP_M(Y)$	$PP_M(Y,S)$	$PP_M(Y)$	$2^{h_M(Y,S)}$
12-state, Random, seed = 1	1.574	1.654	1.580	1.690
12-state, Random, seed = 2	1.595	1.639	1.595	1.694
12-state, Random, seed = 4	1.647	1.786	1.646	1.921
20-state, Random, seed = 1	1.487	1.592	1.497	1.652
20-state, Random, seed = 2	1.492	1.526	1.499	1.570
20-state, Random, seed = 4	1.546	1.683	1.555	1.773
20-state, Expert	1.969	1.985	2.001	2.005

Alternatively, if the model is correct, the same information should be obtained by directly calculating the entropy of the underlying Markov process (Ref.[9], pages 63-69), i.e.,

$$h_M(Y,S) = -\left\langle \log_2 \Pr \begin{pmatrix} & y(t+1) = y_a \\ \text{and} & s(t+1) = s_a \\ \text{given} & s(t) = s_b \end{pmatrix} \right\rangle_{y_a, s_a, s_b}$$

where $\langle\rangle$ indicates expected value. In Table 1 the difference between $PP_M(Y,S)$ and $2^{h_M(Y,S)}$ indicates that the decoded sequences $(\mathbf{y}_O, \mathbf{s}_d)$ found by the Viterbi algorithm are more probable than the sequences that would typically be produced by the "Random" models.

An upper bound on $h(Y)$ is provided by the *KS entropy* $h_\mu(\phi)$. For the Rössler system (Eq. 1), calculations have given $h_\mu(\phi) = 0.190$ bits, or equivalently $PP(Y) \leq 2^{0.190} = 1.140$. The fact that the models evaluated in Table 1 have $PP_M(Y)$ values larger than 1.140 means much of the structure in the data is left unused.

4. Conclusion

Hidden Markov modeling techniques fail to discover much of the structure in chaotic time series. Because the state space of a dynamical system is simple enough to be visualized it is easy to build an intuition for how the modeling process works or how it fails, and since a dynamical system is low dimensional and has a small entropy, it does not take as many observations to determine its statistical structure as it would for a higher dimensional problem. Thus such data provide an ideal test bed for model building methods. While one might test HMM methods (training, decoding, etc.) on data generated from an actual Markov model, data from a dynamical system has the advantage that like speech, it is unlikely that the structure of any proposed HMM is entirely appropriate[3]. In future work I hope to use such synthetic data to test improved modeling schemes.

[3]For instance Peter Brown [10] invokes the inappropriateness of the model to justify using an alternative training criterion in the forward-backward algorithm.

Acknowledgements

Much of this work was done at The University of Texas and was supported by the Department of Energy Office of Basic Energy Sciences. I thank Harry Swinney for encouragement and Michael P. Marder for helpful suggestions.

References

1. G. Benettin, L. Galgani, A. Giorgilli, and J.-M. Strelcyn. Lyapunov characteristic exponents for smooth dynamical systems and for Hamiltonian systems; a method for computing all of them. *Meccanica* **15**, 9 (1980).
2. L.R. Bahl, F. Jelinek and R.L. Mercer. A maximum likelihood approach to continuous speech recognition. *IEEE Trans. on Pattern Analysis and Machine Intelligence* PAMI-**5(2)**, 179–190 (1983).
3. O.E. Rössler. An equation for continuous chaos. *Phys. Lett. A* **57**, 397–398 (1976).
4. J.-P. Eckmann and D. Ruelle. Ergodic theory of chaos and strange attractors. *Rev. Mod. Phys.* **57**, 617–656 (1985).
5. K.F. Lee. *Large-Vocabulary Speaker-Independent Continuous Speech Recognition: The SPHINX System.* PhD thesis, Carnegie Mellon University, Pitsburgh, 1988.
6. L.E. Baum and J.A. Eagon. An inequality with applications to statistical estimation for probabilistic functions of Markov processes and to a model for ecology. *Bulletin of the American Mathematical Society* **73**, 360–363 (1967).
7. A.J. Viterbi. Error bounds for convolutional codes and an asymptotically optimum decoding algorithm. *IEEE Transactions on Information Theory* **13(2)**, 260–269 (1967).
8. A.M. Fraser. Information and entropy in strange attractors. *IEEE Transactions on Information Theory* **35(2)**, 245–262 (1989).
9. R.G. Gallager. *Information Theory and Reliable Communication.* John Wiley and Sons, New York, 1968.
10. P.F. Brown. *The Acoustic-Modeling Problem in Automatic Speech Recognition.* PhD thesis, Carnegie Mellon University, Pitsburgh, 1987.

CAUSALITY AND INFORMATION FLOW

SETH LLOYD

Laboratory for High Energy Physics
California Institute of Technology
Pasadena, CA 91125, USA

Abstract

Correlation does not imply causation. If the outcome of an event A is correlated with the outcome of another event B, then either A is a cause of B, B is a cause of A, or the two events possess some common cause. For more than two variables, however, the pattern of correlations can provide detailed causal information. Since variation in cause produces a correlated variation in effect, the question of which events are causes of which other events is equivalent to the question of how information flows between events. I examine the conditions under which underlying causal structure can be deduced from statistical information.

1. Introduction

Suppose, for a moment, the truth of cigarette companies' assertions that smoking does not cause cancer. Suppose that there exists some as yet unidentified gene that causes its human carrier to be nervous. People that have this gene will be more likely than average to light up to sooth their nerves, and, because of the suppression of the immune system that their nervousness induces, be more susceptible to infection by cancer–causing viruses. In this fanciful story, smoking does not cause cancer; rather, smoking and cancer are common symptoms of an underlying genetic predisposition. The statistical correlation between smoking and cancer — the fact that people who smoke are more likely to get cancer than people who don't smoke — is implied equally by a direct causal link between smoking and cancer, and by an indirect link through such a hidden common cause.

It is straightforward to test for the existence of this hypothetical genetic common cause. One need simply create a population in which smoking is not correlated with this gene, say by selecting two thousand people at random from the population, and paying one thousand to smoke and one thousand not to smoke. If the cigarette companies' tale is true, then within this group people who smoke should be no more likely to get cancer than people who don't smoke. If, on the contrary, smoking has

Information Dynamics, Edited by H. Atmanspacher and
H. Scheingraber, Plenum Press, New York, 1991

a direct effect on getting cancer, then the usual positive correlation between smoking and cancer should hold within the experimental group.

This example illustrates two points. First, different causal scenarios imply different sets of statistical relations, allowing the elimination of causal scenarios inconsistent with the observed statistical evidence. Second, the introduction of independent information (in the example above, the division into two experimental groups independent of genetic proclivity) is helpful for sorting out what's causing what, even in the presence of hidden common causes. Unfortunately, as in the example given, the experiments required to obtain the statistical information needed to distinguish between causal scenarios may be difficult to perform. The question then arises, under what circumstances can one deduce the actual causal relations between events from existing statistical relationships, in the absence of the possibility of direct intervention?

Attempts to answer this question date back to Hume [1]. In this century, there is a considerable philosophical literature on the subject of causality beginning with Reichenbach [2-4]. The present attempt to unravel the relationship between causality and statistics by use of graphical methods has its beginnings in Wright's path analysis [5, 6], and in the methods of inference developed by Pearl *et al.* [7-10].

2. Events and Statistics

Consider a set of events $A, B, C, \ldots$ with possible outcomes $\{a_i\}, \{b_j\}, \{c_k\}, \ldots$, respectively. Suppose that we are given statistical information about a set of events in terms of a joint probability distribution over their outcomes, $\{p(a_i b_j c_k \ldots)\}$, satisfying $\sum_{ijk\ldots} p(a_i b_j c_k \ldots) = 1$. This joint probability distribution gives rise to marginal probability distributions for the outcomes of events taken on their own, e.g., $\{\ p(a_i) \equiv \sum_{jk\ldots} p(a_i b_j c_k \ldots)\ \}$, and to conditional distributions for the outcomes of events given that other events have had particular outcomes, e.g., $\{\ p(b_j|c_k) \equiv p(b_j c_k)/p(c_k)\ \}$, together with the corresponding entropies for each of these probability distributions, e.g., $S(C) \equiv -\sum_k p(c_k) \log_2 p(c_k)$.

The sign of causal influence is the existence of correlation between events. We will say that two events A and B are correlated if and only if the mutual information between A and B, $I(A;B) \equiv S(A) + S(B) - S(AB)$, is greater than zero (mutual information is a more general measure of correlation than covariance or linear correlation coefficients). The basic relation between causal and statistical relations is the rule that two events A and B are correlated if and only if

(1) A is a cause of B,

(2) B is a cause of A, or

(3) A and B have some common cause.

The simple fact of correlation allows us to conclude only that one or more of the above three alternatives is the case. However, when more than two variables are present, various different patterns of correlation arise from different patterns of cause and effect. One example of such a pattern is the phenomenon known as "screening–off" [2, 3]. We say that an event C screens off A from B if A and B exhibit no correlation whenever the outcome of C is fixed, i.e., $I(A;B|C) = S(A|C) + S(B|C) - S(AB|C) = 0$, or equivalently, $p(a_i b_j|c_k) = p(a_i|c_k)\, p(b_j|c_k)$ for all i, j, k. We expect screening–off to occur when all correlation between A and B is a result of their common cause C, as in the cigarette companies' story of the common gene that causes smoking and cancer, or when A's effect on B occurs only through C as an intermediate cause, as when disconnecting the antenna of a television destroys the correlation between what is broadcast and what is seen on the screen.

Screening–off is a pattern of correlation in which fixing the outcome of an event or set of events renders two other sets of events independent. For example, when the only connection between two events is through a common cause, fixing the outcome of that common cause screens off the one event from the other. Fixing the outcome of a common effect of two independent causes can have the opposite effect of rendering the causes dependent. For example, consider an exclusive OR gate with inputs A, B and output C: $C = 0$ if either A or $B = 1$ but not both, $C = 0$ otherwise. Suppose that all inputs AB have the same probability. A and B are then uncorrelated: $I(A; B) = 0$. If the output, C, is fixed, however, then A and B exhibit perfect correlation (when $C = 1$, for example, $A = 0$ implies that $B = 1$ and *vice versa*), and we have $I(A; B|C) = 1$.

Clearly, common causes and common effects induce different patterns of correlation. One can ask whether these different patterns suffice to distinguish between cause and effect in a set of events. The answer to this question is given by the theory of Bayesian networks [7].

3. Bayesian Networks

It is common practice to depict causal relationships by an arrow — "$A \longrightarrow B$" corresponds to "A has an effect on B" — and sets of causal relations by directed graphs. A Bayesian network is a graphical representation of the screening–off relationships between sets of events. In what follows, italic $A, B, \ldots$ will refer to individual events, and bold–faced $\mathbf{A}, \mathbf{B}, \ldots$ will refer to sets of events. A Bayesian network for a joint probability distribution $p(a_i, b_j, \ldots, z_k)$ over the outcomes of a set of events $\mathbf{U} = \{A, B, \ldots, Z\}$ is a directed, acyclic graph $(\mathbf{U}, E)$, whose nodes correspond to the events $A, B, \ldots, Z$, and whose links, when interpreted as representing relations of cause and effect, are consistent with the screening–off relationships present in $p(a_i, b_j, \ldots, z_k)$. That is, a Bayesian network is a model of cause and effect between events, in which the screening–off relationships implied by the pattern of links in the network hold for the actual probability distribution over events. Furthermore, a Bayesian network is a minimal model in the sense that the network obtained by deleting any link from a Bayesian network implies some screening–off relationship that does not hold for the actual probability ditribution.

The way in which a Bayesian network implies screening–off relationships is derived from the two notions discussed above: fixing all common or intermediate causes between two sets of events renders them independent as long as no common effect of the events is also fixed. More formally, consider a particular path between two disjoint sets of events $\mathbf{A}$ and $\mathbf{B}$: if we fix a set of events $\mathbf{Z}$, disjoint from $\mathbf{A}$ and $\mathbf{B}$, to particular outcomes, then the path is capable of generating correlation betwen $\mathbf{A}$ and $\mathbf{B}$ if every node along the path with converging arrows is inside $\mathbf{Z}$ or has a descendant in $\mathbf{Z}$, and every other node is outside $\mathbf{Z}$. In the terminology of Pearl [7] such a path is said to be "active" given $\mathbf{Z}$. Paths that are not active given $\mathbf{Z}$ are said to be "blocked" by $\mathbf{Z}$. Fixing $\mathbf{Z}$ screens off $\mathbf{A}$ from $\mathbf{B}$ if all paths between $\mathbf{A}$ and $\mathbf{B}$ are blocked by $\mathbf{Z}$. For example, in the following network

$$A \longleftarrow B \longrightarrow C \longleftarrow D \longrightarrow E$$

the path between A and E is active given C, since fixing C can induce correlation between B and D, and hence between their descendants A and E. On the other hand, fixing the set $\{C, D\}$ blocks this path, since D is the only route through which E can receive information about A.

In summary, a Bayesian network for a probability distribution is a directed, acyclic graph such that (i) any screening–off relationship implied by the graph holds for the probability distribution, and (ii) if any edge is removed from the graph, condition (i) ceases to hold. A Bayesian network for a probability distribution is thus a minimal causal model whose implied screening–off relations hold for the probability distribution. Note that the probability distribution may exhibit screening–off relationships and statistical independencies above and beyond those implied by a given Bayesian network. The conditions under which the patterns of correlation exhibited by a probability distribution are exactly those implied by a particular network are discussed by Pearl [7].

It is important to note that in the simple form described here, Bayesian networks are incapable of coping with feedback loops, and with the existence of "hidden" causes — events outside the given set of events that are responsible for inducing correlations between events within the given set. For the moment, we will concentrate on scenarios without feedback loops and hidden causes, returning to discuss their treatment in a later section.

A Bayesian network is a model of cause and effect that is consistent with the screening–off relationships — conditional independencies — inherent in a joint probability distribution. The question of whether the screening–off relationships inherent in a probability distribution over a set of events suffice to determine uniquely the pattern of cause and effect between the events is equivalent to the question of whether the probability distribution admits only one Bayesian network. In fact, a given probability distribution admits many different Bayesian networks, each giving a different consistent model of cause and effect. The existence of different, incompatible Bayesian networks for the same probability distribution follows from a result due to Verma [7, 9]:

Theorem: Given a probability distribution $p(x_1, x_2, \ldots, x_n)$ and any ordering d of the events, the directed acyclic graph created by designating as parents of X_i any minimal set $\mathbf{\Pi}_{X_i}$ of predecessors satisfying

$$p(x_i | \mathbf{\Pi}_{X_i}) = p(x_i | x_1, \ldots, x_{i-1}), \qquad \mathbf{\Pi}_{X_i} \subseteq \{X_1, \ldots, X_{i-1}\}$$

is a Bayesian network of p. If p is strictly positive, then all of the parent sets are unique and the Bayesian network is unique given the ordering d.

The nonuniqueness of Bayesian networks arises because *any* ordering d of the events gives rise to a Bayesian network in which the direction of the links respects the ordering d. Verma's result implies that for every direct link in a given Bayesian network, there exists another Bayesian network in which that link is absent or reversed. Therefore, the form of screening–off relationships alone is insufficient to identify uniquely the structure of cause and effect.[1]

---

[1]The form of the screening–off relationships together with the ordering of events in time gives a unique Bayesian network for strictly positive probability distributions. Here, however, we are interested in the question of whether cause and effect can be deduced in the absence of an aforeknown temporal order.

4. Information Flow

Patterns of conditional independence — screening–off relationships — alone do not determine the direction of causal influence. However, there exist additional statistical patterns that can be exploited to distinguish between different possible causal scenarios. We now turn to the analysis of such patterns.

The advantage of discussing correlations in terms of information theory, as we do here, is that causality can be identified with *information flow.* To say that variation in a cause induces a correlated variation in its effect is equivalent to saying that information flows from cause to effect. The arrows in a Bayesian network can be regarded as representing relationships of cause and effect, or as indicating the direction of information flow, and the network itself is a model of where information is flowing that is consistent with screening–off relations.

Let us first consider how information flows in a system that has been constructed exactly to regulate where information is flowing — an electronic logic circuit (Fig. 1). The AND and OR gates in such a circuit are essentially switches: the output is equal to one of the inputs when the other input is equal to 1 (for the AND gate) or 0 (for the OR gate), respectively. In the above circuit, we can say that information flows from A to D when $B = 1$, from A to E when $B = 1$ and $C = 0$, from E to F when $B = 0$, but never from A to F. We can characterize information flow within the circuit by a set of statements of the form, "Information flows from A to B under conditions γ_{AB}," where γ_{AB} is a set of conditions, each one of which lists which variables must be fixed to which values and which variables must be allowed to vary, if information is to flow from A to B. For example, for information to flow from A to E, B must be fixed to 1, and D must be allowed to vary.

The pattern of information flow in a logic circuit obeys the following rules:

(1) If information flows from A to B under conditions γ_{AB}, and from B to C under conditions γ_{BC}, then information flows from A to C under conditions $\gamma_{AB} \cap \gamma_{BC}$.

(2) Information flowing from A to B under conditions γ_{AB}, and from A to C under conditions γ_{AC} induces mutual information between BE and C under conditions $\gamma_{AB} \cap \gamma_{AC}$.

In accordance with the identification of information flow with causality, we could equally well characterize the causal structure of the circuit by statements of the type, "Variation in A induces a correlated variation in B under conditions γ_{AB}," or "A has an effect on B under conditions γ_{AB}," where the conditions γ_{AB} are the same for each of the three ways of talking about causal structure.

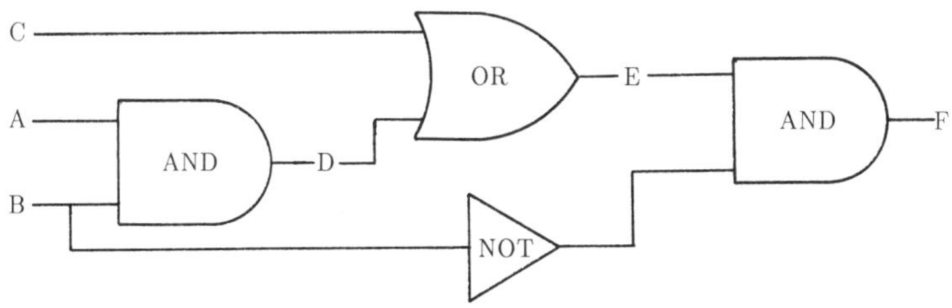

Figure 1: A logic circuit in which information flow is not transitive. Information can flow from A to E and from E to F, but not from A to F.

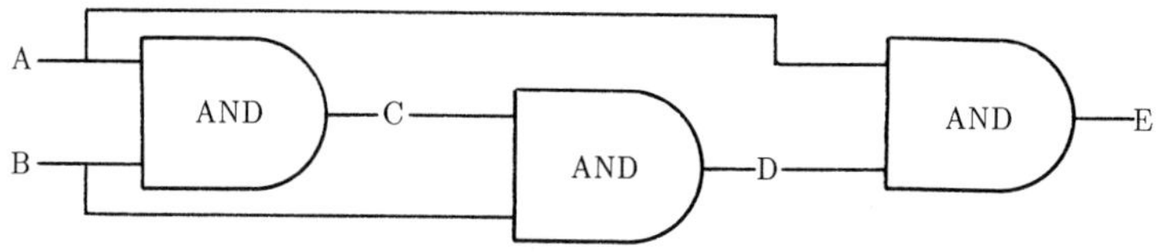

Figure 2: A logic circuit with noisy logic gates. Each gate operates correctly only half the time; the remainder of the time it emits a zero or one at random.

A comprehensive set of statements of the above form about where information is flowing between which events under what conditions gives more information than a Bayesian network for the same set of events: such a set tells not only what events can influence what other events, but under what conditions that influence takes place. One may summarize this difference as follows: if certain variables are fixed to particular values, a Bayesian network tells us between which other sets of variables to expect independence, while a pattern of information flow tells us in addition between which sets of variables to expect mutual information. Thus, while there may be many incompatible Bayesian networks consistent with a particular joint probability distribution over a set of events, there may be many fewer consistent patterns of information flow betweeen the events. In particular, while there are always several Bayesian networks consistent with a particular joint distribution, there may be only one pattern of information flow.

Let us compare Bayesian networks and patterns of information flow for a particular example in which we know exactly where information is flowing. Consider the logic circuit depicted in Figure 2. The set of allowed values for the variables in this circuit is:

$$\begin{array}{ccccc} A & B & C & D & E \\ 0 & 0 & 0 & 0 & 0 \\ 0 & 1 & 0 & 0 & 0 \\ 1 & 0 & 0 & 0 & 0 \\ 1 & 1 & 1 & 1 & 1 \end{array} \tag{1}$$

It is immediately obvious that no causal ordering between C, D, and E can be effected by statistical means alone, as they are perfectly correlated. Suppose, however, that this circuit suffers from dynamical noise, so that each AND gate operates correctly only half the time, and the other half of the time emits a 0 or 1 at random. The probabilities for the outcomes $abcde$ for $ABCDE$ are given by

$$p(abcde) \;=\; p(a)\,p(b)\,p(c|ab)\,p(d|bc)\,p(e|ad) \tag{2}$$

where $p(a), p(b)$ are the input probabilities for the values of A and B, and the three conditional probabilities have values $p(i|jk) = 1/4$ for $i = 0, (j,k) = (1,1)$, or $i = 1, (j,k) = (0,0), (0,1)$, or $(1,0)$; $p(i|jk) = 3/4$ otherwise. If the initial probabilities for A and B are equal for $a, b = 0, 1$, then the actual probabilities are:

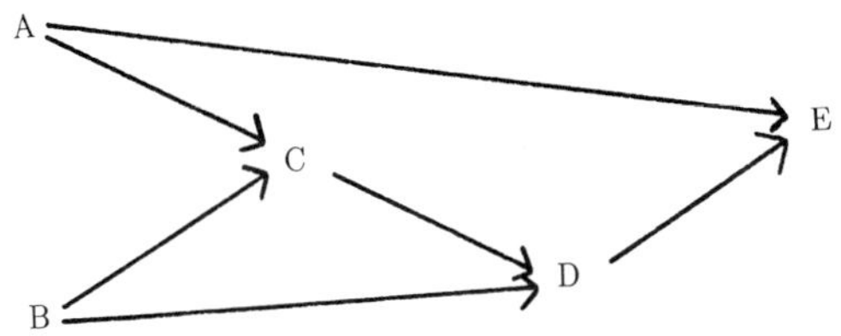

Figure 3: A Bayesian network for the logic circuit in Figure 2 that corresponds to the correct pattern of information flow between the variables.

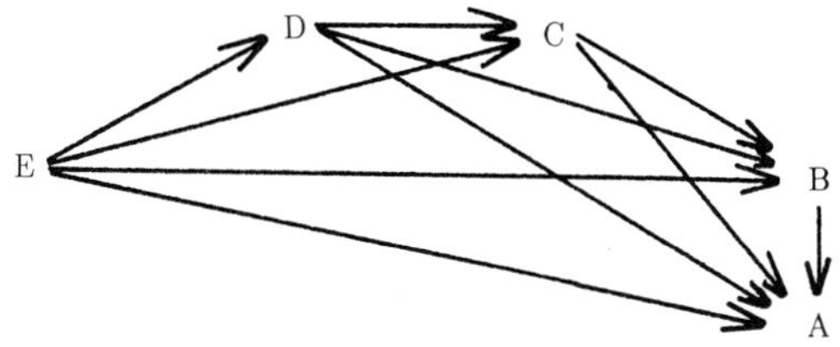

Figure 4: A second Bayesian network for the logic circuit in Figure 2.

$$\begin{array}{llll} p(00000) = 27/256 & p(01000) = 27/256 & p(10000) = 27/256 & p(11000) = 9/256 \\ p(00001) = 9/256 & p(01001) = 9/256 & p(10001) = 9/256 & p(11001) = 3/256 \\ p(00010) = 9/256 & p(01010) = 9/256 & p(10010) = 3/256 & p(11010) = 1/256 \\ p(00011) = 3/256 & p(01011) = 3/256 & p(10011) = 9/256 & p(11011) = 3/256 \\ p(00100) = 9/256 & p(01100) = 3/256 & p(10100) = 9/256 & p(11100) = 9/256 \\ p(00101) = 3/256 & p(01101) = 1/256 & p(10101) = 3/256 & p(11101) = 3/256 \\ p(00110) = 3/256 & p(01110) = 9/256 & p(10110) = 1/256 & p(11110) = 9/256 \\ p(00111) = 1/256 & p(01111) = 3/256 & p(10111) = 3/256 & p(11111) = 27/256 \end{array} \qquad (3)$$

A Bayesian network consistent with this probability distribution can immediately be derived from the actual wiring diagram of the noisy logic circuit shown in Fig. 3. This network implies the factoring of the probabilities in Eq. (1) above. Other orderings of the variables give equally consistent Bayesian networks, however. For example, the Bayesian network obtained by taking the variables in reverse order, $EDCBA$, is given in Figure 4.

This network implies that $p(abcde)$ can be decomposed as follows:

$$p(abcde) \;=\; p(e)\, p(d|e)\, p(c|de)\, p(b|cde)\, p(a|bcde) \qquad (4)$$

Each Bayesian network gives a different model of how information enters the system, and how it is transformed. The network of Figure 3, for example, gives a model in which independent information is injected into the network at A and B, flows from A and B to C under the conditions given by the matrix $p(c|ab)$, flows to D from B and C under the conditions given by $p(d|bc)$, and flows to E from A and D under conditions $p(e|ad)$. More independent information is injected at each point along the way, due to the random firing of the gates: that is, the value of a particular variable is not completely determined by its immediate predecessors. In the model of information flow implied by the network of Figure 4, in contrast, information is injected at E, transformed on its way to D, further transformed and mixed with information from E on its way to C, mixed with information from E and D on its way to B, and mixed with information from E, D, C before arriving at A.

Each Bayesian network implies a model for information flow analogous to a logic circuit. According to a particular network, the way in which information flows through a node, X, is determined by the form of the Markov matrix $p(x|\pi_X)$ that gives the conditional probabilities for the states of X given the states π_X of X's immediate predecessors $\mathbf{\Pi}_X$. One can think of this matrix as determining the function of a generalized "switch" whose inputs are X's immediate predecessors, and whose output is X. The conditions under which information flows along a particular path $X_1 \to X_2 \to \ldots \to X_m$ are obtained by fixing the immediate predecessors of $X_2, \ldots X_m$ other than $X_1, \ldots, X_{m-1}$ to particular values, and seeing whether variation in X_1 propagates along the chain of variables $X_2, \ldots, X_m$. More formally, let $\mathbf{\Pi}'_{X_2 \ldots X_m} = \mathbf{\Pi}_{X_2} \cup \ldots \cup \mathbf{\Pi}_{X_m} - \{X_1, \ldots, X_{m-1}\}$. The set of conditions $\gamma_{X_1 \ldots X_m}$ under which information propagates from X_1 along the path $X_1 \to \ldots \to X_m$ is then the subset of possible outcomes $\pi'_{X_2 \ldots X_m} \in \mathbf{\Pi}'_{X_2 \ldots X_m}$ for which there exist x_1, x_m such that:

$$\sum_{x_2 \ldots x_m} p(x_2|x_1\pi'_{X_2 \ldots X_m})\, p(x_3|x_2\pi'_{X_2 \ldots X_m}) \cdots p(x_m|x_{m-1}\pi'_{X_2 \ldots X_m}) \neq p(x_m) \qquad (5)$$

We call a model of information flow given by a Bayesian network together with the Markov matrices for its internal nodes and the input probabilities for its initial nodes, a *causal model.* Any causal model derived from a Bayesian network for a joint probability distribution suffices to reproduce the probability distribution exactly. Not all causal models are equally good models, however. When one selects a model for the mechanism that produced a particular data set, one is looking for more than fidelity of reproduction: in addition, one desires a concise model — one that doesn't take up too much space — and a robust model — one that reproduces all features of the data without fine–tuning of initial conditions.

5. Data Compression and Occam's Razor

The size of a causal model is equal to the number of parameters required to specify the Markov matrices for the model's internal nodes and the probabilities for its initial nodes. Let $n(X)$ be the number of possible values taken on by the variable X. The number of parameters required to specify the probabilities for an initial node, A, is $n(A) - 1$. The number of parameters required to specify the Markov matrix for an internal node, Y, is:

$$\Pi_{X \in \mathbf{\Pi_Y}} n(X)(n(Y) - 1) \qquad (6)$$

Given an ordering $X_1, \ldots, X_m$ for the variables in a joint probability distribution, the model derived from the resulting Bayesian network requires the maximum number of parameters to specify when each variable has as parents all previous variables in the ordering (as in Figure 4 above); in this case, to specify the causal model requires $n(X_1) - 1 + n(X_1)n(X_2 - 1) + \ldots + n(X_1) \ldots n(X_{m-1})(n(X_m) - 1) = n(X_1) \ldots n(X_m) - 1$ parameters, which is exactly the number of parameters required to specify the original joint probability distribution directly. Any consistent causal model with more sparse connections contains the same information as the original probability distribution, but in a more succinct form. That is, causal models compress data.

Given a set of causal models of different sizes, each one consistent with a given joint probability distribution, Occam's razor often suffices to pick out a unique, most concise model. For example, it is a straightforward albeit tedious task to show that the causal model corresponding to the Bayesian network of Figure 3 is the smallest model consistent with the joint probability distribution of Eq. (3).

Of course, the most concise model is not guaranteed to reflect correctly the mechanism that produced a particular set of data. The probabilities of Eq. (3) *could* have been produced by a logic circuit designed after Figure 4; however, in a world that began from simple boundary conditions, in which the sort of mechanisms with which we are surrounded arose out of a combination of chance and natural selection (broadly construed), the most concise model for a set of data is likely to reflect how that data was produced.

6. Models That Require Fine–Tuning of Initial Conditions

Both Bayesian networks, Figures 3 and 4, imply conditional independence relations that appear in the actual probability distribution. For example, the first network, Figure 3, corresponding to the flow of information in the actual circuit, implies 1) that A and B are independent, 2) that fixing the outcomes of B and C screens off A from D, and 3) that fixing the outcomes of A and D screens off B and C from E. The second network, Figure 4, implies no screening–off relations; hence all screening-off relations that it implies are present in p.

Simply by virtue of its greater predictive power, and by Occam's razor, we would be more inclined to accept the first network, Figure 3, as the correct causal model for the circuit. However, in addition to failing to predict the conditional independencies correctly implied by the first network, the second network, Figure 4, fails as a model of information flow by predicting dependencies where they do not exist, and mutual information where there is none to be found. For example, consider the prediction that the second network gives for how information flows from E to A and B. The dependency of A and B on E is mediated by the Markov matrix, $p(ab|e)$, where:

$$\begin{array}{llll} p(00|0) = 12/41 & p(10|0) = 10/41 & p(00|1) = 4/23 & p(10|1) = 6/23 \\ p(01|0) = 12/41 & p(11|0) = 7/41 & p(01|1) = 4/23 & p(11|1) = 9/23 \end{array} \tag{7}$$

It is easy to verify that almost any variation in E generates mutual information between A and B. In fact, the only set of probabilities for E that fails to generate $I(A;B) > 0$ is $p_E(0) = 41/64$, $p_E(1) = 23/64$.

Similarly, the network, Figure 4, predicts that almost any variation in E should generate mutual information between A and D given that B and C are fixed. The only set of probabilities for E that fails to generate mutual information between A and D with B and C fixed, is $p_E(0) = 41/64$, $p_E(1) = 23/64$ — once again, exactly the probabilities for E in the actual distribution.

Regarded as a model of information flow, the Bayesian network in Figure 4 predicts generically that $I(A;B) > 0$ and that $I(A;D|BC) > 0$. If this network were the true model for the causal structure underlying the probability distribution, the fact that

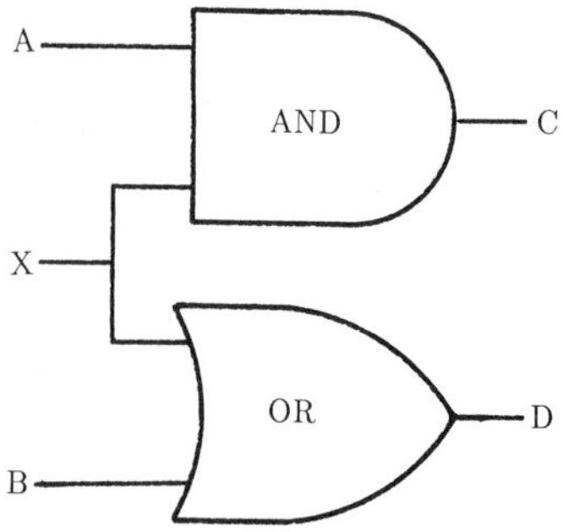

Figure 5: A noisy logic circuit in which one of the variables, X, is inaccessible.

these mutual informations are equal to zero in the actual probability distribution could only be regarded as a highly implausible coincidence. (Just how implausible a coincidence this fine–tuning of probabilities is depends on the size of the sample from which the given probability distribution was constructed.) We thus reject this network in favor of the network of Figure 3 as the correct model of information flow, on the grounds that any model that requires implausible coincidences to explain observed features of reality should be rejected in favor of a model in which all observed features arise naturally, without requiring any fine–tuning of initial conditions.

7. Feedback Loops

In the presence of feedback loops (closed, unidirectional paths), the rules for which paths are active and which are blocked by fixing the values of a given set of variables are the same as for graphs without feedback loops. Therefore, each directed graph implies the existence of certain screening–off relations whether it has feedback loops or not. If a graph with loops implies some screening–off relationship that does not hold for the probability distribution in question, then that graph must be rejected as the basis for a causal model. In addition, the conditions under which information flows along particular paths are still governed by Eq. (5).

The main limitation of graphs with feedback loops for constructing causal models is that the probability distribution $p(x_1 \dots x_m)$ no longer factorizes according to Verma's theorem. That is, even if all the screening–off relations implied by a particular graph hold for the probability distribution in question, it is no longer necessarily the case that $p(x_1 \dots x_m) = p(x_1|\pi_{X_1}) \dots p(x_m|\pi_{X_m})$.

8. Hidden Causes

Consider the noisy logic circuit according to Figure 5, where as before, each gate operates correctly half the time, and fires at random the remainder of the time.

Suppose that we have access only to $ABCD$, and not to X, so that our data comes in the form of a probability distribution $p(abcd)$. The correct model of information flow for this circuit is derived from the Bayesian network (Figure 6). But in the absence of information about X, we can construct consistent models of information flow based on Bayesian networks involving $ABCD$ alone, e.g., shown in Figure 7.

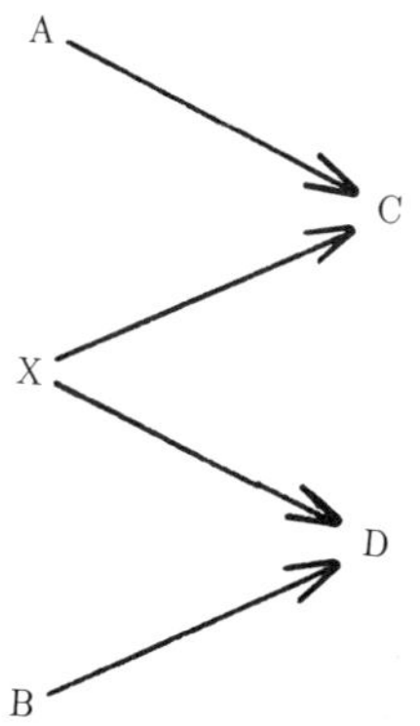

Figure 6: A Bayesian network corresponding to the correct pattern of information flow in the circuit of Fig. 5.

Comparing the size of these different causal models, we see that the model of Figure 6 requires 11 parameters to specify, while that of Figure 7 requires 12 parameters. In fact, all causal models for the mechanism that produced $p(abcd)$ that involve $ABCD$ alone require at least 12 parameters. That is to say, Occam's razor, when applied to the probability distribution $p(abcd)$, suggests the existence of hidden causes.

9. Conclusion

The methods given here for distinguishing between consistent and inconsistent causal models are at their most powerful when some *a priori* information is known about the causal structure of events, for example, when the order of events in time is known, or certain events are known to have a direct effect on other events. In the *Gedanken*experiment proposed on the first page of this article, the introduction of independent information that determines directly whether one smokes or not, but has no direct effect on whether one gets cancer, suffices completely to determine the existence of a direct causal link between smoking and cancer.

In general, there can be many different causal scenarios consistent with a particular statistical data set. Bayesian networks and models of information flow allow us to determine which causal models are consistent with the data, and which are inconsistent. If we appeal to Occam's razor, and reject fine–tuning of boundary conditions, we can often identify a unique, most concise causal model.

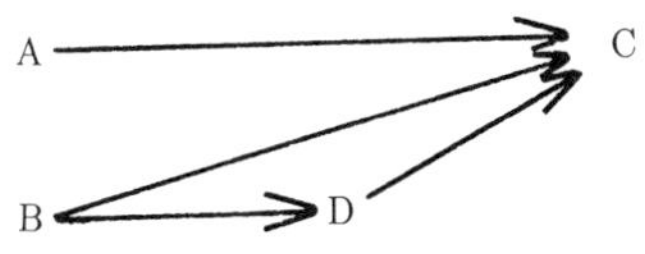

Figure 7: A Bayesian network for the circuit of Fig. 5 that assumes no "hidden" variables.

Acknowledgements

The author would like to thank J.D.Farmer for helpful discussions, and the Institute for Scientific Interchange for its hospitality and support.

References

1. D. Hume. *A Treatise on Human Nature.* Posthumous edition 1777. 2nd edition Clarendon Press, Oxford, 1927.
2. H. Reichenbach. *The Direction of Time.* University of California Press, Berkeley, 1956.
3. J.L. Mackie. *The Cement of the Universe: A Study of Causation.* Clarendon Press, Oxford, 1974.
4. P. Suppes. *A Probabilistic Theory of Causation.* North Holland, Amsterdam, 1970.
5. S. Wright. *Journal of Agricultural Research* **20**, 557–585 (1921).
6. S. Wright. *Ann. Math. Stat.* **5**, 161–215 (1934).
7. J. Pearl. *Probabilistic Reasoning in Intelligent Systems: Networks of Plausible Inference.* Morgan Kaufmann, San Mateo, 1988.
8. J.Pearl. *Artificial Intelligence* **35**(2), 259–271 (1988).
9. T.S. Verma. Technical Report R-65, Cognitive Systems Laboratory, University of California, Los Angeles (1986).
10. J. Pearl and M. Tarsi. *Journal of Complexity* **2**(1), 60–77 (1986).

SOME REMARKS ON THE PROBLEM OF REPEATABILITY OF EXPERIMENTS DEALING WITH COMPLEX SYSTEMS

WALTER VON LUCADOU

WGFP
Hildastr. 64
D-7800 Freiburg, FRG

1. Introduction

It is an empirical fact that experiments which include complex systems show a smaller degree of repeatability than experiments dealing with simple systems. The conventional "explanation" for this fact assumes that complex systems cannot easily be prepared, cannot be fixed or varied in a systematic way.

The *model of pragmatic information* (MPI), which assumes an underlying "quantum mechanical" structure of complex systems, makes the situation even worse. The problem of so–called "negative measurements" will be discussed and also how the repeatability of experiments is affected. The pragmatic information about the complex system which is available to the experimenter plays an important role for the preparation of the system. If the pragmatic information available from a previous experiment is "used" by the experimenter to enhance the transinformation–rate or efficiency of the experimental system, the internal coherence of the system may collapse, leading to a result completely different from the previous experiment. From this viewpoint, the well known experimenter effect in psychology (Rosenthal) appears in a new light.

The consequences for the design of experiments with complex systems will be discussed, and it will be shown that double blind conditions require more constraints than is usually assumed.

2. The Role of Experiments in Natural Sciences

Traditionally an experiment is regarded as a question posed by an experimenter and answered by "nature". From a certain point of view of modern philosophy of science, however, an experiment is a highly artificial system. The conclusion that it describes "nature" cannot be drawn in general [1]. It is true that an experiment may approximate "real life situations" via sufficiently general operationalizations, but it is a well known fact for instance in social sciences that any attempt to adapt an experimental setting to a real life situation results in a certain vagueness of the methodological aspects of the experiment, e.g., the statistical evaluation becomes less efficient, etc. [2, 3, 4].

The constructive theory of action [1, 5] is able to describe experiments in terms of actions of the experimenter and the problem of repeatability in terms of instructions to the experimenter. In the following scheme the development of an experiment is described:

$$!a \;:\; S_0 \rightarrow a \rightarrow S_1 \rightarrow P \rightarrow S_2 \; (\rightarrow i)$$

$!a$: instruction for the experimenter

S_0: initial situation

a: action of the experimenter

S_1: artificially prepared system

P: non–accessible process, dynamics, time evolution

S_2: final situation

i: pragmatic information from the experiment (result)

Any experiment has to start with an initial situation S_0 which includes the available instrumentation, materials, subjects, and so on. In order to be able to perform an experiment with this material, the experimenter needs an instruction $!a$ how to do it. This instruction includes his socialization, education, knowledge of the state of the art of the field, and the so–called "tacit knowledge" which describes the knowledge of the scientific "in–group" for instance of the laboratory. It is important to note that $!a$ is not fully explicit. However, scientific method requires that the description of the initial situation S_0 and the action of the experimenter a to prepare the experimental setting S_1 must be made as explicit as possible. Once the artificial experimental situation is established, the system evolves according to its own rules P and leads to the final situation S_2. S_2 describes not only the situation of the experimental setting after the experiment but also the experimental data and the result i of the evaluation. From this point of view any experiment can be regarded as the production of "meaningful information" i on the basis of an "action" a and the "dynamics" P of an artificially prepared system S_1. The produced information is in principle part of the final system S_2 and the question arises whether the system S_1 could ever be prepared again if the results of the experiment i are implemented in $!a$ or a of a replication study. Conceptually, additional requirements are necessary to prevent an influence of i on $!a$ or a in future experiments.

3. What Does Repeatability Mean?

The usual distinction between identical and conceptual replication is based on the assumption that the process P under investigation is something objective which can be abstracted from the other parts of the experimental scheme above. Additionally, the concept of identical replication assumes that it is possible to replicate at least the steps a and S_1 in an identical way. This implies two main criteria: (1) the accessibility of S_0 and P, and (2) the realizability of a and S_1. The first criterion seems trivial, however, in real experiments the access to resources plays an important pragmatic role (instrumentation, funding), and the accessibility of P is a hypothetical assumption which might not be warranted. Theoretically it is possible that the process depends on "internal" constraints and/or changes absolutely in time (historical events). The second criterion includes explicitness and completeness of a, for instance the preparation of the initial and boundary conditions and their context–independence which is of special interest here. It is assumed that a and S_1 do not depend on a special experimenter or other contextual parameters such as the place or the time where the experiment is performed, or which meaning it may have for society. Furthermore a clear criterion must be given defining the end of a measurement.

It is obvious that even in experiments with classical complex systems these requirements cannot be fulfilled in general. In most cases it is unknown whether all relevant variables of the process P are accessible, and even if this should be the case, then it is not necessarily possible to manipulate them explicitly.

Statistical accumulation is normally regarded as a standard method to solve this problem. It is assumed that the unknown variables are randomly distributed and contribute only to the variance of the experimental result. But again it is assumed that the unknown variables do not depend on contextual parameters. One could even use this assumption to define classical systems in the following way: *A classical system is a system S_1 which allows for a complete separation of all variables (known or unknown), thus distinguishing the process P from the rest of the experimental system, namely $!a$, S_0, and a.*

Since the work of Rosenthal [6, 7] it is known in psychology that the separability of variables of the experimental process P and the preparation a of the system S_1 is not always warranted. However, this does not necessarily include that the system is a non–classical one, because it can be assumed that the action a of the experimenter was not described explicitly enough, thus leading to differently prepared systems S_1 and S_1'. The classical Rosenthal – experiments were able to demonstrate that expectations of different experimenters can influence the results of their experiments. He assumed that the experimenters unconsciously used different experimental actions a which caused the differences in the results i.

As a consequence, in psychology a double blind technique is used to avoid experimenter expectancy effects. The experimenter who exerts the action a to perform the experiment is not allowed to know the distribution of the critical (independent) variables, the influence of which is to be tested by the experiment.

Paradoxically enough, it turned out that the well known "Rosenthal – effect" itself cannot always be replicated [8]. At least with certain types of experiments it shows a typical decline with the number of replication studies. Unfortunately, Ref.[8] gives no detailed information about the decline effect. Only the difference between the studies before and after 1969 is reported. But the data show clearly that the most successful experiments about animal learning and inkblot tests show a considerable decline. In spite of their initial success they were repeated less frequently whilst studies with less impressive results gained more attention since 1969. These experiments do not show a decline and seem to yield more stable results, although not more than 40% reached the significance level. This may indicate that scientists tend to neglect experiments which are not easily replicable in spite of rather strong effects (see also the open peer commentary to Ref.[8]). The inter– and intra–experimenter decline effect in replication studies can be observed in social sciences so often that it requires an additional explanation (see Ref.[9] for an example from personality psychology).

4. Non–Classical Complex Systems

According to the definition given above, at least certain quantum mechanical systems are examples of non–classical systems as defined above. For instance, Bell's theorem [10] and the results of the corresponding experiments [11, 12] demonstrate that any classical local theory fails to describe the experimental results of these EPR – type experiments. [13] This means that an experimental situation can be constructed in such a way that a complete separation of all relevant variables is not possible any more. The problem can be illustrated by means of the so–called "negative result measurement" [14]. The double slit experiment may serve as a very simple example. Quantum theory predicts an interference pattern behind the two slits even in the case that the intensity of the particle beam is so low that only one single particle passes the apparatus at once. Recently this predicition could be demonstrated directly in an experiment [15]. On the other hand, the interference pattern vanishes if a detector at one slit would be able to detect a particle passing it. According to the classical local theory one could argue that the perturbation of this measurement destroys the interference pattern. However, according to quantum theory the interference pattern also vanishes if the particle is not detected by the detector since it passed the other slit. In this case an interaction with the detector does not occur (negative result measurement). The only reasonable interpretation of this fact is the assumption that the mere presence of a "potentiality" to detect the particle at a given (precise) location changes the whole system. (Quantum theory simply states that the two operators X and P do not commute.) Or to put it in other words: the actual possibility for the measurement of a certain observable changes the system even in the case that a factual measurement is not performed. The actual potentiality of a measurement changes so to say the "propensity" [16] of the whole system. As a generalization of the situation of EPR – experiments, we call such non–local changes "non–local correlations".

In spite of the well–known debate about the measurement problem in quantum physics [17] it is rather unlikely that the quantum mechanical "non–locality" should also include that in macroscopic systems variables are non–locally correlated to contextual parameters of $!a$ or a of the experimental scheme described above. However, the situation changes if it is assumed that macroscopic systems may exhibit a quantum mechanical structure on a macroscopical level.

The model of pragmatic information (MPI) [18-24] assumes that the structure of highly complex (self–referential) systems is isomorphic to the structure of quantum theory. This holds especially for psychological systems. In psychology many phenomenological similarities to quantum physics can be found [25-28]:

1. Perturbations caused by a measurement cannot be eliminated in principle.
2. The order of measurements cannot be exchanged.
3. Quantum steps cannot be observed.
4. Uncertainty relations can be formulated.
5. Non–local correlations can be observed.

Especially the last phenomenon is of crucial importance for the MPI, because all the others could also be described by classical models. However, experimental evidence [29-32, 22-24] for such non–local correlations in the realm of psychology is still very controversial and further research is necessary. A recent meta–analysis [33] shows at least that this seemingly anomalous phenomenon cannot be neglected any more.

One of the most important features of the MPI is the assumption that highly complex self–referential systems that are able to establish "organizationally closed units" [34] "use" non–local correlations to establish their unity. As an analogous phenomenon the exchange force in molecular physics [31] might be mentioned. Thus the decomposition of an organizationally closed system "consumes" pragmatic information to cancel the non–local correlations inside the system. Pragmatic information is a concept that can be used to quantify meaningful information [18, 35]. It is measured by the action it exerts on a system. From this point of view the preparation of the experimental system S_1 by means of the action a is described by the amount of pragmatic information which is "transferred" to the system S_0. After the preparation the system S_1 "produces" a certain amount of pragmatic information P which changes the system dynamically until it reaches the "output state" S_2. The analogies to a scattering process (S–matrix formalism, respectively) are obvious.

From this point of view the seemingly paradoxical situation of the "negative result measurement" can be understood in a natural way. Each attempt to get information about the "internal structure" of a non–classical system means a decomposition of the system and thus a "scattering process" of pragmatic information which changes the system as a whole. Or to put it into terms of information theory: "Each attempt to use non–local correlations as a signal transfer will make them vanish". The term "signal transfer" means in this context that it is possible to identify the conditions or scores of individual events of S_1 only on the basis of knowledge of the (non–local) correlations and the output state S_2 (or that S_1 can be "reconstructed" by i and S_2 [36]).

5. "Gedankenexperiment" Involving Non–Local Correlations

Under very special conditions (EPR–situation, singlet state, spin conservation) the classical correlation function $c_0(v_1, v_2)$ and the non–classical correlation function $c_1(v_1, v_2)$ between two measured variables (spin measurements) differ considerably (for details see, e.g., Ref.[37]). However, any attempt to "use" the non–local part of the correlation $c_1 - c_0$ as a "superluminal telegraph" [38] cannot work because it would be necessary to select single particles from an isotropic spin distribution without performing a spin measurement. The mere measurement of the correlation does not include a "superluminal" signal transfer because a coincidence measurement of corresponding particles does not include a selection only on the basis of S_2 but uses classical (local) signal transfer from both detectors.

Let us now assume that in a set of data from an experiment with a sufficiently complex system a correlation can be found which includes non–local correlations within the system. Let us further assume that the experimenter does not know before the experiment which variables show a correlation. In this case he is actually not able to prepare the system S_1 via the action a in such a way that he could select single cases and transfer an amount of information by means of the (non–local) correlations. From the point of view of the MPI his ignorance about the possible correlations is a part of the action a which constitutes the system.

After the experiment, however, the pragmatic information i gained from the experiment does not only change the knowledge of the experimenter but also the "meaning" of the previous action a to a'. If the experimenter would start a (seemingly) "identical replication", he would again use action a to prepare the system S_1. However, since the meaning of a has changed he uses a' in the case of an identical replication, and he could actually use the new pragmatic information i to select cases and to transfer signals via the non–local correlations. Since this is not allowed, the non–local correlation must necessarily decline even in the case that the experimenter does not factually use this possibility. Similar to the negative result experiment in quantum mechanics a factual measurement is not necessary to collapse a non–local correlation.

At first glance the change from an action a to a' seems to be immaterial. From a system theoretical point, however, and also from the point of view of operationalization, there is a clear and distinct change in the system caused by the pragmatic information i of the previous experiment. It seems not very useful to neglect this fact.

6. Replication of Experiments Involving Non–Local Correlations

One should not conclude from the previous paragraph that a direct replication of an experiment involving non–local correlations is not possible. The MPI simply shows that the idea of an "identical replication" is an illusion. On the other hand, the MPI demonstrates that the usual double blind technique which is used in social sciences to prevent an experimenter–expectancy effect is not sufficient to allow a direct replication of a previous experiment.

To clarify this point, let us assume that in an experiment it has been found that a certain "independent" variable v_i correlates with a "dependent" variable v_d. The usual procedure in social sciences requires at least a separation of two groups with high, respectively low, scores in v_i (for instance highly intelligent versus less intelligent mice). Under double blind technique the experimenter is not allowed to know which characteristic belongs to which group, but in most cases (for practical reasons) it is known that two experimental groups with different characteristics exist. In practice the experimenter of the replication study actually knows that a difference between the groups is expected. It is obvious that in this case the experimenter could "use" his knowledge to code a signal with the two groups even if he does not know the actual value of the variable v_i. Thus it would be possible to transfer this coded signal to the dependent variable v_d via the expected correlation. If the correlation is a non–local one, then it must consequently decline.

This is of course a very simple example, and in real replication experiments the conditions under which the experimenter could use the knowledge about the previous experiment can vary considerably. Moreover, such conditions are very often not reported in detail. Therefore it is very often difficult to estimate the size of this "non–classical experimenter–effect". However, it is posible to prevent it rather easily: it has to be warranted that the knowledge of the experimenter about the previous experiment does not enable him or her to operationalize any kind of signal transfer via the measured correlations. For instance, the information from the measurements could be coded in such a way that the experimenter remains "blind" until the final evaluation is performed. It is clear that only the experimenter of the first study fulfills this condition in an ideal way, and therefore it is not surprising that in social sciences a remarkable decline effect between original experimental reports and the effect size in later replication studies is often observed.

Obviously it would be necessary to test this model on the basis of existing replication studies. The main problem at the moment seems that most experimental reports are not very explicit concerning details of the double blind conditions because this is not generally regarded as a problem. However, at least one special study that seems to be a good candidate of measuring non–local correlations yields a direct confirmation of this model. The experiment [39] was a replication of an experiment to test the hypothesis of "formative causation" [40]. It was not a strict identical replication but the methodological approach was very similar. The hypothesis of "formative causation" (Sheldrake [41]) in the context of these experiments assumes that the perception of unknown material is easier for a subject if this material has been in use by many persons beforehand. Sheldrake postulates a "morphogenetic field" or a "morphic resonance" which facilitates the perception of such material in contrast to totally new material. In both experiments subjects had to guess the meaning of words in a language unkown to them, and it was supposed that this would be easier if the words were really meaningful (in that language) instead of meaningless.

The first study used Hebrew words and confirmed Sheldrake's hypothesis. It is reasonable to assume that if the effect should really exist then it could be interpreted as a manifestation of non–local correlations. In this case a direct replication would not be possible as discussed above. Indeed, it turned out that this was the case. The experimenter used Braille instead of Hebrew. Of course, a double blind technique warranted that she did not know which of the two sets of the material contained the meaningful and which the meaningless material. However, she knew that two sets were used. The results showed that the expected effect was reversed in relation to the previous experiment. As a consequence, it is not possible, in this special case, to reconstruct the original experimental groups by means of the results and of the knowledge of the correlations alone. The expected effect, however, would have made this possible. Only in the situation when both groups where mixed together so that the experimenter was actually not able to perform a selection of the independent variable on the basis of the results the previous results could be replicated.

It is clear that a single experiment is not sufficient to confirm the discussed model of pragmatic information and further research is needed. However, the model does not only provide a natural explanation for an effect which has frustrated so many researchers in social science but gives also an instruction how to prevent it.

References

1. P. Janich. Natur und Handlung. Über die methodischen Grundlagen naturwissenschaftlicher Erfahrung. In: *Vernunft, Handlung und Erfahrung. Über die Grundlagen und Ziele der Wissenschaften.* Beck, München, 1981, p. 69.
2. T.X. Barber. *Pitfalls in Human Research – Ten Pivotal Points.* Pergamon, London, 1976.
3. G. Westland. *Current Crises of Psychology.* Heinemann, London, 1978.
4. W. Bungard, ed. *Die 'gute' Versuchsperson denkt nicht. Artefakte in der Sozialpsychologie.* Urban & Schwarzenberg, München, 1980.
5. H. Tetens. Was ist ein Naturgesetz? *Zeitschr. allg. Wissenschaftstheorie* **13**, 70 (1982).
6. R. Rosenthal. On the social psychology of the psychological experiment: The experimenter's hypothesis as unintended determinant of experimental results. *American Scientist* **51**, 268 (1963).

7. R. Rosenthal. Interpersonal expectancies: the effects of the experimenter's hypothesis. In: *Artifacts in Behavioral Research.* Eds. R. Rosenthal and R. Rosnow. Academic, New York, 1969.

8. R. Rosenthal and D.B. Rubin. Interpersonal expectancy effects: The first 345 studies. *Behavioral and Brain Science* **3**, 377 (1978).

9. A.L. Edwards and R.D. Abott. Measurement of personality traits, theory and technique. *Ann. Rev. Psych.* **24**, 241 (1973).

10. J.S. Bell. On the problem of hidden variables in quantum mechanics. *Rev. Mod. Phys.* **38**, 447 (1966).

11. J.F. Clausner and A. Shimony. Bell's theorem: experimental tests and implications. *Rep. Prog. Phys.* **41**, 1881 (1978).

12. A. Aspect, J. Dalibard, and G. Roger. Experimental test of Bell's inequalities using time–varying analyzers. *Phys. Rev. Lett.* **25**, 1804 (1982).

13. A. Einstein, B. Podolsky, and N. Rosen. Can quantum mechanical description of physical reality be considered complete? *Phys. Rev.* **47**, 777 (1935).

14. M. Renninger. Messungen ohne Störungen des Meßobjekts. *Z. Phys.* **158**, 417 (1960).

15. A. Tonomura, J. Endo, T. Matsuda, and H. Ezawa. Demonstration of single–electron buildup of an interference pattern. *Am. J. Phys.* **57**, 117 (1989).

16. K.R. Popper. *Logik der Foschung*, 6. Aufl. J.C.B. Mohr, Tübingen, 1976.

17. F. Selleri. *Die Debatte um die Quantentheorie.* Vieweg, Braunschweig, 1983.

18. E. v. Weizsäcker. Erstmaligkeit und Bestätigung als Komponenten der pragmatischen Information. In *Offene Systeme I.* Ed. E. v. Weizsäcker. Klett, Stuttgart, 1974.

19. K. Kornwachs and W. v. Lucadou. Pragmatic information and nonclassical systems. In *Cybernetics and Systems Research.* Ed. R. Trappl. North Holland, Amsterdam, 1982, p. 191.

20. K. Kornwachs and W. v. Lucadou. Pragmatic information as a nonclassical concept to describe cognitive processes. *Cognitive Systems* **1**, 79 (1985).

21. K. Kornwachs and W. v. Lucadou. Open systems and complexity. In *The Paradigm of Self–Organization.* Ed. G. Dalenoort. Gordon & Breach, New York, 1989, p. 123.

22. W. v. Lucadou. Non–locality in complex systems. In *Self–Steering and Cognition in Complex Systems. Studies in Cybernetics, Vol.22.* Eds. E. Rosseel, F. Heylighen, and F. Demeijere. Gordon & Breach, New York, 1989.

23. W. v. Lucadou. *Chaos und Psyche.* Aurum, Freiburg, 1989.

24. W. v. Lucadou. Complementarity and non–locality in complex systems. In: *Parallelism, Learning, Evolution.* Eds. J. Becker, I. Eisele, and F. Mündemann. Springer, Berlin, 1990.

25. J. Fahrenberg. Das Komplementaritätsprinzip in der psycho – physiologischen Forschung und psychosomatischen Medizin. *Zeitschrift für klinische Psychologie und Psychotherapie* **27**, 151 (1979).

26. O.W. Hill. Further implications of anomalous observations for scientific psychology. *Am. Psych.* 1170 (1986).

27. K. Kornwachs. Cognition and complementarity. In *Nature, Cognition, and Systems I*. Ed. M.E. Carvallo. Kluwer, Amsterdam, 1987, p. 95.

28. D.M. Snyder. The significance of human observation in measurement in quantum mechanics: the nature of the traditional separation between psychology and physics. In *Bell's Theorem, Quantum Theory, and Conceptions of the Universe*. Ed. M. Kafatos. Kluwer, Dordrecht, 1989.

29. R.G. Jahn. The persistent paradox of psychic phenomena: an engineering perspective. *Proc. of the IEEE* **70**(2), 136 (1981).

30. E.C. May, G.S. Hubbard, and B.S. Humphrey. New evidence for interaction between quantum systems and human observers. Paper presented at the Fourth Annual Meeting of the Society for Scientific Exploration. Charlotteville, USA, 1985.

31. R.G. Jahn and B.J. Dunne. On the quantum mechanics of consciousness with applications to anomalous phenomena. *Found. Phys.* **16**, 721 (1986).

32. W. v. Lucadou. Experimentelle Untersuchungen zur Beeinflußbarkeit von stochastischen quantenphysikalischen Systemen durch den Beobachter. H.A. Herchen, Frankfurt, 1986.

33. D.I. Radin and R.D. Nelson. Evidence for consciousness–related anomalies in random physical systems. *Found. Phys.* **19**, 1499 (1989).

34. F.J. Varela. Autonomy and autopoiesis. In: *Self – Organizing Systems*. Eds. G. Roth and H. Schwengler. Campus, Frankfurt, 1981.

35. D. Gernert. Measurement of pragmatic information. *Cognitive Systems* **1**, 169 (1985).

36. W. v. Lucadou. Non–locality in complex systems – a way out of isolation? Paper presented at the Sugarloaf Meeting on *Relationship Between Mind and Matter*. To be published.

37. S.L. Braunstein and C.M. Caves. Chained Bell inequalities. In *Bell's Theorem, Quantum Theory, and Conceptions of the Universe*. Ed. M. Kafatos. Kluwer, Dordrecht, 1989.

38. Costa de Beauregard. Time, the physical magnitude. In *Boston Studies in Philosophy of Science, Vol.99*. Eds. R.S. Cohen and M.W. Wartofsky. Reidel, Dordrecht, 1987, p. 158.

39. H. Kling. Der Einfluß von Wahrnehmungsmodalität und Sinnhaftigkeit auf die Sicherheit in der Bedeutungseinschätzung von Blindenschriftwörtern. Diplomarbeit Universität Mainz, 1988.

40. G.E. Schwartz. Unconscious pattern recognition of true Hebrew in subjects who do not know Hebrew: Evidence for formative causation? Unpublished manuscript, Yale University, 1986.

41. R. Sheldrake. *A New Science of Life*. Blond & Briggs, London, 1981.

NON–BOOLEAN LOGIC OF THE THEORY OF EVOLUTIONARY SCIENCE

FRANZ R. KRUEGER

Ingenieurbureau Dr. Krueger
Messeler Str. 24
D-6100 Darmstadt 12, FRG

1. Introduction

Evolution theory as an empirical science claims to be able to model a gain of information content by which evolving systems are to be described. It is obvious that the propositional structure of such theories cannot be logical in a completely Boolean sense. Namely, Boolean logic conserves information, because those logical transformations are, strictly speaking, tautological, as "deductions" are. "Inductions" are information conserving, if they are done with a closed set of propositions, otherwise senseless. The non-Boolean figure of logical transformation is the "abduction". On the basis of a lattice theoretical description of the structure of evolution theory, abduction has a very rigorous (by no means vague or even feuilletonistic) meaning.

In certain limiting cases (in terms of stability or reversibility of the systems, commutability of the propositions) Boolean structures are valid as subsets of the entire propositional lattice. The non-Boolean lattice structure of evolutionary science has been given by Krueger [1] in a German textbook, and is shortly described here with some of its implications for science theory and ontology.

2. Lattice Structure of Propositions

In the following very shortly the properties of lattices will be given with special emphasis on those items being important for science theory in our very context.

$A, B, C \in V$ are propositions gained in a dialogical sequence, maybe regarding facts (specialities), maybe regarding theories (generalities). V is an algebraic lattice on the set G (normally with group structure) whose elements represent that empirical knowledge the propositions are referring to. Among all lattice-elements A, B, an order relation $A \subseteq B$ defines a directional semi-order on these propositions.

Just for easier handling of a lattice, an identical representation with two two-

Information Dynamics, Edited by H. Atmanspacher and
H. Scheingraber, Plenum Press, New York, 1991

place operators "meet" $\sqcup$ and "join" $\sqcap$, those being commutative, is possible (for the Boolean limiting case "or" $\vee$ and "and" $\wedge$, see below). With this isomorphous representation two consequences come along:

- The two operator representation makes a much more economic handling of propositions possible - in the language of biologists one may state that logic at all is an evolved category (or Kant's *a priori* beyond empiry).
- The two operator representation causes a huge redundancy by duplication of all combined propositional relations.

Namely, all formal lattice axioms are also valid if the $\vee$ and $\wedge$ are just fully exchanged - the meaning of the propositions may be, however, different. (In science theory we shall see, which meaning is conserved if the operators are exchanged.) This property is called "duality", and most of the scholastic ontological interpretations are just trivial redundancy phenomena.

The dual axioms of the lattice and some consequences are as follows:

Commutativity :	$A \wedge B = B \wedge A$	$A \vee B = B \vee A$
Associativity :	$A \wedge (B \wedge C) = (A \wedge B) \wedge C$	$A \vee (B \vee C) = (A \vee B) \vee C$
Amalgamation :	$(A \wedge B) \vee B = B$	$(A \vee B) \wedge B = B$
Idempotency :	$A \wedge A = A$	$A \vee A = A$

The partial order supplies the connection with the unique directional operator:

$$A \subseteq B : A \wedge B = A = \inf(A, B)$$

and dually:

$$A \subseteq B : A \vee B = B = \sup(A, B)$$

A class of lattices (e.g., of Boolean lattices) is modular, as defined by the autodual relation

$$A \wedge (B \vee C) = (A \wedge B) \vee C$$

from which follows:

$$A \wedge (B \vee C) \supseteq (A \wedge B) \vee (A \wedge C)$$

Without proof we state here that all propositional logics are modular!

A class of lattices (e.g., Boolean or evolutionary lattices) is complemented, i.e., to each proposition A there exists at least one complement (negation) $B = \neg A$ with $A \vee B = 1$ (largest sup of V) and $A \wedge B = 0$ (smallest inf of V). In the special case of a Boolean lattice (classical logic) the complement is unique and thus the lattice is distributive:

$$A \wedge (B \vee C) = (A \wedge B) \vee (A \wedge C)$$

It should be pointed out more clearly again that the redundancy due to duality causes an inflation of conceptual terms (into conjugated term-pairs) which do not (!) have an ontological meaning at all, but are the result of the economic lattice structure of human thinking. Examples of these pairs are:

extensions – intensions	extensive vs. intensive variables
object – subject	canonically conjugate quantities
fact – model	causality – finality
system – theory	possibility – necessity
matter – mind	disposition – manifestation

3. Non-Boolean Attributes of Evolutionary Logic and Information Gain

There are some peculiarities with (evolutionary or synergetic) processes the propositions about which do not obey the Boolean lattice rules. As we shall see, consequences arise that the negation of such a proposition is not necessarily unique (as, for instance, known for the orthocomplement in quantum logic), and that even more than one type of complement must be introduced.

3.1. *Catastrophes, Hystereses, Switching, Memory Storage*

If $(A \wedge C) \vee (A \wedge \neg C)$ is valid then also A. *Vice versa* only in Boolean logic the implication is always true. Namely, the subsequent empirical proof of the *vice versa* – implication demands the commutativity of the operators with the propositions A, C. Or, with higher systems: The actual subsequent measurement of the values regarding the propositions A, C (in the formula from right to the left) may disprove distributivity in cases, where catastrophe or hysteresis phenomena are involved, in any case of the non-linear process "switching", as the precedent action may influence inseparably the subsequent one.

As a consequence, at least two types of complements are needed, namely:

- $\neg_1 A$, the – empirically indistinguishable – totally antisymmetric proposition to A, only needed for formal descriptions of conservation of symmetry;
- $\neg_2 A$, the non-unique empirical complement in order to model bifurcations and information gain.

In the limiting case of classical systems both complements merge to the one unique complement.

3.2. *Bifurcations, Symmetry Breaking, Mutation*

Near a bifurcation point the dialogical sequence of testing a proposition concerning the sensitive state variable may exhibit that the negation of that proposition is not unique. Namely, the state trajectory of a non-linear system is not simply the case or not, but may split. Thus a subsequent test of the proposition may reveal multiple answers rather than "yes" or "no". The two complement operator types thus behave differently: Let $H \subset G$ be a subgroup of G with the $V(H)$ acting on H. Then we get:

$$A \wedge \neg_1 A = 0 \qquad A \wedge \neg_2 A = a \in V(H)$$

$$A \vee \neg_1 A = 1 \qquad A \vee \neg_2 A = V/V(H)$$

$$\neg_1 \neg_1 A = A \qquad \neg_2 \neg_2 A = \text{iso } A \neq A \text{ or } = A$$

Observe, that the proposition 0 in the lattice V acts on the element 1 of the group G. Thus propositions about bifurcation construct a wider basis of references by a semi-group (!) operation. Whereas classical Boolean propositions act (in a certain representation) on given symmetry groups, the evolutionary propositions create new symmetry groups by semi-group operation due to their symmetry breaking character. Biological mutation can also be formalized this way.

4. Dual Properties of Facts and Models

4.1. *Incommensurability Relations*

The incommensurability of propositions about certain facts is well known in quantum mechanics, namely of those which refer to not completely separated subspaces in Hilbert space. In evolutionary science the same is found with non-linear behaviour, which indeed is founded on the basis of quantum mechanical effects. In both cases this incommensurability can be written (the proposition about a proposition is again a proposition): "There are pairs of different facts which fulfill both incommensurability relations."

$$\bigvee A, B \in V : A \neq (A \wedge B) \vee (A \wedge \neg B)$$
$$\wedge : B \neq (B \wedge A) \vee (B \wedge \neg A) \qquad \text{(FACTS)}$$

The complete exchange of all $\wedge$ and $\vee$ should again provide a meaningful relation, however, with other propositions. This dual set reads:

$$\bigwedge L, W \in V : W \neq (W \vee L) \wedge (W \vee \neg L)$$
$$\vee : L \neq (L \vee W) \wedge (L \vee \neg W) \qquad \text{(MODELS)}$$

More accurately, in both sets the $\neq$ -signs can be exchanged by the $\subseteq$ and $\supseteq$, respectively. This model incommensurability relation reads: "For all pairs of different models either the one or the other incommensurabilty relation is valid." These fact-model relations are just a short description of the incommensurability concept of Feyerabend [2].

As one can easily realize with these incommensurability relations the non-Boolean truth values are split due to the ambiguity of the complement: "false" is a value in models, and "true" in facts propositions, however, in models the possible ambiguity just allows "approved" instead of "true", and with facts a subsequent test just allows "doubtful" instead of "false".

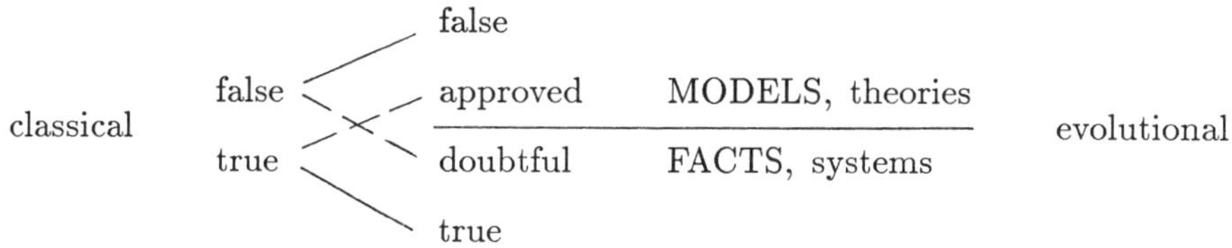

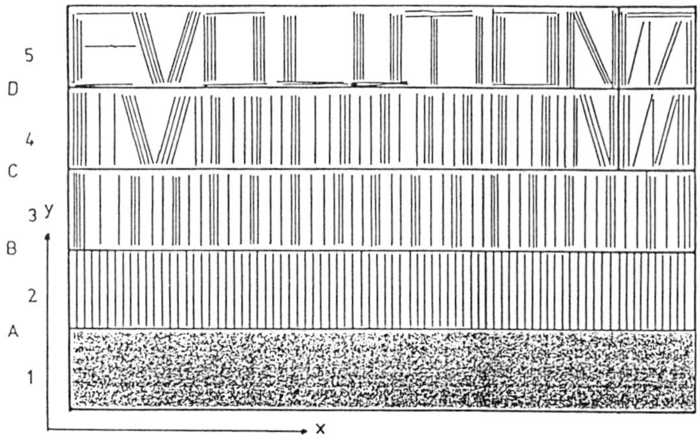

Figure 1: Illustration of a sequence of symmetry breakings, conservation laws, entropy, information creation (from Ref.[1], p.35).

4.2. *Strata of Systems and Theories as Dual Representations of Symmetry Breaking*

Evolution of systems is maintained by subsequent symmetry breaking, giving rise to new system variables and related conservation laws. A system S' being advanced relative to another one, S, possesses at least one more symmetry subgroup than the retarded one. Dual to system evolution is theory progress, which indeed is symmetry healing. A theory T' being advanced relative to another one possesses at least one less symmetry subgroup than the retarded one. Or:

$$T = T' + \text{set of contingent conditions}$$

$$S' = S + \text{set of contingent conditions}$$

Especially the systems are worthwile to realize their stratification in the evolution process (S, S', S'', S''', and so on, with further complexity). Within each level (stratum) all (intra-system) propositions can be cast into Boolean form, however, in between the levels (strata) at least some (inter-system) propositions behave non-Boolean.

The properties of symmetry breakings A, B, C, D are illustrated in Fig.1. These breakings connect the 5 system levels together. (The real course of symmetry breakings by fluctuation and selection is, however, not illustrated here!) Level (1) illustrates the total symmetry (the "Nothing"); it is fully translational invariant in x- and y-direction. A breaks the translational invariance in x-direction into a special invariance x = x+a in level (2). The "grey" of the "Nothing" is periodically split into "white" and "black". The total grey-sum, however, is a conserved entity through all levels (1-5). B breaks again the invariance in x-direction, namely, into a superperiodicity of 5 lines each. However, the number of lines is now a conserved entity through all levels (2-5). The total symmetry in y-direction is still valid in level (3). C now breaks the symmetry in y-direction by tilting several lines in level (4). However, the tilting sum (left - right) is zero and thus a conserved quantity through all levels (3-5). A certain mirror-symmetry for single line elements is now valid. D finally breaks the remaining symmetry by further tilting into right angle. However, there is a new conservation law through levels (4-5), as the number of elements are meanwise symmetric relative to the axis y/2. The result is level (5).

A general law is thus illustrated, that with each symmetry breaking the invariant entities remain conserved, however, one additional invariant entity is born. If one type of order is evolved (due to long-mode survival, not illustrated here), here a certain order of letters, with progressing structuration – level (4,5) – one needs an "outer world" (right hand side box). This "outer world" receives the overflowing disorder (or entropy) which is maintained when an "inner world" information is evolving; here the text EVOLUTION. This total entropy increase is thus equivalent to the total invariance principle together with system formation.

Before we pass over to the quantitative concepts of information, we may state here the three qualitative concepts of information, namely:

- syntactic information: It just governs the formal logical transformations within one level, say S' (intra-system information);
- semantic information: It refers to an evolved code, fixed in the next lower level S, now providing contingent information in the reference level S' (inter-system-bottom information);
- pragmatic information: It refers to a possible meaning, arising in the next higher level S'', which may result from a fixed information in the reference level S' (inter-system-top information).

5. Information Concepts with Tautological vs. Evolutional Propositions

Now it is the task to provide information concepts suitable for quantification, with which the information content at each level, and the information gain (or loss) in each dynamic (evolutionary) process can be precisely given.

The intra-system (syntactic) information can best be modeled by the Shannon-formalism describing the transmission of information (messages) in principle without gain or loss of information. For con- and disjunctions of propositions A, B, and themselves, probabilities P are defined fulfilling the autodual (!) relation:

$$\text{Probability } P: \quad P(A \wedge B) = P(A) + P(B) - P(A \vee B)$$

Identifying an area with the probability, Venn-diagrams can illustrate these conditions of information conservation. That formula is only another representation of saying that A and B are commensurable (commutable) propositions. The kind of reasoning within that lattice structure governed by the Shannon concept of information is still the figure of "deduction".

If information is not necessarily conserved in the dialogical empirical sequence of testing the propositions A, B, then the above autodual relation splits into two dual relations, which contradict the Shannon information definition. Thus P is not any more a measure of probability, but splits dually into "plausibility" and "believability", as it can be seen in the work of Klir [3], referring to the Dempster-Shafer definition of information:

$$\text{Plausibility } Pl: \quad Pl(A \wedge B) \leq Pl(A) + Pl(B) - Pl(A \vee B)$$

$$\text{Believability } Bl: \quad Bl(A \vee B) \geq Bl(A) + Bl(B) - Bl(A \wedge B)$$

It is now obvious that these relations are just the rewritten incommensurability relations for fact and model propositions in evolutionary logic. The kind of reasoning within that (non-Boolean) lattice structure governed by the Dempster-Shafer concept of information now is the thus precisely stated figure of "abduction".

Needless to say here that for the limiting case of conservation of information both above relations merge again to the Shannon concept to be treated by Boolean logic, and deduction is the limiting case of abduction.

6. Me-Ontology as a Consequence of the Retrospective of Evolution

The only irreducible entity describing the world obviously is "information", being transmitted and created by "action". In contrast, "substance" and "matter" are only stabilized sets of information. "Information", on the other hand, is identical with "asymmetry", the measure of which is the distribution function of actuation of symmetry group elements.

The various distinct levels of being are evolved in a sequence of symmetry breakings. As an example one may think about the evolution from elementary particles to atoms, from atoms to molecules, from molecules to biomolecular organization, and so forth. At any step, a symmetry is broken and thus new information is accumulated on the basis of new-type variables and additional related conservation laws.

All these conservation laws persist through all higher levels of organization. Thus a local symmetry breaking (creating a new entity) is always compensated globally. As an example, in the transition from laminar to turbulent flow a symmetry of angular momentum is broken locally thus giving rise to new entities, the whirls, but, of course, the total angular momentum is conserved.

It seems to be worthwile to note that the following conclusion as well as most of the arguments given were already published by Nicolaus Cusanus [4]. One should only exchange his "God" (deus) by the term "Nothing" and His (God's) "word" (verbum) by the term "information".

Namely, if one goes the ladder of symmetry breaking backwards down, one ends up with a total symmetry, i.e., at no action whatsoever. In our empirical sense this very pristine state is characterized as "Nothing". Now taking the fact seriously that all the conservation laws persist throughout all higher levels of being, we must state: "The Nothing is conserved." Or: "All being is the result of subsequent non-linear transforms of the Nothing." It is just a philosophical question, whether "information" is "anything" or not. Besides this, the ontological question is answered by the "$\mu\eta$–$o\nu$". The famous ontological difference is obviously zero.

References

1. F.R. Krueger. *Physik und Evolution.* Paul Parey, Berlin, 1984.
2. P. Feyerabend. *Against Method. Outline of an Anarchistic Theory of Knowledge.* NLB & Suhrkamp, Frankfurt/M., 1975.
3. G.J. Klir and T.A. Folger. *Fuzzy sets, Uncertainty, and Information.* Prentice-Hall, New York, 1988. See also G.J. Klir. This volume.
4. N. Cusanus. *De Possest* (1460). German translation: Reclam, Stuttgart, 1963.

CHAOS AND INFORMATION PROCESSING: MULTIFRACTAL ATTRACTORS MEDIATING PERCEPTION AND COGNITION

JOHN S. NICOLIS

Department of Electrical Engineering
University of Patras
Patras, Greece

Information theory quantifies and optimizes (with respect to given constraints) the efficiency of designing, performing, and interpreting (the results of) single or multiple measurements.

To this end, information theory devises codes (cascades of one–to–one mappings between sets of symbols) in order to compromise two (conflicting) objectives in signal transmission: speed of transfer and reliability of reception. Specifically, the aim is to ensure that the "receiving site" will be able to detect and correct single or multiple, uncorrelated or correlated errors in the presence of (quite often overwhelming) noise, i.e., uncontrolled fluctuations. In short, information theory as originally devised by C. Shannon deals with symbol manipulations (much like the execution of a program in a digital computer) at a purely syntactic level, without any reference to meaning of the messages involved. No "understanding" or "thinking" takes place during this sort of transaction.

After clearing out the received messages from noise, the next step is the "semantic" one, namely the extraction of meaning: The *sine qua non* condition for that is the unambiguous partitioning of the set of undifferentiated "raw" external stimuli onto coexisting asymptotically stable categories ("minimal length algorithms" $\rightarrow$ chaotic attractors) and the establishment of a language of communication (a sequence of jumps) amongst them. The attractors/memories have to be asymptotically stable so that the syndrome of self – reference which characterizes any linguistic scheme does not turn into a paradox. Now, the raw messages that the processor receives may come from three different sources (dynamical systems):

(a) Dynamical systems with intrinsic relaxation times T_i orders of magnitude above the intrinsic relaxation time(s) τ of the processor itself: $T_i \gg \tau$.

(b) Dynamical systems for which $T_i \ll \tau$.

(c) Dynamical systems for which $T_i \approx \tau$.

Information Dynamics, Edited by H. Atmanspacher and
H. Scheingraber, Plenum Press, New York, 1991

It is perhaps not accidental that the two sciences man has developed to the point of perfection are (a) celestial mechanics ($T_i \gg \tau$) and (b) quantum mechanics ($T_i \ll \tau$). In case (a) one takes a lot of variables and/or parameters as practically constant and in case (b) as boundary conditions. In case (c) on the other hand, the processor has no other choice but to "coevolve" with the system under observation or to wage a game amounting essentially to a cascade of decision making under uncertainty and conflict. Such a "game" constitutes the third and last level of the informational edifice we call information hierarchy, namely the pragmatic level.

The algorithmic and dynamical aspects of the ongoings at the three distinct hierarchical levels (syntactic, semantic, and pragmatic) as well as some hints about the way these levels interact are treated in references [1-4].

Synopsis

We first consider a chaotic dissipative flow (e.g., the Rössler attractor) or map (e.g., the logistic equation),

$$\dot{\mathbf{x}} = \mathbf{F}(\mathbf{x}, \xi) \tag{1}$$

with div $\mathbf{F} < 0$. We apply an arbitrary (but Markovian) partition of M subintervals on the x – axis, thereby imposing an "alphabet" so that transitional points between neighbouring cells (symbols) are mapped onto transitional points. Otherwise, i.e., in case of non – Markovian partitions, the alphabet will get distorted along the flow or along the cascade of iterations.

Taking one subinterval as the pause and iterating the flow or the map we generate (after transients subside) artificial texts of words (strings of symbols) separated by the pause. We investigate the "bulk" statistical properties of such a "language", namely the Shannonian entropy, the Kolmogorov – Sinai entropy, the transinformation, and the "Zipf like" behavior of the text, that is the probability of appearance of a "word" in a long text *versus* its ranking order [2, 3]. Figure 1 shows a comparison of Zipf's law for natural languages (Greek prose), genetic "texts" (RNA polymerase), and numerically generated chaotic symbol sequences. We also investigate the Markovian memory of the language (usually of fifth order) as well as the grammatical and syntactic rules which allow only a small fraction of all M^k possible words of length k made from M symbols to be realized. So we deduce that the chaotic dissipative flow acts as an efficient "symbolic filter" allowing variety and reliability to coexist, and mimicking the evolutionary processes – whatever they are – which have shaped the formation of natural languages.

All these deliberations characterize a language at the syntactic level, as they essentially refer to word dynamics or plain information theory. Moving next to deliberations at the semantic level (e.g., at the hierarchical level where pattern classification and recognition takes place) we deal with correspondences or mappings (not one–to–one) amongst subsets (sub–basins) of external objects and functions (stimuli) and the coexisting categories/attractors of the processor. The empirical study of the ongoings at the semantic level is of course the object of cognitive psychology.

At the semantic level the processor/classifier is simulated by a chaotic dynamics as given above (Eq. 1) possessing multiple coexisting chaotic attractors separated, in general, by fractal basin boundaries. Each attractor is a "cognitive receptor" which far from behaving like a "photographic memory" – as in the case of a steady

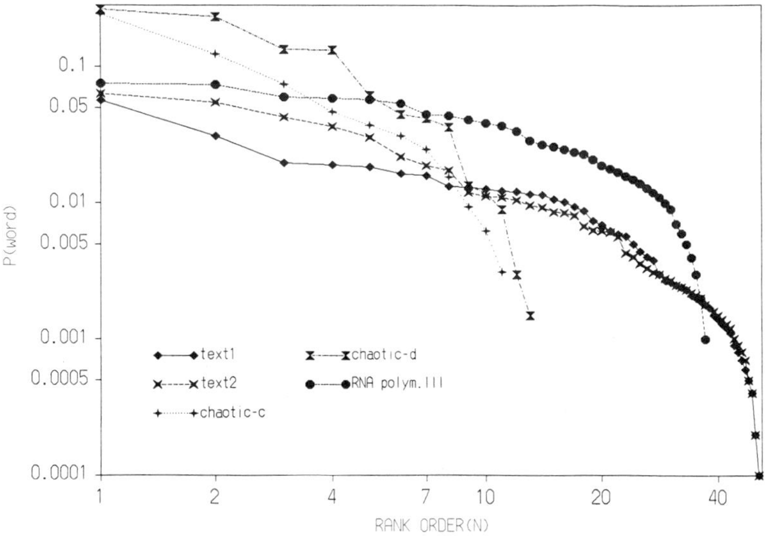

Figure 1: Comparison of Zipf's law – like behavior for natural language (text "Antilegomena" by I.M. Panayiotopoulos, two different versions), genetic "text" (RNA polymerase III), and chaotic symbol sequences generated by the logistic map at $r = 4$.

state – improvises and evolves "from within". The partitioning of a large number of environmental stimuli (time series of length N) onto a set of few chaotic attractors of information dimension $D_i \ll N$ amounts to compression of the raw stimuli and the consruction of minimal length algorithms (programs) whose length is proportional to the information dimensions of the attractors involved. Due to the fractal character of the basin boundaries, the above categorization takes place mostly under conditions of uncertainty and conflict, that is it involves an intermittent jumping of a given stimulus from attractor to attractor (see Fig.2).

In case where coexisting attractors are stable (not repellers) external noise is needed in order to make the different categories to establish a language amongst themselves. In (human) information processing the thalamocortical pacemaker (whose one–dimensional trace is the EEG from a lead on the scalp) is responsible for permutating the attractors on a time–division–multiplexing basis.

This is accomplished in a fashion which far from being a "random" process possesses all the characteristics of a metalanguage. Indeed we know that the thalamocortical pacemaker itself possesses a chaotic attractor which is more or less uniform during epilepsy, drowsiness or sleep but becomes a multifractal under the demanding regime of execution of mental tasks such as pattern recognition or calculations. Since a multifractal attractor is highly inhomogeneous (it spends practically all the time in the vicinity of few "hot spots" in state space) we can understand the parsimonial reason behind information selection which underlies, e.g., recognition of shapes and faces, that is the reason why very few (information rich) details (edges, sharp turns) are enough for the construction of the whole image in the processor, essentially from a contour.

Finally we come to the deliberations taking place at the pragmatic level where games are involved. Here we study the correspondence or mapping amongst the psychological motivations or strategies of the partners involved and their objectives

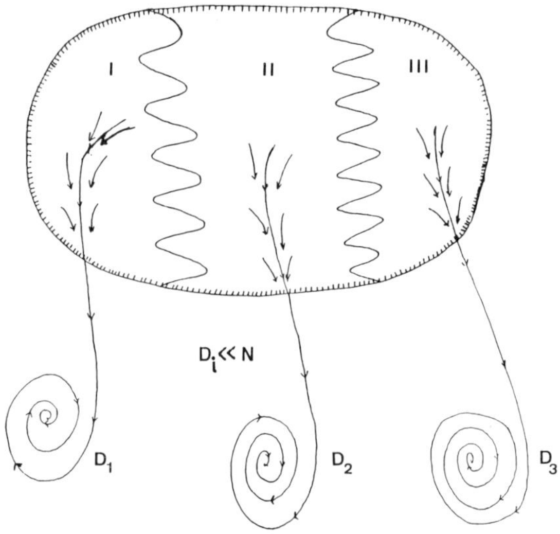

Figure 2: The partition of an undifferentiated set of impinging "raw" stimuli (basins I, II, III) on the processor in a set of coexisting semi–stable chaotic attractors (D_i) is isomorphic with a decision making procedure under uncertainty and conflict (e.g., game–like behavior) whenever the basin boundaries are fractals.

(attractors) under uncertainty and conflict. We show in some carefully chosen examples from paradoxical games (like the game of blackmail or "chicken"; see Ref.[4]) that allowing the payoffs of the game to vary with time we come up with an intermittent process between a stable attractor (the "synergetic" state) where the partners' strategies converge asymptotically and stay there, and an inconclusive regime in state space where the exchange of "blows" never ends. In some respects then, non–trivial pattern recognition and paradoxical games share an important feature which makes them isomorphic: They both involve decision making under uncertainty and conflict.

Let us mention that the parameter "attention" is related to the degree of excitation of the ascending branch of the reticular formation of the brain and directly influences the polarization of the specific thalamic nuclei which are the generators of the thalamocortical pacemaking activity. "Attention" then deeply determines (and modifies) the topology of the chaotic attractor of the thalamocortical pacemaker thereby dictating the policy of intermittent (multifractal) jumps amongst the coexisting attractors/categories of the cognitive processor.

Conclusions

We propose a new selection principle which acting via inhomogeneous intermittency in phase space (e.g., via a multifractal chaotic attractor: the thalamocortical pacemaker) is responsible for a highly nonlinear linguistic filtering process, limiting drastically:

- at the syntactic level: the grammatically legitimate words;
- at the semantic level: the "interesting" key features of a pattern;

- at the pragmatic (game–playing) level: the set of possible "moves".

This is an *a priori* selection principle, whereas "natural selection" is an *a posteriori* selection principle acting on phenotypes (e.g., on attractors) via shifting control parameters and resulting in generation/elimination/fusion ("crises") of coexisting attractors/categories.

Multifractal chaotic dynamics applied to information processing devices may thus be instrumental:

(a) in drastically simplifying hardware without compromising the breadth of the functional repertoire (software);

(b) in reducing or compressing memory capacity: instead of storing via "pixelization" of a pattern in all details one limits attention to very few key features (associated with small radii of curvature) coding for disproportionally high amounts of stored information. So, through an inhomogeneous intermittent dynamics in state space we filter out only key words, key features, and key moves.

The principle of natural selection in linguistics involves further scrutinization on those small, *a priori* selected subsets.

References

1. J.S. Nicolis. *Chaos and Information Processing.* World Scientific Publishers, Singapore, 1990. See also references therein.
2. J.S. Nicolis and I. Tsuda. *Prog. Theor. Phys.* **82**, 254 (1989).
3. A. Katsikas and J.S. Nicolis. *Nuovo Cim.* **D**. In press.
4. L. Drossos, T. Bountis, and J.S. Nicolis. *Nuovo Cim.* **12 D**, 155 (1990).

RECONSTRUCTING COMPLEXITY: INFORMATION DYNAMICS IN ACOUSTIC PERCEPTION

MANFRED EULER

Fachbereich Elektrotechnik
FH Hannover
D-3000 Hannover 91, FRG

Wär' nicht das Auge sonnenhaft,
die Sonne könnt' es nie erblicken!
J.W. v. Goethe

1. Concepts of Reality Mediated by Sensory Perception

The above quote from Goethe, that our eyes must be sun–like in order to perceive the sun is only tenable in a very indirect evolutionistic interpretation: The operating principles of the eye have to be in accord with the physical principles governing the behavior of light. Biological evolution must have learnt the principles of optics. Of course, there is no direct sun–likeness in the sense that the eye is an active light source. Instead, if Goethe had referred to the acoustic sense organ, likening the ear to an orchestra, he would have been correct even in a straightforward interpretation! In fact, if the ear did not work actively similar to a certain class of musical instruments, we were unable to hear as well as we do! The receptor organ ear is a sounding organ. Biological evolution uses principles from active dynamical systems to improve ear performance. One aim of the present article is to work out the rationale of this unexpected activeness in terms of performance, and to relate it qualitatively with general principles of information dynamics in a biological context, such as information reduction and the emergence of meaning.

Moreover, hearing conveys a concept of reality that seemingly goes deeper beneath the surface of phenomena. The acoustic objects are of a strange ethereal qualitity, transcending the realistic facade of visual objects. In the collectice awareness of mankind hearing has always played a certain mystical role. This is for instance shown by the archetypical concept of harmony that emerged from the acoustic world, providing a paradigm for an explanation of the working together of systems, creating well ordered behavior that evades straightforward explanation in mechanistical terms.

Information Dynamics, Edited by H. Atmanspacher and
H. Scheingraber, Plenum Press, New York, 1991

There is a strange analogy between the perception of acoustical objects and the measurement of quantum objects, giving rise to some perception paradoxa with a strong quantum-like flavour. The second aim of the present contribution is to substantiate this admittedly rather mystical parallelism between the internal and external world. Such an analysis is facilitated by a two level access to the system. It can be studied from the outside, at least at more peripheral stages of information processing, e.g., in the inner ear, and it is accessible from within by introspection and by psychophysical experimentation.

2. Inner Ear Dynamics: Cognitive Processes at Mechanical Level

2.1 *Psychophysical Data of Ear Performance*

The ear viewed as a naturally evolved measuring device for sound shows some highly astonishing features [1, 2, 3]. The most remarkable technical data being:

- threshold sensitivity,
- dynamical range,
- frequency discrimination.

The minimum audible sound intensity $I_{th} = 10^{-16}$ W/cm^2 is close to the physical limitations. A further increase would let us "hear the grass grow". The threshold is only slightly above thermal noise power density at the eardrum in the audible frequency range. This corresponds to an audibility of a current of 600 electrons per second, each with an energy of 1 eV, if they could be coupled in properly. Transformed into basilar membrane vibration, the incredibly minute amplitude of 10^{-11} m is sufficient to produce a sensation. We macroscopic beings are coupled to mechanical vibrations in the atomic range! Hearing must involve effective amplification and noise suppression mechanisms. The physiological range of hearing covers 12 orders of magnitude in intensity from hearing level (0 dB) to the threshold of pain (120 dB). Such a broad operating range requires data compression. Besides amplification for low power signals (expansion), limitation processes in the high power regions are necessary (compression). The ear as frequency analyzer has a remarkable resolving power. In discriminating the frequency difference δf of successive tones two different operating modes occur. In the low frequency region (< 500 Hz) it works with an absolutely constant frequency resolution $\delta f \approx 1$ Hz, whereas in the high frequency mode (> 1 kHz) the relative frequency resolution is constant, $\delta f / f \approx 0.0035$.

One might suppose, that all these features, such as nonlinear transformation of the input signal or frequency discrimination, are performed by some clever neural machinery. Yet, a good deal of the performance is already encoded on a mechanical level in the dynamics of the inner ear, a small fluid-filled snail-like receptor organ, the cochlea. Stated in technical terms, the cochlea is an intelligent sensor which does a lot of adaptive preprocessing on its own. How can all these different tasks of intelligent sensing, of vibration amplification and vibratory pattern analysis be accomplished in the inner ear on a mostly mechanical basis? What is the rationale of this "mechanical" intelligence? For fully appreciating the performance of the sensory organ we have to consider a hierarchy of signal transformation steps which mix biological fitness with adaptive power in a physical and technical sense.

2.2 *Pattern Analysis by Passive Wave Propagation*

No unnecessary energy loss of the incoming sound signal is a prerequisite of sensitivity. As a direct coupling of soundwaves to sensory cells is ineffective because of impedance mismatch, the first step to improve energy flow is to locate the cells in the cochlear fluid with roughly the same density. This requires an impedance matching procedure, performed by the middle ear acoustical transformer system. Within the cochlea a second transformation process occurs, changing the compression wave in the cochlear fluid to a surface wave along the basilar membrane. This step represents an important velocity matching process at the mechano-neural interface, slowing down the group velocity of wave packets along the basilar membrane to the signal processing velocities of the neural substrate. The first comprehensive explanation of inner ear function is given by the travelling wave theory of Bekesy [4], a passive model of the dispersive action of the basilar membrane. The basilar membrane is a 35 mm long elastic plate dividing the cochlear canal along the middle. In the basal part it is narrow and stiff, getting broader and less stiff towards the cochlear apex. The graded mechanical properties along the cochlear partition together with longitudinal coupling (directly and via the surrounding fluid) turn the system into a strongly dispersive waveguide. Because of high membrane stiffness and little effective mass waves travel fastest in the basal part and continuously slow down towards the apical region. The smooth decrease is important to secure unidirectional energy flow and to prevent standing waves.

Let us consider the basilar membrane excitation pattern of a sine-tone. As wavelength decreases towards the apex, the amplitude must increase by energy conservation. An unlimited steepening of wave crests is counteracted by damping mainly because of the surrounding cochlear fluid. A proper tuning of dispersion and dissipation produces a maximum in the envelope of travelling wave pattern. The location of the maximum depends on the driving frequency and moves towards the apex for decreasing frequencies. This is the place principle of hearing. It is one important mechanism of auditory frequency analysis. The frequency components of the incoming sound are transformed to certain positions of maximum vibration on the basilar membrane. This analysis mode is characterized by the mentioned high frequency scaling ratio $\delta f/f = const$, with an underlying logarithmic frequency–place mapping function, and a sampling of the pattern with constant spatial resolution by the neural system. In addition to the place principle, the neural system can transmit the periodic structure of the sound signal with a frequency up to 2 kHz. This is the periodicity principle of hearing, which provides a second channel of acoustic information processing, complementary to the place principle (see below). The working of the periodicity principle is characterized by the constant frequency resolution mode $\delta f = const$ in the low frequency range.

Though different in phase, the mechanical analysis of sound by travelling waves is closely related to Fourier–analysis and represents essentially a linear and passive paradigm for mechanical pattern analysis. The interplay of dispersion and dissipation must not necessarily involve nonlinearities to produce the desired frequency–place transformation. Of course, any realistic cochlear model must be nonlinear to include hydrodynamical effects especially at high signal levels. Yet the frequency–place transformation effect is already shown in one dimensional models of linear dispersive media with velocity dependent damping. Consider for instance a chain of coupled pendula with variable lengths as a simple mechanical model.

However, it turns out that the linear and passive structure principle in mechanical pattern analysis is not sufficient to explain the observed frequency discrimination power of the ear. Because of strong dissipation the mechanical resonance curves of basilar membrane sections are broad. Consequently, the quality of the basilar membrane as a purely passive mechanical filter is only poor. Two perceptually different neighbouring tones produce practically identical mechanical excitation patterns. Additional mechanisms must be present, sharpening the excitation patterns so that they can be discriminated. A neural sharpening mechanism suggests itself [5]. It was hypothesized that after a first mechanical filter a second neural filter was to enhance contrast. The so–called second filter hypothesis is in close agreement with optical signal processing where sharpening of contrast is achieved by neural means. In vision a neural network is already present at receptor organ level, producing contrast enhancement by the interplay of activation and inhibition. However, in audition attempts to locate a lateral inhibition neural network were unsuccessful. There is no such anatomical structure in the inner ear. As the tuning curves of auditory nerve fibers exhibit sharp peaks, contrast enhancement has to occur already more peripherically before that stage. Thus, the hypothetical second filter must be operating on a substrate different from a neural network. The experimental evidence of the Bekesy travelling wave model including heavy damping and broad tuning is based on measurements using dead material [4]. Mechanical basilar membrane tuning curves obtained by Mössbauer–spectroscopy in animal experiments with physiologically intact inner ears reveal a different result and exhibit fine tuning tips at low levels of the driving signals, changing into the expected broad tuning only at high excitation levels [6, 7]. Mechanical sharp tuning heavily depends on the functional integrity of inner ears and disappears, when inner ear metabolism is disrupted.

Obviously, there seems to be a difference in mechanical behavior between living and dead biological tissues, showing much less friction and therefore strong resonances in the living state. Is this a manifestation of the mysterious *"élan vital"*, a hypothetical living force that counteracts ubiquituous dissipation and the tendency to approach thermal equilibrium, giving life to inanimate matter? *Cum grano salis* the answer is yes, taking into account that biological systems exhibit coherent, well coordinated behavior. Yet we do not have to postulate new biology–specific principles to explain the effect of mechanical basilar membrane fine tuning circumventing dissipation. Though surprising, the effect is in agreement with the principles of physics, showing all the ingredients of the phenomenon of self–organization, a general principle underlying structure formation in different field such as physics, chemistry or biology [8].

The travelling wave maximum can only be sharpened by reducing the effective damping. The level dependent mechanical fine tuning suggests a nonlinear damping. This might at least partially be accomplished passively by cochlear micromechanics, the combined action of fluid dynamics with the complicated mechanical arrangements of sensory cells and the tectorial membrane in the Corti–organ. In addition to that, Nature counteracts damping in a very surprising way. The basilar membrane is actively undamped, by pumping metabolic energy into mechanical vibration. The basilar membrane is not only a passive vibratory system, similar to a microphone membrane driven by the sound signal. There is an additional active autonomous process operating at low signal levels and reaching saturation 40 dB above threshold. At high signal levels, the ear is set into passive resonance, at low levels it actively locks to the driving signal. Details of the feedback mechanism are still far from being understood. The process must be fast enough to account for sharp tuning in some mammal ears up to 100 kHz (bats). Certainly outer haircells are involved. A parametric amplification process seems possible. The tiny hairs (cilia) of the sensory cells consist of contactile proteins and can act both ways as sensors and as actuators. Bending the cilia causes a flow of ions, giving rise to a change in the receptor potential [9]. The resulting electrical signals, the cochlear microphonics, are phaselocked to the driving signal. Their magnitude is in the order of some 10 mV, strong enough to be measured in the far field with scalp electrodes [10]. These potential gradients could become mechanically relevant by modulating the mechanical parameters of the system, e.g., the stiffness of the haircells.

In view of mechanical feedback the performance of the sound receptor organ becomes plausible:

- Hearing threshold is lowered by the amplification process.
- The dynamical range is enlarged by 40 dB.
- Reduction of damping provides an increase of fine tuning, narrowing receptor bandwidth and reducing receptor noise.

Feedback produces a pattern recognition task by mechanical means with a result similar to lateral inhibition. Contrary to the discrete logic of neural networks, the present mechanism operates on a mechanical continuum, the basilar membrane. In both cases feedback and nonlinearities are important. The lateral inhibition mechanism can be viewed as implemented on a neural network, a special purpose parallel computer, dedicated to pattern enhancement. In this case the program consists in interconnecting the cells for short distance activation and long distance inhibition. This principle can be regarded as the universal program that governs both pattern formation in the outside world and pattern recognition in our brains [11]. Active basilar membrane pattern formation operates on a more basic mechanical level by purely hardware means. The reason for its implementation is the much higher limiting frequency compared to neural analysis. The "programs" of vibration analysis working behind the scene are the same principles that govern the generation in self sustained oscillations. Thus the accordance between pattern formation and pattern recognition principles also holds at the mechanical level in inner ear dynamics. Nature exploits the very same principles that human intuition uses to generate self sustained oscillations in organ pipes or violins. Related principles were also applied to improve the performance of radio sets and their quantum mechanical generalisation underlies the generation of coherent electromagnetic vibrations in masers and lasers.

The results of mechanical activity can be measured. There must be a critical parameter for the onset of self sustained oscillations. In many subjects narrow band acoustical emissions can be detected in the ear canal [12]. Usually, these signals are below the threshold of perception. Apart from these spontaneous emissions acoustically stimulated signals occur in response to an incoming sound signal [13]. As a consequence of amplification the impedance match in the active region is lost and a small amount of energy flows backward giving rise to echos. Their group delays reflect cochlear dispersion.

3. Mechanical Self–Organization, Biological Fitness, and the Emergence of Meaning

In order to improve its performance as a sound receiver which goes in parallel with the improvement of biological fitness, the ear has turned into a sound producing organ. Perception can be considered as an active modelling process. Hearing reveals that this is more than psychological parlance, adequate only say for high level mental models. It is even true in a strictly basic mechanical sense. Perception mechanisms actively rebuild reality by using model systems isomorphous to reality. Information dynamics on biological substrates mirrors information dynamics on the material level. The shift from passive to active reception, though of minor importance at a first glance, creates new phenomena alien to the linear and the passive system. The above raised question of mechanical intelligence can now be specified further. There is a bunch of new effects in the mechanical self–organization paradigm that transcends a purely mechanistical view, defining new levels of organization, creating relevance and meaning, and introducing a teleological component at least in our interpretation of the interplay of components. There is a powerful constructive potential in the underlying nonlinear dynamics. In effect, Nature received much more than just an increase in sensitivity, selectivity, and biological fitness by turning the receptor organ mechanically active.

3.1 *Meaning via Threshold–Logic and Discrete States*

The most primitive concept of meaning is embedded in the threshold concept. The information processing device considers only those signals exceeding a certain critical level as relevant and as valuable enough to inform subsequent stages about their presence. Subthreshold signals are discarded as meaningless, preventing an upward flow of irrelevant signals (e.g., thermal noise). Being essential to technical information processing systems like computers, threshold logic is also ubiqituous in all levels of the nervous system, from the firing of a single neuron to the working of neural networks. The present paradigm of active perception creates a threshold logic already on receptor level. As a descriptive model of basilar membrane action at low signal levels we consider an active oscillator, e.g., an organ pipe driven by an external signal (Fig. 1). Such a driven autonomous oscillator exhibits the well known frequency pulling effect. Its basic behavior can be modelled by the van der Pol oscillator. Given a certain frequency difference between external and internal oscillator, the driving signal power must exceed a critical value before it is able to pull over the internal oscillator. Signals below threshold pass unnoticed. They are not "perceived" by the internal oscillator, because their power is insufficient to produce locking. In this picture the peripheral auditory perception mechanism is viewed as an interaction between the perceiving system and the system perceived, producing phase coherence between external and internal oscillators.

Apart from generating a threshold logic further cognition–like functions arise from a mechanical context. These meaning generating processes can be studied psychophysically by two tone interaction experiments. A nonlinear interaction effect, well known from everyday experience, is the masking phenomenon. In the presence of a narrowband noise a sine tone is perceived only if the distance between signal and noise exceeds a certain sharp limit with respect to level and frequency difference. Though masking is common to neural information processing, it is already inherent in inner

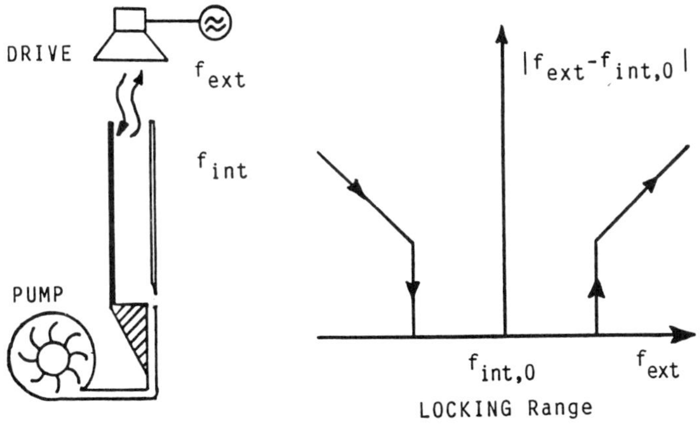

Figure 1: Driving an organ pipe as a model for acoustic perception. The internal oscillation is locked to the external signal at the fundamental frequency or at subharmonics. Models like this describe mechanoreception in haircells as well as neural periodicity.

ear dynamics. The above model of "perception by locking" provides an adequate mechanical explanation. In order to be perceived, i.e., to produce locking, the signal must exceed a critical distance to the masking noise. This seemingly unsocial suppression of the weak by the strong is a highly constructive feature of intelligent peripheral signal preprocessing. It performs data compression by drastically reducing the number of independent basilar membrane modes. Theoretically, the basilar membrane has an infinite number of degress of freedom, which is reduced by mechanical feedback to about 30 independent localized modes. These correspond to the critical bands known from psychophysics. About 30 basilar membrane sections, each about 1 mm long, can vibrate independently at low signal levels [2]. The bands can be viewed as natural coarse graining imposed on receptor state space, providing an upper limit to information flow. Assuming that a switching rate of 5 s^{-1} in each band can be perceived, a data transfer rate of 150 bit s^{-1} can be estimated.

3.2 *Meaning via Generation of New Objects*

On this level of description, every signal that is "understood" via state change of the receptor organ carries meaning. This rather primitive mechanistical concept of meaning provides already a rudimentary form of context dependence. An isolated signal can be meaningful, but its meaning may disappear in the presence of another masking signal, or its meaning may be changed. Such a change is brought about by nonlinear interaction of the vibratory pattern of sound signals. Two periodic signals with fundamental frequencies f_1 and f_2 create new combination products of the form:

$$f_{comb} = mf_1 \pm nf_2 \quad (m, n \in N)$$

Two types of nonlinear distortion products occur. Regular distortion products such as the quadratic combination tone $f_1 - f_2$ decrease with decreasing level of the primary tones, i.e., the underlying nonlinearity can be linearized. This is the regular behavior that one would expect for an amplitude limiting nonlinearity. However, there is another distortion product with an irregular nonmonotonous level dependence. The cubic difference tone $2f_1 - f_2$ cannot be linearized and is present even close to hearing threshold [14]. The cubic combination tone can be viewed as the "sound of self–organization", because essential nonlinearities of at least cubic degree are prerequisite of self–organization. The essential nonlinearity in active systems, e.g., in the damping term of the van der Pol system, creates combination products with such a behavior.

Receptor nonlinearities both in the amplitude limiting passive and in the active case create new frequency components with informational relevance. If the ratio of fundamental frequencies of the primary tones corresponds to the ratio of small natural numbers, the tones are consonant:

$$f_1/f_2 = m/n \quad (m, n \in N)$$

In case of consonance, the resulting products due to nonlinear combination coincide with the superposed patterns and do not create any new spectral features. Because of fusion of spectral components, the complexity of the mechanical excitation pattern does not increase compared to linear superposition. In case of dissonance, a more complex excitation pattern, temporally as well as spatially, is created by the combination products. Especially, the low periodicity components are perceived as a more or less pleasant roughness. Though roughness requires an adequate temporal detection in the subsequent neural stages, it is justified to say that the concept of consonance is already founded mechanically. Nonlinear receptor dynamics imposes a structure on the continuum of frequencies giving preference to small numbers in frequency intervals. The subsequent neural periodicity detection uses closely related schemes, e.g., phase locking also creates consonant intervals, so that the all pervading logic of nonlinear organization underlies our ability to create objects, numbers and even to detect ratios of small numbers. The nonlinear detection scheme is highly adequate to analyze natural sounds that have been created by effectively one–dimensional oscillators, e.g., the human voice or most sounds of musical instruments, which can be characterized by one fundamental frequency. In these cases, the interplay of nonlinearities and periodicity detection even allows the recovery of missing spectral structures, the so called residual pitch phenomenon. In more–dimensional sounds, e.g., in sounding bells or plates, the different frequency components are not necessarily rational, giving rise to complex sounds with a virtual pitch that does not correspond to the physical spectrum.

From an evolutionistic view an increase in sensitivity and selectivity is the primary reason for a mechanically active inner ear. Activeness goes together with additional nonlinear effects that generate meaning and may well reach beyond that peripheral level of organization. In a wider sense, the ability to appreciate music can be regarded as an epiphenomenon to evolution. It can be traced down to nonlinear effects in the receptor organ, improving biological fitness, and to closely related nonlinear effects in neural dynamics, yet transcending a purely evolutionistic explanation the higher we climb up in the information processing hierarchy towards conscious perception. Leibniz considered music as a hidden unconscious mathematical exercise of our soul. As will be shown in the following sections, our mind computes by properly combining linear and nonlinear dynamical principles in the information chain from the sensory organ to conscious perception.

4. Complementarity by Complex Information Dynamics

4.1 *Neural Volley Principle: Linearizing a Nonlinear System*

The above mentioned periodicity principle of hearing raises the question, how a system can transmit signals up to a limiting frequency of 2 kHz although single nerve cells as basic transmission elements have maximum firing rates of 500 s^{-1}. How can an extremely nonlinear element, working in the all–or–nothing–way, be used to transmit signals beyond its upper frequency limit? While in inner ear dynamics the basic trick to improve performance is a change from linear to nonlinear, the converse holds for certain levels of neural dynamics. Neural periodicity is based on linearizing a system of extremely nonlinear transmission elements, the recipe being a mixture of linear and nonlinear principles, of superposition, phase locking, and noise. An ensemble of acoustic nerve fibers, sampling neighbouring basilar membrane places, is driven coherently. Each fiber responds with a pulse sequence, phaselocked with the driving signal at its fundamental frequency or at subharmonics. As the phases of the individual fibers differ, the superposed spike pattern consists of phaselocked volleys of spikes, which are smeared to a certain extent, because of noise (phase jitter). Summation in subsequent neural stages can recover the temporal structure of the signal with some distortion similar to rectification of an ac–signal [15].

Neural periodicity is essential for pitch perception in the low frequency region. The periodicity principle of pitch is complementary to the place principle in various aspects. It processes information in a delocalized way in the neural system, with the temporal structure of the signal being the important clue. Conversely, in the place principle the spectral features are processed in a localized mode as geometric patterns according to frequency–place–transformation, with local vibration amplitudes corresponding to spectral power. Neural periodicity invalidates to some extent this Fourier–analyzer paradigm of hearing, and allows the detection of phase effects. Signals with the same amplitude spectrum but different phase spectra, which have different temporal patterns, will also sound somewhat different.

Periodicity plays an important role in binaural signal processing. The simple yet impressive experiment of binaural beats demonstrates, that the acoustic nerve is able to transmit low frequency signals similar to a telephone cable. The ears are stimulated separately by stereo headphones with two tones of slightly different frequencies. Though the sounds cannot interfere physically, beats from neural interference are clearly perceived that seem to originate from somewhere within the head [16]. Contrary to monaural superposition of two tones, no new frequency components are created by binaural interaction. On that level of information processing obviously linear principles prevail. Our two ears are the base of an acoustical interferometer, a linear analysis arrangement, which uses phase differences to locate the sound source. Apart from phase or binaural time delay, which can be used only in the low frequency range where neural periodicity is in action, other clues to lateralisation are binaural intensity differences and the directional sensitivity of the outer ear transfer function.

At low signal levels close to threshold, a different mechanism of sound localization seems to be relevant, directly involving the active tuning processes of the basilar membrane. The presence of a signal in one ear influences the mechanical fine tuning of the opposite ear in the same frequency band. The mechanical tuning is reduced which is shown by a decrease of group delay in contralateral cochlear echos in a slowly sweeping tone stimulation [17]. The tuning of one ear diminishes the discriminative power of the opposite ear. Again, as in pattern formation, this is a nonlinear inhibition scheme. Both ears interact in a flip–flop manner to produce symmetry breaking. This low–level localization scheme might be superior to conventional localization in view of noise limitations.

4.2 *Interplay of Linear and Nonlinear Structure Principles: Mind from Matter?*

The emergence of complex patterns in the outside world is governed by linear and nonlinear structure principles, by superposition and by self–similarity [18]. The first is an additive, arithmetic structure principle, the second a multiplicative, geometric one. In a way, the isolated action of each principle is simple, but the interplay of (+) and (*), of linearity and nonlinearity, causes complexity. The reconstruction of reality uses isomorphous dynamical principles, and both (+)– and (*)–processes can be uncovered in auditory perception phenomena. The (+)–principle governs monaural and binaural beats. (*)–effects can be perceived in two–tone interaction, where cochlear nonlinearity as well as the locking patterns of neural periodicity allow the detection of consonant intervals. Both principles complement each other and create a dualistic description of reality. Passive pattern formation by travelling waves, an essentially linear structure principle is enhanced by active amplification, invalidating the superposition principle. By phase locking the perception system is informed and a primitive form of meaning generates itself. Receptor states are created that behave according to the logic of mechanical objects. Two mechanical objects have never been observed in the same place simultaneously. The logic of our internal acoustic space–time is in accordance with that everyday experience (at least, as long as periodicity is not involved). The oneness of objects and the "either–or but not both" of facts in classical logics is already created in the first (*)–process (mechanical amplification). However, this strict "either–or" is relaxed by complementary (+)–processes in the next level.

The subsequent stage of neural processing recovers (+) from (*). The transition from mechanical to neural code is essentially nonlinear, creating neural spikes as discrete elementary signals. Out of that discreteness, the continuum of our sensations is built up. This involves summation (+) and discretisation (*) on various stages of neural processing. The first of these linearization processes can be perceived directly (binaural beats). Obviously, the interplay of (+) and (*), of superposition and nonlinear pattern formation governs the whole information chain from periphery to the central, from receptor organ via nervous system to conscious perception. Though the last step is unaccessible to objective experimentation, it can be studied by introspection and reveals dynamical features typical for nonlinear systems, with phenomena like fusion, switching, jumps, and hystereses effects. We mention two acoustical experiments:

- One can voluntarily focus the attention to certain regions of the auditory space, e.g., to objects on the left or right, to high– or low–frequency sources, to foreground and background. This voluntary symmetry breaking can easily be shown by interleaving tunes, with two separate melodies mixed together. If the interleaved tunes are in the same frequency range, it is impossible to separate their frequency–time pattern. The two tunes fuse to a single new one. If their frequency difference is big enough, we can switch between the high– and low–frequency tune. Again, discreteness appears in that ultimate stage of conscious switching. Only one object of consciousnes emerges, the rest is interpreted as background.

- One can try to focus attention to a rising tone sequence that crosses a falling sequence. In the neighbourhood of the crossing point, the ability to follow the tune is lost, which goes together with a loss of individuality of the object of attention. The object perceived is now a certain rhythmic pattern of tones. After that, the attention is locked to the wrong, falling tone. It takes some time until the wrong attracting state is realized and a switch to the proper tone sequence takes place [19]. This is an example for a built in bias (prejudice), and even musically trained people are unable to perform the task properly.

Both experiments convincingly demonstrate the oneness of conscious perception. Voluntary switching reduces the incoming information to effectively one single object of attention. The remaining incoming signals are merely background, a superposition of potential objects of attention. Even in pure introspection the switching to a certain thought fits into the general scheme of an interplay of (+) and (*). One specific pattern emerges from the background of superposed unconscious thoughts: oneness of conscious perception as the ultimate super–pattern. As the states of consciousness behave qualitatively like certain attracting states created in more peripheral stages of the perception system or even in simple nonlinear mechanical systems, no additional structure principles are obviously necessary. Is conscious perception merely a matter of complexity?

This brings up the central mind–body problem. Although perception–like functions are present in simple nonlinear mechanical or electronical devices, nobody can seriously claim that they do consciously perceive. In the present scheme, the problem is encoded in infinite regression. Each stage of the perception system performs pattern formation as an input to pattern recognition in the subsequent stage. We have to cut open the infinite hierarchy of layers perceiving each other, by saying that somewhere in our finite brain, mind consciously selects a pattern. Ultimately, our perception system defines what is regarded as meaningful. It actively selects what it assumes as relevant information. This conscious super–pattern formation remains mysterious to us, as we are part of the system that we try to understand. Though not analyzable as a whole, parts of the system may be modelled by fully mechanistic means. Stated differently, the analysis of sound signals into primitive features (spectral components, periodicity, phase etc.) is completely mechanistic, the re–synthesis to tunes as the objects of perception space remains a mystery (for a related discussion of visual objects see [20]).

5. Perception Dynamics and Quantum Magic: The Micro–Macro Connection

In the preceeding discussions of perception dynamics, several issues with a strong quantum mechanical flavor appeared. Are these mere superficial similarities or is there more to the relation between thoughts and quantum processes, which Bohr always considered as a deep analogy [21]. As he stressed, the sensitivity of our sense organs touches the quantum realm. The threshold of the eye amounts to a few light quanta. Although no single quantum processes are involved in hearing because of the minute quantum energies in the acoustic frequency range, the inner ear is even more interesting as a dynamical system, because it amplifies mechanical vibrations smaller than the diameter of an hydrogen atom to macroscopic yes–no alternatives. There have been arguments, that quantum noise could be the ultimate limiting factor of ear sensitivity [22]. Although this view has been questioned [23], it is clear that such a performance cannot be achieved with a passive detector in thermal equilibrium. The receptor must become mechanically active to reduce bandwidth and to lower noise temperature. Though in perception a conclusive evidence of quantum coherence percolating to a macroscopic level is missing, it cannot be completely ruled out. Maybe, Nature still hides surprises similar to high–temperature superconductivity.

While it is unsettled that real quantum effects are necessary to understand conscious perception, could it be that quantum–like effects are utilized by way of simulation? There are striking parallels between acoustic perception and the measurement process in the conventional Copenhagen interpretation. The "irreversible act of amplification" is essential in creating both perception and quantum phenomena. The establishment of coherence is important in mechanoreception, neural transmission as well as in macroscopic quantum effects. The analogy holds more or less closely for the peripheral as well as the central levels of perception. The ultimate step creating oneness of perception is similar to the reduction of the wave function. It changes potential information, inherent in the superposition of yet unobserved states, to actual information. Active perception with a nonlinear coupling between the perceiving system and the system perceived, formally described as a $(*)$–process, corresponds to nonunitary evolution of the wavefunction in the quantum mechanical measurement process. Quasi–linear $(+)$–processing in the intermediate stages corresponds in quantum mechanics to the unitary evolution of the unobserved states. We have a clear–cut distinction between the dynamics creating a phenomenon (perception or quantum) and the sub–phenomenal processes. Although these analogies may seem superficial in view of the still missing quantum link, they require a closer consideration. As an exemplification of perception phenomena with nonclassical features, we consider the emergence of a quantum–like logic and the existence of quantum–like topological effects.

5.1 *Binaural Interactions and Nonclassical Logic*

Though, at a first glance, the logic conveyed by sensory perceptions appears identical with the classical Boolean logic, a closer look reveals certain non–Boolean features. We restrict ourselves to the phenomenon of unmasking a signal in binaural perception, where the interplay of $(+)$– and $(*)$–processes on different levels creates a non–Boolean logic similar to quantum logic, for which a breakdown of the distributive law is characteristic [24]. Let A be a sine tone and B a narrow–band noise with the center frequency of the sine tone. Both signals are presented monaurally, and the level of B is adjusted such that it is just sufficient to mask A completely. If additionally the same noise signal is presented to the contralateral ear, the subject can uncover the signal A from noise via binaural correlation. The signal recovery works also, if the noise signal is inverted or presented with a certain time delay. In both cases audibility goes together with additional information. Binaurally, signal and noise background appear well separated in their direction. In monaural presentation, the $(*)$–processes create an incompatibility between two signals in the same critical band: B masks A. The $(+)$–process in the subsequent level of binaural neural periodicity processing can recover the masked signal by correlation because of its "hidden" temporal structure.

To show that the binaural unmasking paradigm hides nondistributivity, we consider the set of propositions, concerning the perception of combinations of the signals

A and B and their inverted versions –A and –B. An ordering relation ($\leq, \geq$) can be introduced in the set of propositions through the perceptibility of the signals. If signal B masks signal A then it is obvious to define:

$$A \leq B$$

Let meet ($\wedge$) define the coming together of two signals in one signal channel (i.e., monaural superposition) and join ($\vee$) define the union of signals from the two separate left and right channels giving rise to neural binaural superposition. The inversion of each signal has the properties of a complement. In monaural presentation, both signals completely interfere:

$$A \wedge -A = 0$$

Binaural perception yields:

$$A \vee -A = 1$$

The inversion has not quite the properties as the orthocomplement in quantum logics. Apart from the above relations it satisfies:

$$-(-A) = A$$

However, orthocomplementation requires the validity of a third relation [25]:

$$A \leq B \;\Rightarrow\; -B \leq -A$$

This relation is not satisfied in the present case, because, if a signal is masked by noise, the inverted signal is also masked by the inverted noise! Though analogies to quantum mechanics will show up, complementation makes an essential difference. The above use of meet and join in a one or two channel setup, is similar to the way optical filters or polarizers are used to demonstrate the outcome of classical and nonclassical logics optically [26]. According to the masking paradigm, only the most relevant signal sets the stage for its own analysis, covering less dominant signals. Hence, the presence of a signal corresponds to a certain setting of an adaptive filter, blocking other signals in the same critical band. We consider the truth of the proposition that the sine–tone is perceived by the subject. Let "1" denote the perception of the sine signal irrespective of any additional, e.g., directional, clue. This yields noise as another complement, because:

$$A \wedge B = 0 \text{ and } A \vee B = 1$$

These equations also hold for –B instead of B. Now we can state a violation of the distributive law because formally (see Fig. 2):

$$\begin{aligned} &A \wedge (B \vee -B) = A \wedge 1 = A \\ &\qquad > 0 = (A \wedge B) = (A \wedge -B) = (A \wedge B) \vee (A \wedge -B) = 0 \vee 0 \end{aligned}$$

It must be admitted, that a consistent interpretation of the inequality on one single semantic level is not possible, because different levels of information processing or concepts of reality are mixed. Similar semantic problems arise with respect to the truth values of conjunction and disjunction of incompatible propositions in the corresponding quantum mechanical problem (see Jammer [27], p. 349 for a discussion). The present binaural "paradox" reduces to saying, that a separate evaluation of $A \wedge B$

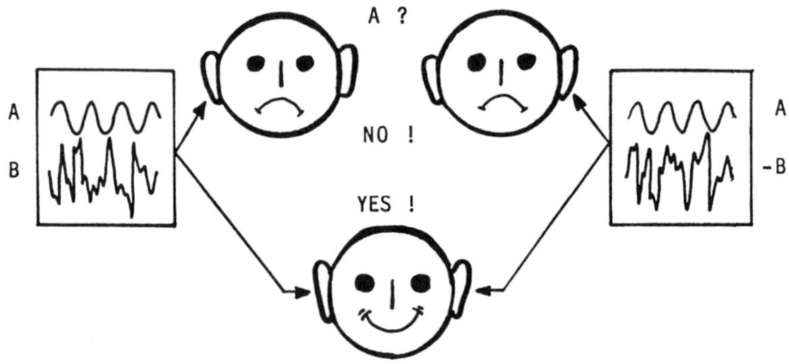

Figure 2: Invalidation of the distributive law in binaural hearing. In separate monaural presentation signal A is masked by noise B respectively –B. The joint binaural evaluation recovers the signal from noise.

and $A \wedge -B$ yields a result different from the combined evaluation. The whole is more than the sum of the separate parts. This can be viewed as a clear–cut violation of distributivity, though one should keep in mind, that two different experimental situations are involved. Both expresssions are factual only in the respective one channel experiment. In the two channel experiment, new potential information is contained in the hidden dynamical variable (=periodicity). It causes interference, allowing to separate the signal from noise. The new information, created in the next level of the information chain, corresponds to lateralisation of the sources.

It seems, that perception simulates quantum logics by invalidating the distributive law. What is the rationale behind the violation of classical Boolean logic at the peripheral stage of perception? How poorly is Boolean logic established in our intuition, when its obvious violation in perception has passed unnoticed? The breakdown of the distributive law is necessary but not sufficient to yield quantum logic. There are other ways to non–unique complementation which lead to a violation of distributivity. As has been shown recently, also in self–organizing classical systems the distributive law no longer holds, because of non–unique complementation by inner and outer complements [28]. These ideas have been shown to be relevant in the context of information dynamics [29]. From the above discussion it should have become clear, why nondistributivity is essential for perception as well as for quantum measurements. As the classical distributive law is a mere restatement of facts, it cannot be used to describe the production of any new information. Hence its breakdown indicates information production from one level to the next, with the levels corresponding to different concepts or hypotheses of reality (e.g., the monaural view: "yes or no" or the binaural view: "left or right").

5.2 *Infinite Tunes and Topological Perception Effects*

So far we have only considered the complexity of perception dynamics. An equally interesting problem is the complexity of the perception space. Here again analogies to quantum–systems appear. One of the most impressive acoustical illusions is the infinitely rising tune (Shepard–scale [30]), which is most elegantly created by an acoustic output of a self–similar waveform [31]. Different ways of generating such a scale on home computers are described in [32, 33]. A stepwise increase of the fundamental frequency of a compound sound signal generates a tone sequence. Before the fundamental frequency completes one octave cycle, it jumps back to the first value and the cycle starts again. As the sound signal is built up from octave tones, no jump is perceived, although its fundamental frequency is varied in a sawtooth fashion, and pitch goes up like a virtually never ending spiral. One period of the sawtooth can be interpreted as a cycle in a two–dimensional perception space built up from an appropriate averaging according to the place and the periodicity principle. After one cycle the external variables have returned to their original values, while a change of the internal variable (i.e., the pitch) persists.

A similar situation in quantum mechanics refers to the Aharonov–Bohm effect, where changes in the internal state variables of the electron are accumulated by encircling a trapped magnetic flux. This quantum mechanical effect is of topological origin, because the region outside the shielded volume is doubly connected, with single valuedness of the wavefunction as boundary condition [34]. Gauge theory offers an adequate explanation. What causes a similar complicated geometry of auditory space? The answer is completely open. Is this an indication for quantum effects in perception? The answer is no, because, as shown recently, geometrical phases [35] can also occur in classical systems [36]. The experiment demonstrates that ideas from gauge theory and concepts from differential geometry, such as connections of fiber bundles, are equally relevant for external and internal theories. Is there any difference between internal and external?

6. Perception, Thoughts and Quantum Processes an Eternal Program?

The description of principles of information dynamics in acoustic perception has revealed numerous parallels between the unfolding of physical structures in the outside world and their reconstruction in our heads. Surprising features show up, that superficially one would ascribe to the quantum realm, but on a second look, they also appear on the level of complex classical systems. The utilization of quantum–like effects by our brain may be the key to our "unreasonable" ability to understand the world. It allows switching between different local concepts of reality, producing new information in this creative jump. Is the inherently local and therefore incomplete simulation of quantum effects by an effectively classical cognitive system sufficient? The old question of Bohr [21] and Bohm [37] on definite relations between quantum processes and thought processes has only recently been reanalyzed on a broader scope [38]. The convergence in the epistemic background of various intractable phenomena like consciousness, measurement process and irreversibility may well point to an ultimate connection.

The correspondence of external and internal dynamical principles has an ambiguous epistemological taste. Numerous abstract principles, that were discovered only recently, have since ever been implemented in our heads. Are we doomed to uncover only those structures, to which our perception system has been optimized during evolution, or is the logic inherent in self–organization sufficient to comprehend self–organized structure principles, operating on different scales in the micro– and macrocosmos? How will scientific progress go on in view of the self–referential background of cognitive processes? All these are ultimate questions, for which no final answers exist. Yet an acoustical answer can be given! Play the above mentioned infinitely rising tune and listen to a parable of scientific progress: The progress goes on, but the central epistemological questions remain invariant.

References

1. W.D. Keidel and W.D. Neff, eds. *Auditory System, Handbook of Sensory Physiology*, Vol.**5**, 1–3. Springer, Berlin, 1974.
2. E. Zwicker and R. Feldtkeller. *Das Ohr als Nachrichtenempfänger.* Hirzel, Stuttgart, 1967.
3. E. Zwicker. *Psychoakustik.* Springer, Berlin, 1982.
4. G. v. Bekesy. *Experiments in Hearing.* McGraw–Hill, New York, 1960.
5. G. v. Bekesy. *Sensory Inhibition.* Princeton Univ. Press, Princeton, 1967.
6. B.M. Johnstone and J.J. Boyle. Basilar membrane vibrations examined with the Mössbauer technique. *Science* **158**, 390 (1967).
7. W.S. Rhode. Observations of the vibration of the basilar membrane in squirrel monkeys using the Mössbauer technique. *J. Acoust. Soc. Amer.* **49**, *1218 (1971).*
8. H. Haken. *Synergetics, an Introduction.* Springer, Berlin, 1983.
9. A.J. Hudspeth and D.P. Corey. Sensitivity, polarity, and conductance change in the response of vertebrate hair cells to controlled mechanical stimuli. *Proc. Nat. Acad. Sci. (USA)* **74**, 2407 (1977).
10. M. Euler and J. Kiessling. Far–field cochlear microphonics in man and their relation to cochlear integrity. *EEG Clin. Neurophysiol.* **56**, 86 (1983).
11. A. Gierer. *Die Physik, das Leben und die Seele.* Piper, München, 1985.
12. P.M. Zurek. Spontaneous narrowband acoustic signals emitted by human ears. *J. Acoust. Soc. Amer.* **69**, 514 (1981).
13. D.T. Kemp. Stimulated acoustic emissions from within the human auditory system. *J. Acoust. Soc. Amer.* **64**, 1386 (1978).
14. E. Zwicker. A model describing nonlinearities in hearing by active processes with saturation at 40 dB. *Biol. Cyb.* **35**, 243 (1979).
15. H. Spekreijse and H. Oosting. Linearizing, a method for analyzing and synthesing nonlinear systems. *Kybernetik* **7**, 22 (1970).
16. G. Oster. Auditory beats in the brain. *Sci. Amer.* **229**, 94 (1973).
17. M. Euler. In preparation.
18. S. Großmann. Selbstähnlichkeit: das Strukturgesetz im und vor dem Chaos. *Phys. Blätter* **45**, 172 (1989).
19. L. van Noorden. Temporal coherence in the perception of tone sequences. Dissertation, Eindhoven, 1975.
20. J. Eccles. *Das Rätsel Mensch.* Reinhardt, München, 1982.
21. N. Bohr. *Atomic Physics and Human Knowledge.* Wiley, New York, 1963.
22. W. Bialek and A. Schweitzer. Quantum noise and the threshold of hearing. *Phys. Rev. Lett.* **54**, 725 (1985).
23. W. Denk and W.W. Webb. Thermal–noise–limited transduction in mechano–sensory receptors of the inner ear. *Phys. Rev. Lett.* **63**, 207 (1989).
24. G. Birkhoff and J. v. Neumann. The logic of quantum mechanics. *Ann. Math.* **37**, 823 (1936).
25. E.G. Beltrametti and G. Cassinelli. *The Logic of Quantum Mechanics.* Addison–Wesley, Reading, 1981.

26. C.G. Adler and J.F. Wirth. Quantum logic. *Am. J. Phys.* **51**, 412 (1983).

27. M. Jammer. *The Philosophy of Quantum Mechanics.* Wiley, New York, 1974.

28. F.R. Krueger. *Physik und Evolution.* Parey, Berlin, 1984.

29. H. Atmanspacher. Incommensurability of Liouvillean dynamics and information dynamics. In *Parallelism, Learning, Evolution.* Eds. J. Becker, I. Eisele, F. Mündemann. Springer, Berlin, 1990.

30. R.N. Shepard. Circularity in judgements of pitch. *J. Acoust. Soc. Amer.* **36**, 2346 (1964).

31. M.R. Schroeder. Auditory paradox based on fractal waveform. *J. Acoust. Soc. Amer.* **79**, 186 (1986).

32. M. Euler. Biophysik und Erkenntnistheorie des Hörens unter dem Paradigma der Selbstorganisation. In *Grundprinzipien der Selbstorganisation.* Eds. K.W. Kratky and F. Wallner. Wiss. Buchgesellschaft, Darmstadt, 1990.

33. M. Euler. Physical principles of information processing in biological systems. In *Proceedings Europhysics Study Conference on "Interdisciplinary Aspects of Physics Education".* Ed. H. Kühnelt. World Scientific, Singapore, 1990.

34. C.N. Yang. Gauge fields, electromagnetism and the Bohm–Aharonov effect. In *Proceedings Int. Symp. Foundation of Quantum Mechanics in the Light of New Technology.* Ed. S. Kamefuchi. Phys. Soc. Japan, Tokyo, 1983.

35. M.V. Berry. Quantal phase factors accompanying adiabatic changes. *Proc. Roy. Soc. Lond.* **A 392**, 45 (1984).

36. J.H. Hannay. Angle variable holonomy in the adiabatic excursion of an integrable Hamiltonian. *J. Phys.* **A 18**, 221 (1985).

37. D. Bohm. *Quantum Mechanics.* Prentice–Hall, Englewood Cliffs, 1951.

38. R. Penrose. *The Emperor's New Mind.* Oxford Univ. Press, Oxford, 1989.

MEASURING CORRELATIONS IN ENERGY LANDSCAPES AND WHY IT MATTERS

ED WEINBERGER

Max-Planck-Institut für biophysikalische Chemie
Am Faßberg
D-3400 Göttingen, FRG

Abstract

The properties of multi-peaked "energy landscapes" have attracted attention in a wide variety of fields, including evolutionary biology, spin glass physics, and computer science. However, relatively little attention has been paid to the properties of the landscapes themselves. Herein, we suggest a framework for the mathematical treatment of such landscapes. A central role in this discussion is played by the autocorrelation of energies obtained from a random walk on the landscape. Our ideas about average autocorrelations allow us to formulate a condition (satisfied by a wide class of landscapes we call AR(1) landscapes) under which the average autocorrelation approximates a decaying exponential. We then show how our mathematical model can be used to estimate both the globally optimal energies of AR(1) landscapes and their local structure.

The notion of an "energy landscape", that is, the Hamiltonian of a discrete system, has attracted wide attention both inside and outside the physical community. Indeed, if one assumes that biological mutations can sweep through a population instantaneously relative to the time scale of interest, the simplest model of an evolving biological population is the geneology of an ever fitter sequence of representative individuals, and thus a walk "uphill" in fitness, which is formally the same as a walk "downhill" in energy [1]; in computer science, such walks are produced by the widely used "hill climbing" optimization algorithm. The goal of the present study is to characterize such landscapes.

Formally, an energy landscape is a real valued function of the vertices of a finite graph. Although this function can be completely characterized in principle by tabulating the "value" of each vertex, typical landscapes are so large that such an enumeration is impractical. As a first example, consider the Sherrington-Kirkpatrick

Information Dynamics, Edited by H. Atmanspacher and
H. Scheingraber, Plenum Press, New York, 1991

spin glass (SK model [2]), which is the Hamiltonian:

$$H(\sigma) = \sum_{1 \le i < j \le N} J_{ij}\sigma_i\sigma_j$$

Here, $\sigma = (\sigma_1, \sigma_2, \ldots, \sigma_N)$ is a vector of N "spins"; that is $\sigma_i \in \{-1, +1\}$. The J_{ij}'s are independent, identically distributed random variables chosen, for example, from a Gaussian distribution with zero mean and variance $1/N$. Here, a particular choice of the J_{ij}'s defines the landscape. Another example of a landscape is the Travelling Salesman Problem (TSP) with a particular choice of the inter-city distances. In the first case, the "value" of a spin configuration is its energy; in the second, the value is the tour length. In the first case, the number of spin configurations grows exponentially with the number of spins because the configurations are the vertices of the Boolean hypercube. In the second case, the number of travelling salesman tours grows even faster; namely, as $(N-1)!$ for an N city tour.

A more tractable approach to characterizing landscapes assumes that the energies are random variables, and that the landscape is drawn from an ensemble of possible landscapes, much as a particular realization of the SK model is drawn from the ensemble of possible realizations. In general, the energies at each point will be correlated with the energies of "nearby" points on the landscape. The distribution of landscape energies is almost always roughly Gaussian, except possibly near the tails, because each energy is the sum of many statistically independent contributions of relatively equal magnitude, and thus the Central Limit Theorem [3] applies. Regardless of the theoretical justification, there is also the experimental fact that Gaussian landscapes arise ubiquitously: See Stein [4] for a discussion of a Gaussian landscape model of a protein; see White [5] for a Gaussian landscape that arises in computer design; see Kauffman & Weinberger [1] for a Gaussian landscape model of the response of the immune system to pathogens. In these cases, then, the problem has been reduced to inferring the correlation matrix of a joint Gaussian distribution.

The distance between two points on the landscape can be defined in a very general way as the smallest number of graph edges between them. Often, the graph is sufficiently regular that all transitions from one vertex to another are combinations of a relatively few "generators", such as single spin flips in the SK model or the interchange of pairs of cities in the TSP. If one set of generators exists, then many such sets — such as $m > 1$ spin flips for the SK model — exist. Furthermore, if all generators in one such set commute, then all generators in all such sets commute, implying the corresponding distinction between the underlying landscapes. It is clear that commutative landscapes, such as the SK model, have a particularly simple structure that is not shared by, for example, the TSP.

Both the SK model and the TSP landscape are "statistically isotropic" in that ensemble averages, such as the mean energy of each point, and correlations between pairs, triples, etc. of points do not depend on their absolute location. The joint Gaussian distribution of energies on a statistically isotropic landscape is thus determined by the common (unconditional) mean energy for all points on the landscape, subsequently taken to be zero, and the pair correlation function, $R(s)$, of two points separated by s edges.

One way to estimate $R(s)$ is to make a random walk on the landscape. Although we will see later that useful results can be obtained even with a biased walk, we assume for now that the walk is an unbiased walk that starts at a randomly chosen vertex and proceeds to successive, randomly chosen neighboring vertices (i.e., an infinite temperature walk, ignoring energy barriers in configuration space or a selectively neutral walk, ignoring the phenotypic superiority of certain genotypes). If the number of neighbors at each vertex is sufficiently large, the probability of returning to a previously visited vertex is small, and a reasonable estimate of $R(s)$ can be obtained from the autocorrelation function of the random walk. We also observe that our assumption of statistical isotropy implies that the sequence of energies generated from the random walk must form a stationary stochastic process.

Suppose we assume further that no two points on the landscape have been assigned exactly the same energy. The sequence of energies, $\{E_s\}$, generated from the (Markovian) random walk must then also be Markov, because each point can be uniquely identified via its energy. It is easily shown [3] that the only possible stochastic process which is simultaneously Markov, Gaussian, and stationary is the so-called $AR(1)$ process. Energies $\{E_s\}$ that comprise this process are determined by the prescription

$$E_s = \rho E_{s-1} + \Delta E_s \tag{1}$$

where ΔE_s is a sequence of independent, identically distributed Gaussian random variables,

$$0 \leq \rho = \frac{\mathcal{E}[E_s E_{s-1}] - \mathcal{E}[E_s]\mathcal{E}[E_{s-1}]}{\sqrt{\sigma_s^2 \sigma_{s-1}^2}} \leq 1$$

and where $\sigma_s^2 = \mathrm{Var}(E_s)$. Because of statistical isotropy, ρ is independent of s.

A heuristic justification for this model is based on the observation noted above; namely, that the energies of many landscapes are determined by summing many, essentially independent components, and motion from one configuration to a neighboring one results in the replacement of some of these components with others. A more careful justification is obviously dependent on the details of the landscape being approximated. (See Weinberger [6] for a rigorous justification of the "$AR(1)$ness" of the N-k model, a simple class of landscapes that has been used in the study of evolutionary biology. This model is described more fully below.) The primary difficulty here is that, although E_s and ΔE_s are uncorrelated by explicit construction, E_s and ΔE_t for $t > s$ need not be uncorrelated, if only because the walk can self-intersect. However, the model does seem to predict the statistics of certain important classes of landscapes, such as the free energies of a relatively realistic simulation of self-replicating RNA, (Peter Stadler, private communication; for the model, see Fontana *et al* [7]). Sorkin [8] reports similar results for a problem in VLSI layout, and notes that these results are likely to be typical of a wide class of similar combinatorial optimization problems.

Even if the landscape is not statistically isotropic, we can still write

$$E_s = \frac{\sigma_s}{\sigma_{s-1}} \rho_s E_{s-1} + \Delta E_s \tag{2}$$

where ρ_s is the correlation between E_s and E_{s-1}, which now depends on s, as does $\sigma_s^2 = \mathrm{Var}(E_s)$. Although the coefficient of E_{s-1} was chosen to insure that ΔE_s and E_{s-1} remain uncorrelated, the explicit s dependence of all quantities in (2) implies that this coefficient is a local, rather than a global characterization of the landscape. Nevertheless, one can still define an "effective" value of ρ as:

$$\rho_{\text{eff}} = \lim_{t\to\infty} \prod_{s=0}^{t} \left(\frac{\sigma_s}{\sigma_{s-1}} \rho_s \right)^{1/t}$$

Clearly, this limit will be zero if either of the random variables ρ_S or σ_S can assume the value zero with positive probability. Otherwise, the existence of the limit is guaranteed by a simple application of the strong ergodic theorem [9], which also allows us to evaluate its logarithm:

$$\ln \rho_{\text{eff}} = \sum_{(i,j)} \pi_i P_{ij} \ln \rho_{ij} \tag{3}$$

where the sum is taken over all points i in the landscape and all points j that can be reached from i in a single step of the random walk. π_i is the probability of being at point i after the random walk has reached equilibrium, P_{ij} is the probability of the transition from i to j, and ρ_{ij} is the correlation between the energies at points i and j.

The result (3) also applies to biased random walks in which transitions to some neighbors are more probable than transitions to others, such as walks at finite temperature in configuration space or non-selectively neutral walks in genotypic sequence space. This observation has three important consequences. First, if we return to the statistically isotropic case in which $\rho_{ij} = \rho$, independent of i and j, $\rho_{\text{eff}} = \rho$, regardless of the choice of stationary and transition probabilities. In other words, we can use data from a biased random walk to estimate ρ for the statistically isotropic case. In the non-isotropic case, we can estimate the degree of anisotropy by observing the variability of ρ_{eff} as the bias of the random walk varies. We could even, in principle, solve (3) for the ρ_{ij}'s given the function $\rho_{\text{eff}}(\beta)$. (We use β as the bias parameter, reflecting the possibility of biasing a walk in configuration space by assuming a finite, rather than an infinite temperature. This thermodynamic analogy also motivates the assignment $\beta = \infty$ for an unbiased walk and $\beta = 0$ for a purely "downhill" walk with maximum possible bias.) The third consequence of (3) is that even biased walks of infinitely long length are asymptotically $AR(1)$, even for anisotropic landscapes.

In any case, it is clear that the value of ρ_{eff} obtained above is a fundamental property of the corresponding energy landscape. If the landscape is significantly anisotropic, we can "tune" β to obtain information about the distribution of energies at various energy levels: $\rho(\beta)$ for $\beta \approx 0$ gives us information about the distribution of energies surrounding global minima; $\rho(\beta)$ for a β value allowing transitions into and out of a local minimum, but not out of the global minimum, gives information about the distribution of energies surrounding the local minimum, etc. If the landscape is nearly statistically isotropic, then either the landscape is sufficiently smooth that "typical" pairs of points on the landscape have correlations that are $O(1)$, or many pairs of points are effectively uncorrelated, and, if the landscape is Gaussian, effectively independent. In the former case, which includes both the SK model and the TSP examples given above, low lying energy minima are near one another, so that the landscape has a single *massif central* [10]. In the latter case, rough estimates of the ground energy of Gaussian landscapes can sometimes be made via the theory of large deviations. For example, Weinberger [6] considers the case of landscapes on the N dimensional Boolean hypercube in which the energy of each point is composed of N approximately independent Gaussian components, each with mean 0 and variance σ^2/N, so that the energy of each point also has a Gaussian distribution with mean 0 and variance σ^2. Different points on the landscape may include some of the components of other points, thus giving rise to correlations between points. If the number of components shared between neighboring points is the same for any pair of neighbors, the landscape thus generated will be $AR(1)$. We introduce the correlation length τ, with $\rho = e^{-1/\tau}$. We then have $2^{N-c\tau}$ approximately independent points, for some constant c, and

$$Pr\left\{\sqrt{2\sigma^2 \ln 2\left(\frac{N-c\tau}{N}\right)} \leq \max_{n \leq 2^{N-c\tau}} E_n \leq 2\sqrt{\sigma^2 \ln 2\left(\frac{N-c\tau}{N}\right)}\right\} = 1 \tag{4}$$

for sufficiently large N. Although this result was proven under the assumption that the E's are Gaussian, essentially the same results obtain if the Gaussian distribution is replaced by a distribution that is nearly Gaussian, such as the binomial distribution. Additional, surprisingly general results can be obtained via von Mises' theorem (See, for example, [11]).

It is also possible to infer local properties of $AR(1)$ landscapes from a knowledge of ρ. For these landscapes, the distribution of energies of the neighbors of a given point, conditioned on the assumption that the given point has energy E_0, is also Gaussian, but with mean

$$\mu_C = \mu + \rho(E_0 - \mu)$$

and variance:

$$\sigma_C^2 = (1 - \rho^2)\sigma^2$$

This allows an estimate of the probability, $\phi(E_0)$, that the energy E of a specific neighbor is less than E_0. When this probability is $o(N^{-1})$, the starting point has, with high probability, less energy than any of its N neighbors, and is thus a local minimum. We can therefore crudely estimate the typical energy, E_m, of a local minimum as the solution to:

$$\frac{1}{N} = \int_{-\infty}^{E_m} \frac{e^{-(x-\mu_C)^2/2\sigma_C^2}}{\sqrt{2\pi\sigma_C^2}}\, dx \tag{5}$$

This integral is, apart from a polynomial pre-factor that has a positive exponent for small E_m and a negative exponent for large E_m, proportional to the integrand. Ignoring this pre-factor, we have the estimate:

$$E_m \approx \mu_C + \sqrt{2\sigma_C^2 \ln N}$$

If $\sigma_C^2 \ln N \gtrsim 2$, corresponding to relatively low correlation, the integral in (5) changes very slowly with E_m, so that the estimate given above is relatively good, reflecting the fact that the local minima are rather tightly bunched about an average value. However, this is not the case with relatively correlated landscapes, such as the SK model (for which ρ is easily calculated to be $1 - 3/N + 2/N^2$), thus confirming the fact [12] that local minima in the SK model can be found at a large range of energies.

When the above approximation is good, we also have a reasonable approximation for the probability that a randomly chosen site is a local minimum, namely the unconditional probability that a vertex has energy E_m. Because of the tight bunching of energy minima, E_m is, in this case, also close to the average value of local minima found by "down hill" walks from random starts.

The above techniques have elsewhere been applied to a simple model of evolutionary adaptation known as the N-k model [1, 6, 13, 14], which also has interest as a model of a complex energy surface. This model assigns an energy to a vertex on the N dimensional hypercube (N bit string) by first assigning a real valued "energy contribution", e_i, to the ith bit in the string. Each such assignment depends, not just on the value of the ith bit, but also of $k < N$ other bits, such as the $k/2$ neighboring bits on either side of bit i in the N bit string. Thus, the energy contribution of each site, a function of a $k+1$ bit substring of the N bit string, can have 2^{k+1} possible values, one for each of the possible bit patterns in the substring. The energy contributions associated with each of these combinations is assigned by selecting an independent random variable from some distribution $p(x)$, such as the uniform or Gaussian distribution. This generates an "energy table" for the ith bit. There is a different, independently generated table for each of the N bits. Then, given any string of N bits, the total energy of the string, E, is defined as the average of the energy contributions of each bit. The use of a probability distribution in assigning the energy contributions can be interpreted as either an admission of ignorance of the true nature of the complex couplings between the bits or as an attempt to capture the typical statistical properties of a wide class of landscapes with k interconnections per bit. Rather surprisingly, local features of the landscape — the height of local optima, and the length of "downhill" walks from randomly chosen starting points — were remarkably insensitive to the details of how the $k+1$ bit substrings were chosen in computer simulations.

For N-k landscapes, estimates obtained from (5) were found to be in rough agreement with computer simulations [13]. The technique given above for estimating the number of less energetic neighbors of a point with given energy was also shown to be in rough agreement with the computer simulations in [6]. The average energy of a local optimum in the N-k model for large k and the probability that a randomly chosen site is a local optimum have also been computed rigorously [14], thus allowing direct verification of the above claim about the relationship between them. If the site fitness distribution, $p(x)$, in the N-k model is assumed to be Gaussian with zero mean and unit variance, the bounds for $\psi(N,k)$, the probability that a randomly chosen site is a local optimum, predicted by the rigorous calculation, is

$$\frac{1}{2}\exp\left[-\frac{N\ln(k+1)}{k+1}\left(1+\frac{1}{\sqrt{4\pi\ln(k+1)}}\right)\right] < \psi(N,k) < \exp\left[-\frac{N\ln(k+1)}{k+1}\right]$$

and the average energy, $E_m(N,k)$, of these local optima is

$$\sqrt{\frac{(2-\delta)\ln(k+1)}{k+1}} < E_m(N,k) < \sqrt{\frac{2\ln(k+1)}{k+1}}$$

for arbitrarily small δ independent of k and N. In other words, the probability of being at a local optimum is, to first order in the exponent, equal to

$$\sqrt{\frac{N}{2\pi}}e^{-NE_m^2/2} = e^{-NE_m^2/2+O(\log N)}$$

as claimed.

There are two generalizations to the above approach. The first is to consider random walks on non-Gaussian landscapes. Many of our previous results used the fact that the correlation between two *Gaussian* random variables is zero if and only if they are statistically independent. This is not always true for non-Gaussian random variables. However, the "mutual information" between energies E_s and E_{s+t} along the walk is indeed zero if and only if the two energies are independent. It is defined as

$$I(E_{s+t}; E_s) = \int_{-\infty}^{\infty} p_{E_{s+t}E_s}(x,y) \log_2 \left[\frac{p_{E_{s+t}E_s}(x,y)}{p_{E_{s+t}}(y) p_{E_s}(x)} \right] dx\, dy$$
$$= \left\langle \log_2 \left[\frac{p_{E_{s+t}E_s}(x,y)}{p_{E_{s+t}}(y) p_{E_s}(x)} \right] \right\rangle \tag{6}$$

where $p_{E_{s+t}E_s}(x,y)$ is the joint probability density of E_s and E_{s+t} and $p_{E_{s+t}}(y)$ and $p_{E_s}(x)$ are the densities of E_{s+t} and E_s, respectively, and the angle brackets denote averages with respect to $p_{E_{s+t}E_s}(x,y)$. This quantity measures the amount of Shannonian information one has about E_{s+t}, given a knowledge of E_s, as noted in Fraser and Swinney [15]. These same authors describe an efficent algorithm for computing I, assuming that the average in (6) is replaced by an average along the walk. If one assumes statistical isotropy, $I(E_{s+t}; E_s)$ can depend only on t, the distance between the points. The same arguments used to derive (3) can also be used to show that

$$\ln I_{\text{eff}} = \sum_{(i,j)} \pi_i P_{ij} \ln I(E_i; E_j),$$

so that we have a test for statistical isotropy in this more general case. Furthermore, we can use (6) to estimate the number of independent points in the landscape, but, because the landscape is no longer Gaussian, we must use the more general results of von Mises' theorem, rather than (4).

It is also possible to go beyond the "mean square theory" implied by a study of the autocorrelation function to a "pointwise theory", as has already been done in the theory of stationary stochastic processes. In one such approach, the Karhunen-Loeve expansion (see, for example, [16]), a set of basis functions is found for a given stationary process such that the spectral resolution of the process (e.g., expansion in "Fourier series") consists of these basis functions multiplied by uncorrelated random variables. The same Hilbert space framework applies for landscapes, but, in contrast to the Karhunen-Loeve expansion, in which the basis functions are the solutions to an integral equation, the basis functions are vectors of trigonometric polynomials. The theory, described in more detail in [17], is outlined below:

We choose a vertex V_0 as the "origin", and consider transformations that map V_0 into some other vertex, say, V. It is easy to check that these transformations form a group, to be called G, which may or may not be commutative. For example, the relevant group for the spin glass is $\mathcal{Z}_2^N$, the (commutative) group of modulo 2 addition of Boolean vectors of length N. For the TSP, the relevant group is S_N, the (non-commutative) group of permutations of the N cities. Whether or not G is commutative, the energy, now regarded as a functional of elements $g \in G$, can be expanded into a generalization of Fourier series. We treat only the simpler commutative case, and defer the non-commutative case to later work. The generalized Fourier series is then given by the

Fourier Decomposition Theorem for Commutative Groups. Any function f mapping the elements g of a commutative group G into the complex numbers can be written as:

$$f(g) \equiv f(p_1, p_2, \ldots, p_N)$$
$$= \sum_{q_1, q_2, \ldots, q_N} \hat{f}_{q_1 q_2 \cdots q_N} \exp\left[2\pi i \left(\frac{p_1 q_1}{|G_1|} + \frac{p_2 q_2}{|G_2|} + \cdots + \frac{p_N q_N}{|G_N|} \right) \right] \tag{7}$$

Here,

$$G = G_1 \times G_2 \times \cdots \times G_N$$

where the cyclic groups G_k contain $|G_k|$ elements. An elementary theorem of group theory guarantees that a unique decomposition of this form always exists for finite commutative groups. p_k and q_k are integers such that $0 \leq p_k, q_k \leq |G_k|$. The Fourier coefficients $\hat{f}$ are then given by the formula:

$$\hat{f}_{q_1 q_2 \cdots q_N} = \frac{1}{|G|} \sum_{p_1, p_2, \cdots, p_N} f(p_1, p_2, \ldots, p_N) \exp\left[2\pi i \left(\frac{p_1 q_1}{|G_1|} + \frac{p_2 q_2}{|G_2|} + \cdots + \frac{p_N q_N}{|G_N|}\right)\right] \tag{8}$$

The above series is essentially the finite Fourier transform in N dimensions, so that the proof of this theorem parallels the proof in that more familiar case. See Dym & McKean [18] for details.

Because bounded energy functionals form a Hilbert space, many of the standard results in that theory have important things to say about landscapes. One such result states that t_d, the trigonometric polynomial obtained from taking the first d terms of the Fourier series of a real function ϕ, is better than any other d^{th} degree trigonometric polynomial in approximating ϕ in the mean square sense. The analogue here is to trigonometric polynomials, t, defined on a subgroup H of G. We first note that the decomposition mentioned previously implies an isomorphism between elements of G and vectors of integers in which the "product" $g_i g_j$ for $g_i, g_j \in G$ is formed by modular addition of each component of the corresponding integer vectors. (The modulus for the addition of the kth components is $|G_k|$.) It is then easy to show that all subgroups of G correspond to projections of the integer vectors into smaller dimensional subspaces. We can thus extend the definition of t to every $g \in G$ via $t(g) = t(\text{proj}_H(g))$, where $\text{proj}_H(g)$ is the projection of g to subgroup H. The relevant version of the approximation theorem then states that the trigonometric polynomial of this class that best approximates the function f is the Fourier series for f, restricted to H. This allows the approximation of a large landscape by a smaller landscape defined on a subgraph of the original one. Error estimates for this approximation are provided by the analogue of Parseval's Theorem, which states that:

$$\frac{1}{|G|} \sum_{\mathbf{p}} |f(\mathbf{p})|^2 = \sum_{\mathbf{q}} |\hat{f}_{\mathbf{q}}|^2 \tag{9}$$

In many cases, the left side of the above equation can be estimated by sampling points that are sufficiently further than τ steps apart and are thus effectively independent. The quality of the approximation via the smaller landscape can then be assessed by comparing the right side of (9) to the sum of the Fourier coefficients over the relevant subgroup. Another result with a similar flavor states that the Fourier coefficients for a statistically isotropic landscape are statistically uncorrelated, with variance

$$\text{Var}(\hat{f}_{\mathbf{q}}) = \sum_{p} R(|\mathbf{p}|) \exp\left[2\pi i \left(\frac{p_1 q_1}{|G_1|} + \frac{p_2 q_2}{|G_2|} + \cdots + \frac{p_N q_N}{|G_N|}\right)\right]$$

where $|\mathbf{p}| = \sum_i p_i$. (See [17] for a proof.) Because these coefficients will also be Gaussian for a Gaussian landscape, they will, in that case, also be statistically independent. The SK model is thus the simplest non-trivial Gaussian landscape defined on the Boolean hypercube.

It is also shown in [17] that the variance of $\hat{f}_{\mathbf{q}}$ for an $AR(1)$ landscape on the Boolean hypercube is given by:

$$\mathrm{Var}(\hat{f}_{\mathbf{q}}) = \frac{1}{2^N}(1+\rho)^N \left(\frac{1-\rho}{1+\rho}\right)^{|\mathbf{q}|}$$

The contribution from successively higher Fourier coefficients is thus a decaying exponential in $|\mathbf{q}|$. The base of the exponent is a monotone decreasing function for $\rho < 1$, corresponding to the intuition that highly correlated landscapes require little contribution from "high frequency" subgroups, but, as $\rho \to 0$ and $\left(\frac{1-\rho}{1+\rho}\right) \to 1$, these subgroups play a more and more important role. When $\rho = 0$ and the landscape is completely uncorrelated, all subgroups contribute equally, much as the power spectral density of white noise is constant over all frequencies.

References

1. S. Kauffman and E. Weinberger. *J. Theoret. Bio.* **141**, No. 2, 211 (1989).
2. D. Sherrington and S. Kirkpatrick. *Phys. Rev. Lett.* **35**, 1792 (1975).
3. W. Feller. *An Introduction to Probability Theory and Its Applications*, Vol. 2, 2nd ed. Wiley, New York, 1972.
4. D. Stein. *Proc. Nat. Acad. of Sci. USA* **82**, 3670 (June 1985).
5. S.R. White. Concepts of scale in simulated annealing. *Proc. ICCD* (1984).
6. E. Weinberger. *Biological Cybernetics* **63**, 325 (1990).
7. W. Fontana, W. Schnabel and P. Schuster. *Phys. Rev. A* **40**, No. 6., 3301 (1989).
8. G. Sorkin. Combinatorial optimization, simulated annealing, and fractals. IBM Research Report, RC 13674 (1988).
9. S. Karlin and H. Taylor. *A First Course in Stochastic Processes.* Academic Press, New York, 1975.
10. S. Kauffman. *Complex Systems (Proceedings of the 1988 Summer School on Complex Systems in Santa Fe, NM). Santa Fe Institute Studies in the Sciences of Complexity.* Ed. D.L. Stein. Addison-Wesley, Reading, MA, 1989.
11. H. David. *Order Statistics.* Wiley, New York, 1970.
12. B. Derrida and E. Gardiner. *J. Physique* **47**, 959 (1986).
13. S. Kauffman, E. Weinberger and A. Perelson. Maturation of the immune response via adaptive walks on affinity landscapes. *Theoretical Immunology, Part I. Santa Fe Institute Studies in the Sciences of Complexity.* Ed. A.S. Perelson. Addison-Wesley, Reading, MA, 1988.
14. E. Weinberger. *Phys. Rev. A.* To be submitted (1990).
15. A. Fraser and H. Swinney. *Phys. Rev. A* **33**, 1134 (1986).
16. A. Papoulis. *Probability, Random Variables, and Stochastic Processes.* McGraw-Hill, New York, 1965.
17. E. Weinberger. *Biological Cybernetics.* To be submitted (1990).
18. H. Dym and H. McKean. *Fourier Series and Integrals.* Academic Press, San Diego, CA, 1972.

ASPECTS OF INFORMATION PROCESSING: INFORMATION FLOW IN SYNERGETIC COMPUTERS

HERMANN HAKEN

Institut für Theoretische Physik
und Synergetik
Universität Stuttgart
D-7000 Stuttgart 80, FRG

Abstract

Synergetic computers are a new class of highly parallel computers which are constructed on the basic principles of synergetics and which use the analogy between pattern recognition and pattern formation. We show that the information flow in such computers can be determined explicitly. In this way the channel capacity, needed in such parallel computers, can be determined.

1. An Overview

The lectures I presented at the workshop that forms the basis of the present proceedings dealt with the following topics:

1. The various aspects and interpretations of information,
2. How to implement semantics into the scientific concept of information,
3. Pattern recognition by synergetic computers,
4. Information flow in synergetic computers.

Since the material of (1) and (2) is dealt with in a recent book of mine [1], I may refer the reader to that book. Topic (3) is covered in the book [2] which will come out at the time these proceedings appear so that there is no reason to duplicate the results. This leaves us with the results of (4) which are reported here for the first time and which will form, I believe, an important contribution to the understanding of information transfer not only in synergetic computers, but also in synergetic systems, i.e., systems which can spontaneously form structures by self-organization.

Information Dynamics, Edited by H. Atmanspacher and
H. Scheingraber, Plenum Press, New York, 1991

2. Neural Nets and Synergetic Computers: A Reminder

Over the past decade there is a rapidly growing interest in the study of highly parallel computers which may exhibit some properties of neural nets. After McCulloch and Pitts [3] had shown that the logical functions of a Boolean algebra can be realized by a parallel network composed of two-level threshold elements, first attempts at realizing such a network were made by Rosenblatt [4]. Because of an influential book by Minsky and Papert entitled "Perceptrons" [5], where it was shown that Rosenblatt's "Perceptron" cannot perform important logical operations, such as the XOR operation, these efforts found some end. With respect to the physics community it is important to note that Hopfield [6] showed that the network conceived by McCulloch and Pitts can be realized by spin glasses which are systems in thermal equilibrium and which may form specific kinds of patterns. Such a spin glass has the property of an associative memory. If only part of a pattern is provided, the spin glass may reestablish the whole pattern once that pattern had been stored before. The spin glass can do this restoration even in the presence of other stored patterns. There are, however, still some difficulties because the network does not always work properly, i.e. it may get trapped by so–called spurious states which do not correspond to any of the stored patterns. Sophisticated methods, such as simulated annealing have been developed to overcome these difficulties over the past years.

Some time ago I proposed a quite different approach [7] to the construction of highly parallel computers which is based on the concepts of synergetics [8], i.e., of systems far from thermal equilibrium, which may acquire spatial or temporal patterns via self-organization. The basic principles on which the construction of such a synergetic computer is based can be summarized as follows:

1. The concept of associative memory [9], i.e., once part of a set of data is given, the associative memory is capable of supplementing the lacking data. For instance a telephone dictionary is an associative memory, namely it provides the reader with a telephone number once the name of the person is known.
2. The second ingredient is the assumption of a gradient dynamics, i.e., it is assumed that the patterns can be described by means of order parameters which obey a dynamics in a potential field [8].
3. A third ingredient finally is the idea that pattern recognition can be considered as pattern formation [10].

Just recall a typical example of synergetics, namely a liquid layer heated from below. It is known, according to Bénard, that such a liquid layer may form specific patterns, e.g., hexagons or rolls. Consider a circular pattern in which the orientation of the horizontal rolls is not determined uniquely so that the system can show multistability. One may then show that the whole roll pattern in a specific direction is formed if only an individual roll was prescribed as initial state in that specific direction. Thus the liquid may act as an associative memory, namely the single roll pattern is supplemented by the other rolls in the same direction. As I have shown, this same mechanism can be applied to formulate the synergetic computer [7].

Let us assume that we have a number of cells labelled by an index j which may be interpreted as neurons. The neuron with label j has an activity q_j. I have formulated a dynamics so that the set of activities q_j is driven into a final state if an appropriate initial state is given. For instance we may store a number of images of faces as prototype patterns in the computer and then offer the computer a test pattern which consists only of part of the image of the face. It is now the task of the computer to supplement this face and to identify the name of the person shown on this image. This complementation can be achieved by the following evolution equations,

$$\dot{q}_j = \sum_l \lambda_{jl}\, q_l + \sum_{lmn} \lambda_{jlmn}\, q_l\, q_m\, q_n + F_j(t) \tag{1}$$

where λ_{jl}, λ_{jlmn} can be interpreted as synaptic strengths. $F_j(t)$ are fluctuating forces which are assumed to have the property:

$$< F_j(t) F_l(t') >= Q\, \delta_{jl}\, \delta(t - t') \tag{2}$$

In this way, (1) is a set of coupled Langevin equations. The equations (1) can be written in the form

$$\dot{q}_j = K_j + F_j \tag{3}$$

where the forces K_j are assumed to be derivable from a potential function V

$$K_j = -\frac{\partial V(\underline{q})}{\partial q_j} \tag{4}$$

where $\underline{q}$ is the state vector of the system:

$$\underline{q} = (q_1, \ldots, q_N) \tag{5}$$

With the Langevin equations (1) a Fokker-Planck equation for the distribution function $f(\underline{q}, t)$ is connected. This Fokker-Planck equation has the form:

$$\dot{f} = -\sum_j \frac{\partial}{\partial q_j}(K_j f) + \frac{Q}{2} \sum_j \frac{\partial^2}{\partial q_j^2} f \tag{6}$$

In the following we shall be mainly interested in the stationary state which is characterized by:

$$\dot{f}_{\text{stat}} = 0 \tag{7}$$

Because of the properties (4) and (2), the stationary solution of (6) can be determined explicitly (cf. [11]). It reads:

$$f_{\text{stat}} = N \exp(-2V/Q) \tag{8}$$

It is a major task to determine the coefficients λ_{jl} and λ_{jlmn}. As we have shown previously [12], this task can be accomplished by the application of the maximum entropy principle (or maximum information principle) in which the information

$$i = -\int \log_2 f\; d^N q \tag{9}$$

is maximized under the constraints that the second and fourth moments of the activities of the prototype patterns are given:

$$< q_j\, q_l >, < q_j\, q_l\, q_m\, q_n > \tag{10}$$

We assume for simplicity that the moments of odd order vanish:

$$< q_j >=< q_j\, q_l\, q_m >= 0 \tag{11}$$

In this way the problem (1) is well-defined, namely after the moments are given experimentally, the coefficients λ are determined, and as one may show the λs determine the network. We now wish to study which information flows in the connections between different cells occur.

3. Information Flow

According to equation (1) the activity of cell j changes in time because of the inputs from the other cells with label j. As it transpires from equation (1), λ_{jl} has the dimension:

$$[\lambda_{jl}] = \sec^{-1} \tag{12}$$

For this reason we shall write

$$\lambda_{jl}\, q_l = \tilde{\lambda}_j(\sec^{-1} q_l) \tag{13}$$

where

$$\sec^{-1} q_l = \text{signal flow} \tag{14}$$

can be interpreted as signal flow. It will be our task to derive the information flow connected with (14) so that in practical applications the channel capacity can be chosen adequately. To apply Shannon's formula for the information and information flow, we need the distribution function for the random variable q_j. (Note that we changed the label l to j.) Denoting the value which the random variable q_j can acquire by ξ_j, this distribution function is given by:

$$f_j(\xi_j, t) = \int \delta(q_j - \xi_j)\, f(\underline{q}, t)\, d^N q \tag{15}$$

In the general case it is difficult to determine the distribution function $f(\underline{q}, t)$ as a solution of the time–dependent Fokker-Planck equation. We note, however, that we have to determine f_j under the assumption that it starts from an initial distribution function which coincides with the stationary distribution function:

$$f(\underline{q}, 0) = f_{\text{stat}}(\underline{q}) \tag{16}$$

But because of the property of the stationary distribution function, we immediately obtain:

$$f(\underline{q}, t) \equiv f_{\text{stat}}(\underline{q}) \tag{17}$$

Since in any physical system signals are discrete, for instance in a digital computer, we introduce a connection between a discrete probability p_j and the distribution function f_j via

$$p_j(\xi_{j,\sigma}, t) = \epsilon f_j(\xi_{j,\sigma}, t) \tag{18}$$

where ϵ is an accuracy interval. We assume that q runs according to

$$-\frac{Z}{2}\epsilon < q_j < \frac{Z}{2}\epsilon \tag{19}$$

and that we define an interval with label σ by means of:

$$q_{j,\sigma} - \frac{\epsilon}{2} \leq q_j < q_{j,\sigma} + \frac{\epsilon}{2} \tag{20}$$

The information referring to the variable q_j is given as usual by

$$i_j \equiv \frac{I_j}{Z} = -\sum_{\sigma} p_j(\xi_{j,\sigma}) \log_2 p_j(\xi_{j,\sigma}) \tag{21}$$

i.e., the original total information divided by the number Z of symbols which is, in this case, $q_{j,\sigma}$. The logarithm is taken to the basis 2, because then the information is directly expressed in bits. Our task consists in the explicit evaluation of i_j. Before we do this, we mention that one may not only consider the information flow in the channel from one cell to another one, but also within one cell after the inputs have been summed up over the incoming information flows according to the right-hand side of equation (3). According to

$$\eta_j = K_j(\underline{q}) \tag{22}$$

where η_j is just the signal flow coming in at the internal side of the cell j, the probability distribution for the discretized random variable $\eta_{j,\sigma}$ is given by

$$P_j(\eta_{j,\sigma}) = \int \delta(\eta_{j,\sigma} - K_j(\underline{q}))\ \epsilon f_{\text{stat}}(\underline{q})\ d^N \underline{q} \tag{23}$$

and the information flow connected with this variable is expressed by:

$$\tilde{i}_j \equiv \frac{\tilde{I}}{Z} = -\sum_{\sigma} P_j(\eta_{j,\sigma}) \log_2 P_j(\eta_{j,\sigma}) \tag{24}$$

4. Evaluation of the Information Flow

At first sight the evaluation of the integral (15) and the expression (21) is a major task because the distribution function in (15) is an exponential function which depends on powers up to the fourth order of the components of the vector $\underline{q}$. However, we may make the following observations: The stationary distribution function f was determined by means of the maximum information principle under the constraints (10). We may now form the second and fourth order moments according to

$$< q_j^2 > = \int q_j^2\ f_j(q_j)\ dq_j \tag{25}$$

$$< q_j^4 > = \int q_j^4\ f_j(q_j)\ dq_j \tag{26}$$

which must give the experimentally observed results. This suggests to determine f_j again by the maximum entropy principle with respect to f_j under the constraints (25) and (26). This leads immediately to (cf. [10])

$$f_j(q_j) = N_j \exp\left(\alpha_j q_j^2 - \beta_j q_j^4\right) \tag{27}$$

where the normalization factor is given by

$$N_j^{-1} = \int_{-\infty}^{+\infty} f_j(\underline{q}) dq \tag{28}$$

and the Lagrange parameters of the maximum information principle are expressed explicitly as α_j and β_j. Using tabulated values of (27), these Lagrange parameters can be obtained by means of the fixed values (25), (26). (27) can now be inserted into

$$i_j = -\sum_{\sigma} \epsilon f_j(q_{j,\sigma}) \log_2 \left(\epsilon f_j(q_{j,\sigma})\right) \tag{29}$$

which can be cast into the form

$$
\begin{aligned}
i_j &= -\int f_j(q)\log_2 f_j(q)\; dq - \int f_j(q)\log_2 \epsilon\; dq \\
&= -\int f_j(q)\log_2 f_j(q)\; dq - \log_2 \epsilon
\end{aligned}
\tag{30}
$$

when we use the continuity of f_j. By means of tabulated values of the integral, the right-hand side of (30) can now be evaluated so that our task of determining the information and thus the information flow is accomplished, at least within the approximation described above. Otherwise the integral (15) must be evaluated numerically.

References

1. H. Haken. *Information and Self–Organization.* Springer, Berlin, 1988.
2. H. Haken. *Synergetic Computers and Cognition.* Springer, Berlin, 1991.
3. W.S. McCulloch and W.H. Pitts. A logical calculus of the ideas immanent in nervous activity. *Bulletin of Mathematical Biophysics* **5**, 115-133 (1943).
4. F. Rosenblatt. The Perceptron: A probabilistic model for information storage and organization in the brain. *Psychological Review* **65**(6), 386-408 (1958).
5. M.L. Minsky and S.A. Papert. *Perceptrons* (expanded edition). MIT Press, Cambridge, Ma., 1988.
6. J.J. Hopfield. Neural networks and physical systems with emergent collective computational abilities. *Proc. Natl. Acad. Sci. USA* **79**, 2554-2558 (1982).
7. H. Haken. Synergetic computers for pattern recognition and associative memory. In: *Computational Systems, Natural and Artificial.* Ed. H. Haken. Springer, Berlin, 1987.
8. H. Haken. *Synergetics, An Introduction*, 3rd ed. Springer, Berlin, 1983.
9. T. Kohonen. *Associative Memory and Self–Organization.* 2nd ed. Springer, Berlin, 1987.
10. H. Haken. Pattern formation and pattern recognition – an attempt at a synthesis. In: *Pattern Formation by Dynamical Systems and Pattern Recognition.* Springer, Berlin, 1979.
11. H. Haken. *Advanced Synergetics, Instability Hierarchies of Self– Organizing Systems and Devices.* Springer, Berlin, 1983.

THE QUANTUM MECHANICAL COMPUTER

Arnold Rosenblum

Institute of Theoretical Physics
Utah State University
Logan, UT 84322, USA

Abstract

It is shown that it is possible to construct an optical parallel computer that has quantum mechanics built into its hardware. This is done by using the multivalued logic formulation of quantum mechanics.

1. Introduction

As vector processing computers begin to approach the limits of computational speed imposed by technology and physics, major computer manufacturers, and the scientific community in general, have begun to realize that further substantial gains in processing speed (factors of greater than 10) could only be achieved by linking a number of processors together, so that the sequential power of individual processors could be harnessed into a single, parallel multiprocessor computer. Each processor could work in parallel on independent tasks contributing to the fast solution of large problems. For maximal effect each individual processor should be a digital optical computer. Some advantages of optical computers are the following: Unlike electronic ones, optical signals, when superimposed, are decoupled from one another and create no cross interference. The optical signal is a two dimensional wavefront, lending itself naturally to parallel processing obviating the need for ever more accurate clock synchronization required of a serial computer. The absence of inhibiting inductance of electronic components lends itself to higher operation speeds. Signal connections are made without actual physical or electronic contacts solving the so called "von Neumann bottleneck", i.e., the crowding of wires which escalates with an increase in the number of switching elements.

For the reasons given above it is relatively easy to implement multivalued logic on an optical parallel computer. One formulation of quantum mechanics is the so called "quantum logic" approach which is outlined below and which shows that it is possible to formulate quantum mechanical measurements in terms of multivalued logic.

Information Dynamics, Edited by H. Atmanspacher and
H. Scheingraber, Plenum Press, New York, 1991

2. Quantum Mechanics and Quantum Logic

The general mathematical structure of quantum mechanics consists of operators and wave functions in a Hilbert space [1]. We have operators like position, spin, and momentum and wave functions which are solutions of the Schrödinger equation. A Hilbert space has a geometrical content in that it is a linear vector space with an inner product with complex numbers as coefficients. In particular, there are subspaces of Hilbert spaces. At this point, we must define a lattice and relate it to subspaces of Hilbert spaces (i.e., quantum mechanics) and logic. A lattice is a partially ordered set L in which each pair of elements has a greatest lower bound. A partial order relation in set P is a relation which is symbolized by $\leq$ and is assumed to have the following properties:

$$\begin{aligned} \text{Reflexivity}: &\quad X \leq X \quad \forall X \\ \text{Antisymmetry}: &\quad X \leq Y \quad \text{and} \quad Y \leq X \quad \Rightarrow \quad X = Y \\ \text{Transitivity}: &\quad X \leq Y \quad \text{and} \quad Y \leq Z \quad \Rightarrow \quad X \leq Z \end{aligned} \tag{1}$$

Let A be a non–empty subset of a partially ordered set P. An element $X \in P$ is called a lower bound of A if $X \leq a$ for each $a \in A$; and a lower bound of A is called a greatest lower bound of A if it is greater than or equal to every lower bound of A. Similarly an element $Y \in P$ is said to be an upper bound of A if $a \leq Y$ for every $a \in A$; and a least upper bound of A is an upper bound of A which is less than or equal to every upper bound of A. If X and Y are two elements in L, we denote their greatest lower bound and least upper bound by $X \vee Y$ and $X \wedge Y$. A sublattice of L is a non–empty subset L_1 of L with the property that if X and Y are in L_1 then $X \vee Y$ and $X \wedge Y$ are in L_1. If we go back to our Hilbert space, the lattice can be described as the lattice of subspaces of the Hilbert space ordered by inclusion with:

$$\begin{aligned} &M \leq N \Leftrightarrow M \subseteq N \\ &M \wedge N = M \cap N \\ &M \vee N = M + N \end{aligned} \tag{2}$$

where $M + N = \{U + V : U \in M, V \in N\}$, and M and N are subspaces of a vector space.

Lattices also occur in the context of set theory and logic. The set of all subsets of a set can form a lattice if:

$$\begin{aligned} &S \leq T \Leftrightarrow S \subseteq T \\ &S \wedge T = S \cap T \\ &S \vee T = S \cup T \end{aligned} \tag{3}$$

A set of propositions, ordered by implication, forms a lattice if:

$$\begin{aligned} &P \leq Q \Leftrightarrow P \text{ implies } Q \\ &P \wedge Q = P \text{ and } Q \\ &P \vee Q = P \text{ or } Q \end{aligned} \tag{4}$$

Table 1: Truth values for the propositions X, U, $X \wedge U$, and $X \vee U$ (t = true, f = false, u = undecided)

X	t	t	t	f	f	f	u	u	u
U	t	f	u	t	f	u	t	f	u
X ∧ U	t	f	u	f	f	f	u	f	u
X ∨ U	t	t	t	t	f	u	t	u	u

The lattices (3) and (4) differ from the subspace lattice (2) in that they satisfy the distributive law:

$$X \wedge (Y \vee Z) = (X \wedge Z) \vee (Y \wedge Z) \tag{5}$$

The lattice of subsets of propositions both have elements 0 and 1 satisfying $0 \leq X$ and $X \leq 1$ for all X; for subsets of X the empty set is 0 and the whole set X is 1, while for propositions false is 0 and true is 1. In these lattices, every element X is associated with an element X' such that $(X')' = X$, $X \vee X' = 1$, $X \leq Y \Leftrightarrow Y' \leq X'$ (for subsets of X, S' is the complement $X - S$, for propositions P' is the negation, not P). Such a lattice is said to be orthocomplemented. The lattice of subspaces of a vector space is orthocomplemented if the vector space has an inner product: M' is the orthogonal complement $M' = \{u :< u|v >= 0 \; \forall v \in M\}$.

Propositions can be associated with subspaces of the state space of a quantum system by considering for any subspace M_1 the orthogonal projection onto M. This is a Hermitian operator P_m with eigenvalues 0 and 1. We associate M with the proposition: P_m takes the value 1. This proposition is true when the system is in a state belonging to the subspace M. If N is another subspace, the intersection $M \cap N$ contains simultaneous eigenstates of P_m and P_n for which the propositions $P_m = 1$ and $P_n = 1$ are both true. The subspace $M + N$ corresponds to the proposition $P_m = 1$ or $P_n = 1$. The failure of the distributive law shows that the propositions in quantum mechanics do not obey Boolean logic. In addition it is possible to show that it is impossible with only two valued logic to assign truth values to the lattice of subspaces of a state space. This means that if a system is in a state which is not an eigenstate of an observable A, then a statement "A has the value α" is neither true nor false. This calls for the addition of a third truth value "undecided", giving rise to a truth table as given in Tab.1.

3. The Physics of the Hardware of the Quantum Mechanical Computer

It is possible to build a quantum mechanical computer on the macroscopic level by using strained superlattices [2] in a Fabry – Perot interferometer for optical computing. The switching is done by a small change of an incident beam of laser radiation. An increase in the intensity of light causes a change in the strained superlattice refractive index. This in turn changes the intensity of transmitted radiation with only a small change in the intensity of the incident beam. With a superlattice system with multiple bistability each transmission level could stand for values of a multivalued logical system.

On the microscopic level, it is possible by using the optical bistability of indium antimonide to build an optical computer. The natural reflectance of the crystal faces is enough to form a Fabry – Perot cavity. There are optical bistable devices with switching times as low as a few picoseconds which of course could be used in parallel.

4. Conclusion

Optical computing is a field with an exciting future. By combining optical computing with quantum logic, it should be possible to build a computer system that handles quantum mechanics in its hardware.

Acknowledgement

I would like to thank the Department of Energy for partial support of the project. Grant number C85-160-659.

References

1. A. Sudbery. *Quantum Mechanics and the Particles of Nature.* Cambridge University Press, Cambridge, 1986.
2. J. Wilson and J.F.B. Hawkes. *Lasers, Principles and Applications.* Prentice Hall, New York, 1987.

COMPLEMENTARITY OF STRUCTURE AND DYNAMICS

HARALD ATMANSPACHER

Max Planck Institut für Physik und Astrophysik
Institut für extraterrestrische Physik
D-8046 Garching, FRG

1. Introduction

It is well-known for a long time that quantum theory, although being extremely successful with respect to the correct description of experimental results, still involves conceptual problems of different kinds. The following three issues seem to be outstanding: they contain the basic problems to which numerous discussions in the field may be reduced.

- Interpretational problems which have lead to a variety of approaches presented and discussed, e.g., in Ref.[1].
- Quantum theory of gravity, which is expected to provide the missing link between quantum theory and general relativity [2].
- Quantum theory of measurement, an issue closely related to the question of irreversibility and the arrow of time [3].

The present article touches these problems. It is based on a logical approach toward the formulation of knowledge in the framework of quantum theory. The historical step into that direction has been made in the seminal paper by Birkhoff and von Neumann [4] who found that the incommensurability of quantum theoretical operators can be interpreted in terms of non–distributive complemented lattices. More details on this approach and on its applicability to operators (basically position and momentum) will be reviewed in Sec.2.

The problem of irreversibility in quantum systems is related to the existence of a quantum theoretical time operator. Pauli realized already in 1933 [5] that such an operator T cannot exist if it is to provide an incommensurability with the Hamiltonian H. The reason is that the spectrum of H is bounded from below, while the spectrum of an energy operator incommensurable with T would have to be unbounded (and continuous). As it is the case for H, a corresponding operator must be of simultaneous relevance as a dynamical operator. Section 3 introduces candidates for incommensurable operators corresponding to energy and time and derives an incommensurability among them for classical systems. A lattice theoretical interpretation

Information Dynamics, Edited by H. Atmanspacher and
H. Scheingraber, Plenum Press, New York, 1991

of this incommensurability analogous to that of position and momentum in quantum theory will be given. Following ideas of Prigogine and collaborators [6] a transfer of the formalism onto the regime of quantum systems will be suggested.

Section 4 compares the properties of the propositional lattices according to the incommensurabilities discussed in Secs.2 and 3. It is found that the non–distributivity relations in both formulations are dual to each other. Identifying traditional quantum logic as a logic of structure in position space and identifying temporal logic as described in Sec.3 as a logic of dynamics in time, yields a mathematical formulation of a complementarity principle of structure and dynamics. The conceptual level of this complementarity supersedes the levels of incommensurabilities concerning momentum/position and energy/time. As a lattice theoretical duality, it points toward a unified understanding of structure and dynamics, of space and time, and of further apparent dichotomies.

The complementarity of structure and dynamics as a lattice theoretical duality is an algebraical feature. It is therefore interesting to study the properties of a corresponding geometrical duality. This idea has already been indicated in Ref.[4]. It has been picked up and elaborated by Finkelstein *et al.* [7] in 1962. The inclusion of specific notions about dynamics and information flow which had not been available at the time of those papers might perhaps provide more insight into the relevance of the approaches discussed that early. Section 5 will be devoted to this issue. In particular, the geometrical viewpoint seems to be of crucial importance concerning the quest for empirically testable predictions.

2. Structure and Quantum Logic

The development of quantum logic represents an attempt to understand the meaning of the incommensurabilities resulting from their mathematical formulation. (As such, quantum logic represents an interpretational issue and falls under category (1) mentioned in Sec.1.) The "basic" quantum incommensurability is

$$i[P, Q] = h\mathbf{I} \tag{1}$$

where P and Q are momentum and position operators, $i = \sqrt{-1}$, and h is Planck's action. The operators P and Q act on states ψ defined in a complex Hilbert space $\mathcal{H}$, and ψ obeys the Schrödinger equation:

$$i\hbar\partial_t\psi = H\psi \tag{2}$$

How can the incommensurability (1) be understood? A possible approach to answer this question has been suggested by Birkhoff and von Neumann [4] who tried to analyze the meaning of (1) using lattice theory. The relevance of lattice theory for logics on the other hand is responsible for the notion of quantum logic, which has turned out as a key subject at the borderline between quantum physics and philosophy of science.

Any measurement of the position or momentum of a system (particle) can formally be described by a projection of its state (ψ) onto a subspace of its entire Hilbert space. For the projected state, it is then possible to investigate the truth value of propositions like "The projected state is in subspace $\mathcal{H}_q$" etc., which simply correspond to "The eigenvalue of the position operator Q is q". The investigation of those propositions provides the result that they are not either definitely true or definitely false, but that a third contingent truth value, e.g., "undecided", is required to characterize them. This contingent truth value is particularly relevant for the case that the eigenvalue of an operator is to be inferred, if the eigenvalue of the corresponding incommensurable operator has been measured. In other words: if the state vector ψ has been projected onto a subspace of $\mathcal{H}$, then a proposition about ψ's projection onto another (orthogonal) subspace is neither true nor false, but remains undecided (contingent).

This result is far from being evident within our usual experience, which is based on observations of classical (not quantum) systems. But even for quantum systems it is quite difficult to illustrate in which sense a contingent truth value is to be understood. (The most intelligible example I found is given in Ref.[1], in a footnote on page 349.)

The insufficiency of a two–valued logic for quantum observations can formally be shown using lattice theory, a mathematical discipline of very general relevance which still deserves to be discovered for numerous fundamental physical problems. In the framework of lattice theory, the set of subspaces of the Hilbert space of the system as well as the set of propositions about eigenvalues of incommensurable (non–commuting) operators constitutes a complemented non–distributive lattice. It is the non–distributive property which accounts for the non–uniqueness of the complement, appearing in the necessity of a contingent truth value. If the complement of a positive measurement outcome were uniquely negative, then the corresponding lattice would be distributive, and the corresponding logic would be of Boolean type. The non–distributive property will be formally shown in Sec.4, where it will be compared with a second type of non–Boolean logic to be introduced in Sec.3. Nevertheless, some further remarks in context with Eqs.(1) and (2) are appropriate here.

The Schrödinger equation (2) relates the state ψ of a system to its Hamiltonian H. The Hamiltonian H is an evolution operator giving rise to a unitary, time–reversible evolution of the state ψ. At the same time, however, H is the energy operator of the system, and its eigenvalue spectrum is bounded from below ($E > 0$). It is exactly this double function of H which prevents the definition of a quantum mechanical time operator T which would be incommensurable with H. This remarkable point is the cornerstone of the vivid debate about energy–time–uncertainty since the 20's.

This debate is substantial for two reasons:

- Experimental evidence for an uncertainty relation of energy and time exists (e.g., energy bandwidth of atomic levels and their lifetimes are related by $\Delta E \Delta t = h$.)
- Energy and time provide the four–dimensional completion of momentum and position in general relativity. It is therefore not easy to understand why quantum theory should generate an obvious (or apparent?) asymmetry between momentum/position and energy/time.

The gap between momentum/position and energy/time with respect to the second point justifies the distinction of structure and dynamics as it is addressed in the title of this contribution. Here, structure is to be understood as the spatial appearance of a system in terms of the positions of its constituents. Dynamics characterizes the functional behavior of a system in terms of the evolution of its constituents. Even if this distinction appears as an artefact from the viewpoint of general relativity, it is an unresolved problem in quantum theory. Its solution requires a properly defined incommensurability of energy and time which could be used for an interpretation in terms of lattices as for quantum logic.

3. Dynamics and Temporal Logic

In 1978, B. Misra [8] proved that an incommensurability of time and energy can be derived if both are considered to be superoperators (operators acting on operators), and if the energy operator has an unbounded and – at least piecewise – continuous spectrum. Such an energy superoperator can be defined: It has originally been introduced [9] as the Liouville operator for classical systems and is given by:

$$\mathcal{L} = \sum \left(-i \frac{\partial H}{\partial p} \frac{\partial}{\partial q} + i \frac{\partial H}{\partial q} \frac{\partial}{\partial p} \right) \tag{3}$$

As an important point, $\mathcal{L}$ is generally defined for a system described by a density matrix (density operator) ρ instead of a state vector ψ. The latter one represents a pure state which is a point in Hilbert space. Such a description assumes complete knowledge about the state of the system, i.e., this state is considered to be exactly known. However, already the uncertainty relation resulting from (1) indicates that the eigenvalues of P and Q together can never be known exactly. This inconsistency (among other arguments) motivates an incomplete (inexhaustive) description of the state of a system by a density matrix, resembling the idea of a distribution function with non–vanishing higher order moments (correlations). (A state function ψ corresponds to a δ – function in state space, or to vanishing off-diagonal (correlation) terms in the density matrix.)

It is the decay of those correlation terms which enables the definition of a time superoperator $\mathcal{T}$ with eigenvalues t characterizing the age of a system described by ρ. Systems providing such a decay are mixing and satisfy the so–called K–property. This class of systems indeed turns out to provide an unbounded and continuous spectrum of $\mathcal{L}$ as well. The commutation relation of $\mathcal{L}$ and $\mathcal{T}$ is given by:[1]

$$i[\mathcal{L}, \mathcal{T}] = \mathbf{I} \tag{4}$$

with:

$$\mathcal{L}\rho = i\partial_t \rho \tag{5}$$

$$\mathcal{T}\rho = t\rho \tag{6}$$

In classical systems, K–flows are systems with positive K-S-entropy K. Although the notion of K–flows is usually restricted to Hamiltonian systems, the condition $K > 0$ can also be satisfied in dissipative cases. This condition therefore offers an opportunity to classify classical dynamical systems in a very basic way. It has been suggested [11] to understand systems with $K > 0$ as information generators or information processors. The idea is easily intelligible by considering the information flow along the unstable manifolds of such a system with respect to a given (initial) resolution ϵ. This resolution limits the amount of information an observer can gain by

[1] Both $\mathcal{L}$ and $\mathcal{T}$ are superoperators acting on the density operator ρ in "superspace". This superspace is defined as the direct product of $\mathcal{H}$ and its dual. Balian *et al.* [10] interpret both dual spaces as state space and space of observables, reflecting the duality of Schrödinger and Heisenberg representation, respectively.

measuring the system. Since the uncertainty increases along the unstable manifolds, the knowledge of the observer about the state of a system after measurement decreases as a function of time. Ascribing this loss of observer information to an internal information production within the system, generated by its own dynamics, provides:

$$
\begin{aligned}
I(t) &= I(0) + Kt && (7)\\
&= D \log \frac{1}{\epsilon} + Kt && (8)
\end{aligned}
$$

where $I(0)$ is the information obtained by measurement at $t = 0$, $I(t)$ is the grown information at time t, D is the information dimension of the attractor of the system, and K is its K-S-entropy [12]. Now we can (similar to the definition of $\mathcal{L}$) define an information operator $\mathcal{M}$, acting on a distribution function ρ, with eigenvalues according to (7):

$$\mathcal{M}\rho = (I(0) + Kt)\rho \tag{9}$$

This definition, including K as an empirically accessible quantity, has first been given in Ref. [13]. It has been essentially influenced by the idea of an entropy superoperator as suggested in earlier work of Prigogine and his group [6, 8, 14]. From Eq.(9) follows the commutation relation for $\mathcal{L}$ and $\mathcal{M}$ [13]:

$$i[\mathcal{L}, \mathcal{M}] = K\mathbf{I} \tag{10}$$

It represents an incommensurability of different dynamical concepts, which might be called Liouvillean dynamics and information dynamics. In particular, it contains the following aspects:

- It is related to Eq.(4) in providing an empirically accessible version of it. Since K can be obtained for theoretical systems as well as from experiment, it gives a nice and transparent visualization of the existence of a time operator. Depending on the physical meaning assigned to ρ (density operator or distribution function), (4) and (10) are relevant to quantum and classical systems.
- It illustrates the difference between the unitary (time – reversible) evolution based on $\mathcal{L}$ and the non–unitary (time – directed) semigroup evolution based on $\mathcal{M}$. The direction of time has implicitly been defined by Eqs.(7) and (8), based on increasing uncertainty along the unstable manifolds. This choice is in principle arbitrary, but it is motivated by the phenomenological relevance of the second law of thermodynamics (and other phenomenological time arrows).
- It provides the equivalence of the dynamical concepts given by $\mathcal{L}$ and $\mathcal{M}$ in case of information conserving systems ($K = 0$). This point makes clear that the concepts given by $\mathcal{L}$ and $\mathcal{M}$ are hierarchical: whenever $\mathcal{L}$ is applicable, then $\mathcal{M}$ is applicable as well (although in its least interesting special case). Otherwise, if $\mathcal{M}$ is applicable, then $\mathcal{L}$ is in general not applicable.

Although (10) has been derived for classical systems, both (10) and (4) should be relevant to quantum mechanical cases as well, if the superspace formalism is adequate. However, the classical relevance of (10) should enable more direct insight into its meaning than this was possible in the quantum incommensurability (1). Therefore, an analysis of (10) along the lines described in Sec.2 suggests itself. Such an analysis in terms of projectors acting on states ρ in superspace has not yet been worked out in detail. Instead, another possibility to visualize the lattice structure of Eq.(10) has been proposed [15, 16], studying a set of propositions about the state of a system with respect to the issue of prediction. For instance, consider a system with unknown K, such that the set of all possible values of K is given by definition of K: $\{0 \leq K < \infty\}$. The question is whether the system can be described by $\mathcal{L}$ ($K = 0$) or by $\mathcal{M}$ ($K \geq 0$).

Let us first assume that the claim $K = 0$ exists, which means that the accuracy of any prediction about the state of the system be maintained for ever. (There is no unstable manifold being able to generate increasing uncertainty.) This claim may be checked using a prediction about a future state of the system after a time interval Δt, necessarily finite. If the system is not found in the predicted state, then $K = 0$ is definitely false. Otherwise, if the system is found in the predicted state, then $K = 0$ is not definitely true, but "corroborated". For example, it is still possible that $0 \leq K < \Delta t^{-1}$ which does not definitely exclude $K > 0$. This example shows that a proof for $K = 0$ for a specific system would require an infinite time interval between initial observation and test of prediction. Such a definite proof is operationally impossible. Only corroborations (corresponding to a contingent truth value) can be carried out.

The other possibility is to claim $K \geq 0$ (no matter how big K may be). Any prediction about a future state would then *a priori* be known as potentially irrelevant. Thus the claim $K \geq 0$ would trivially be corroborated, even if K vanishes.

In contrast to the structural aspect of quantum logic, the preceding discussion clearly stresses the dynamical aspect of a temporal logic.[2] Formalization of these deliberations in a lattice theoretical way provides a non–distributive propositional lattice as in case of the quantum incommensurability (1). The complement of a dynamical model falsified by a checked prediction is not a verified dynamical model, but merely a corroborated model.

4. Algebraical Unification of Structure and Dynamics

This section describes how the incommensurabilities (1) and (10) can be interpreted in terms of lattices. As mentioned above, we shall not use the lattice of subspaces of $\mathcal{H}$ or of superspace $\mathcal{S}$ to do so, but we shall study the lattice as a propositional lattice. Such a *procedere* can be regarded to be appropriate because our main goal here is to explain the non–distributivity of temporal logic (corresponding to (10)) which is relevant in a classical as well as quantum context. Hence, it must be intelligible in the framework of propositions about classical systems, an issue to which the human mind (at least to some extent) is better adapted than to quantum systems. (Remember that an understanding of quantum logic in terms of propositions

[2]It should be noted that dynamical and structural incommensurabilities are settled on different levels of physical descriptions. While structural quantum logic is clearly restricted to the case of quantum systems (where actions are not large enough to neglect h) and its Hilbert space representation, the temporal logic of dynamics applies in superspace. This implies that temporal logic is relevant in a more general manner than this is the case for quantum logic. These deliberations support the view that quantum theory is (by virtue of Planck's action h) an overdetermined classical theory. This overdetermination enables a description in terms of one of two possible dual representations, those in state space and in observable space (as long as pure states are considered). Superspace representations as they are required for temporal logic include both quantum mechanical representations. They are therefore natural candidates for tracing the transition from quantum to classical systems, including features like non–local correlations, inseparability of composite systems, and irreversibility.

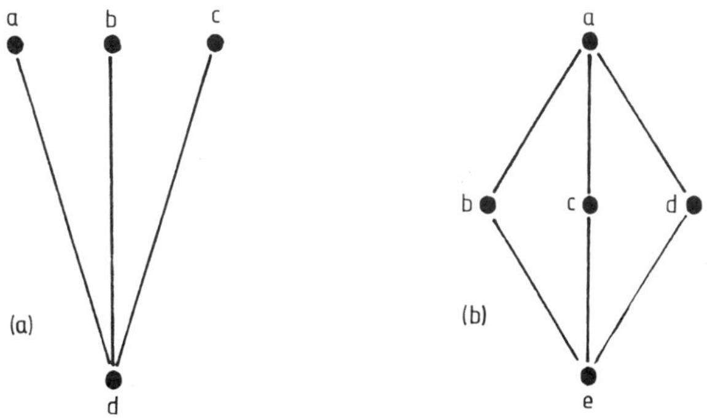

Figure 1: Hasse diagram for two posets. Upper elements imply lower elements of the diagram. Are both posets lattices? a) Poset with greatest lower bound d, but without common least upper bound. This poset is no lattice. b) Poset with greatest lower bound e and least upper bound a. This poset is a lattice (which is even non–distributive; see below).

describing specific situations is a rather difficult subject.)

Propositional logic requires a set G of propositions $\{a, b, \ldots\}$ and relations among them. For instance, $\sqcap$ and $\sqcup$ symbolize conjunction (logical "and") and disjunction (logical "or"). An implication is represented by $a \rightarrow b$, meaning a implies b. If both $a \rightarrow b$ and $b \rightarrow a$ are valid, then one has an equivalence relation, $a \leftrightarrow b$. If elements of G (propositions) can be ordered by implications, then G is a partially ordered set (poset). It is then possible to arrange the elements of G in an order diagram (Hasse diagram) as shown and explained in Figure 1. If there is a common least upper bound $x \sqcup y$ and if there is a common greatest lower bound $x \sqcap y$ for all pairs $(x, y) \in G$, then $(G, \rightarrow)$ is a lattice.

Using set theoretical symbols, $\wedge$ and $\vee$ denote the intersection ("meet") and the union ("join") of elements. Implication and equivalence translate into greater ($>$) and equal ($=$). For reasons which will become clear below we shall use this notation for the subsequent discussion.

Basic algebraical properties of any lattice are:

$$
\begin{aligned}
&\text{Commutativity}: && a \wedge b = b \wedge a \\
&\text{Associativity}: && a \wedge (b \wedge c) = (a \wedge b) \wedge c \\
&\text{Absorption}: && (a \wedge b) \vee b = b \\
&\text{Idempotency}: && a \wedge a = a
\end{aligned}
\tag{11}
$$

together with the corresponding dual properties which are obtained by interchanging $\wedge$ and $\vee$.

To check these relations, propositions constituting G have to be defined. In the context of quantum theory and the associated incommensurability (1), these propositions are propositions about measurements (or more general, observations). Within the Hilbert space description of the state of a system, measurement corresponds to a projection of the state vector onto that subspace characterizing the measured variable. So, propositions about the measured position and momentum of a system have to be taken as elements $a, b \in G$, e.g., $(a) : q = q_o$.

Concerning the dynamical incommensurability (10), we deal with temporal predictions instead of spatial observations. They are projections of the present state of a system onto that subspace of superspace which characterizes the predicted state. Propositions in this predictive sense are clearly different from those described for the quantum context: They are dynamical instead of structural. Relevant propositions for the superoperators $\mathcal{L}$ and $\mathcal{M}$ as introduced in Sec.3 are: $(a) : K = 0$ and $(b) : K \geq 0$. While (a) corresponds to systems obeying $\mathcal{L}$, (b) represents the general case of systems obeying $\mathcal{M}$ including the special case $K = 0$.

The characterization of $\mathcal{L}$ and $\mathcal{M}$ by a set of values assigned to K makes it possible to use the set theoretical notation for propositions in the framework of temporal logic. The set of all values of K which are allowed using the description with $\mathcal{M}$, namely $\{K|K \geq 0\}$, may thus be taken to replace the proposition "$K \geq 0$", and the set $\{K|K = 0\}$ correspondingly replaces "$K = 0$". In this manner it is possible to check conditions (11) for predictions made according to the superoperators $\mathcal{L}$ and $\mathcal{M}$.

Now, it is well known for quantum logic [4] that in addition to (11) a complement of a proposition exists which is not unique. A lattice is complemented if for each element a its complement a' exists such that

$$\begin{aligned} a \wedge a' &= \mathbf{0} \\ a \vee a' &= \mathbf{I} \end{aligned} \tag{12}$$

are satisfied. Both relations are again dual to each other. $\mathbf{I}$ (unity) and $\mathbf{0}$ (null) represent universal upper and lower bounds in G. Complements a' and b' are unique if the lattice is distributive with respect to these complements. This is the case if for each pair $(a, b) \in G$ the relations

$$a = (a \wedge b) \vee (a \wedge b') \tag{13a}$$

$$\vee \quad b = (b \wedge a) \vee (b \wedge a') \tag{13b}$$

and (dually)

$$a = (a \vee b) \wedge (a \vee b') \tag{14a}$$

$$\wedge \quad b = (b \vee a) \wedge (b \vee a') \tag{14b}$$

are satisfied. If the lattice is non–distributive, then (13) and (14) have still to satisfy the distributive inequalities, obtained by replacing $=$ by $>$ in (13), and replacing $=$ by $<$ in (14). Non–distributivity expresses non–uniqueness of the complement, implying a non–Boolean logical structure of the propositional lattice considered. Figure 1b gives an example of a non–distributive lattice: the complement of b, e.g., is not unique, since both c and d are possible "candidates". (In fact, any lattice with (1b) as a sublattice is non–distributive, hence non–Boolean.)

An essential part of the work of Birkhoff and von Neumann [4] was to show that quantum logic does not satisfy (13), i.e., is non–Boolean. Since (13a) and (13b) are connected by a set theoretical union (logical "or"), a falsification of (13) has to provide that both (13a) and (13b) are not true. Indeed, quantum logical non–distributivity manifests itself in:

$$a > (a \wedge b) \vee (a \wedge b') \tag{15a}$$

$$\wedge \quad b > (b \wedge a) \vee (b \wedge a') \tag{15b}$$

As it is the case for all dualities in the context of lattice theory, non–distributivity (i.e. non–(13)) according to (15) implies that the equality (14) is violated, too. The corresponding pair of inequalities reads:

$$a < (a \vee b) \wedge (a \vee b') \tag{16a}$$
$$\vee \quad b < (b \vee a) \wedge (b \vee a') \tag{16b}$$

where the union of both relations (14) means that (14) is violated if only one of both inequalities (16a) or (16b) is found to be true. The dual formulation of non–distributivity in (16) has never been discussed to correspond to any quantum theoretical incommensurability. Since it is nevertheless relevant for mathematical reasons, one might wonder what its physical content consists of.

In the following we shall conjecture that (16) is satisfied as the non–distributive property of temporal logic, thus providing a worthwile hint toward its unification with quantum logic. This can be seen as already indicated at the end of Sec.3. Expressing propositions a and b as sets $\{K|K=0\}$ and $\{K|K \geq 0\}$ according to $\mathcal{L}$ and $\mathcal{M}$, we can write the rhs of (16a) as [15]:

$$\{K|K=0 \vee K \geq 0\} \wedge \{K|K=0 \vee K < (\Delta t)^{-1}\} = \{K|0 \leq K < (\Delta t)^{-1}\} \tag{17a}$$

Here we have used that the initial assumption was $K=0$ and that a corresponding prediction has been checked after a time interval Δt. Since $\{K|0 \leq K < (\Delta t)^{-1}\} > \{K|K=0\}$, the inequality (16a) is satisfied. (Note that $\{K|K=0\}$ is indeed *smaller* than the rhs of (16a) as given in (17a).)

For (16b)

$$\{K|K \geq 0 \vee K=0\} \wedge \{K|K \geq 0 \vee K \neq 0\} = \{K|K \geq 0\} \tag{17b}$$

still provides a distributive equality. Since the inequality (16a) is sufficient for the non–distributivity of the lattice, this does not provide any contradictive argument. In contrast, it indicates a type of asymmetry which reflects an important difference from the "symmetric" non–distributivity of structural quantum logic. The formal reason for this asymmetry is that (10) includes the case $K=0$ of a vanishing commutator as a special case (in which $\mathcal{L}$ and $\mathcal{M}$ provide equivalent representations), while the commutator h in (1) is a non–vanishing universal constant.

This difference is not a defect or a weak point. It accounts for the duality of observation/measurement and modelling/prediction as objective and subjective components of our world view. The truth values associated with both components provide an additional clue to this point. In the framework of quantum logic, a measurement with positive result provides a definitely true proposition about the result, e.g., $q=q_o$. A negative outcome, instead, does always correspond to a contingent truth value, e.g., doubtful or undecided (potentially false). *Vice versa*, temporal logic provides a definitely false proposition, if a prediction leads to a negative outcome (is falsified). Otherwise, if a prediction is not falsified, then it is contingent, i.e., corroborated (potentially true) [16].

All the dualities discussed so far, be they expressed as complementarities, dichotomies, or ambivalences, appear as mathematically dual properties of a non–distributive lattice. (They may be visualized by a rotation of the corresponding Hasse diagram by π.) It is thus tempting to interpret the lattice as an algebraical tool to identify a fundamental unity of nature, on the basis of which all the dualities emerge.

One obvious question arising is whether this unity may be transferred from algebra to geometry. If such a transfer can be made, the next question would be in which way the resulting geometrical dualities correspond to their algebraical counterparts. We started with an abstract algebraical model for the complementarity of structure and dynamics. Is it then possible to infer concrete empirical consequences on the side of geometry? These questions will be addressed in the following section.

5. From Algebra to Geometry

The complementarity of structure and dynamics which expresses itself in the duality of the propositional lattice constituting quantum logic and temporal logic calls literally for a transfer to geometry. The reason is that structure and dynamics as they have been discussed so far, are distinguished by the notions of space and time, the basic categories of the geometrical theory of general relativity. (It should also be recalled that this point was one of the historical motivations to look for a mathematically sound formulation of energy–time–uncertainty relations.)

Now, is it possible to derive knowledge about the geometry of spacetime from the algebra associated to events in space and time? The answer is yes: it has been known for a long time that the algebraical properties of a lattice can be mapped onto the geometrical properties of a vector space under certain conditions. In fact, Ref.[4] used this correspondence to infer logico–algebraical properties from the geometry of complex Hilbert spaces. Already this reference quotes that the resulting non–distributivity requires a skew–symmetric geometry as the appropriate geometrical counterpart. The least complicated skew–symmetric geometry we know is one based on the number system of quaternions. The basis vectors of a quaternionic vector space are ± 1, $\pm i$, $\pm j$, and $\pm k$, where $i = j = k = \sqrt{-1}$, and $ij = k$, $ji = -k$, etc. With $e = \{+1, -1\}$, $b = \{\pm 1, \pm i\}$, $c = \{\pm 1, \pm j\}$, and $d = \{\pm 1, \pm k\}$, it is easy to see that these basis vectors form a lattice as given in Figure 1b.

The first detailed exploration of the consequences a quaternionic Hilbert space would have for quantum theory has been carried out by Finkelstein *et al.* [7]. One of their basic results was that quantum theory seems to be sufficiently represented in complex Hilbert spaces, although the associated logic indicates the relevance of a higher dimensional quaternionic space. Their investigations revealed a particularly interesting point with respect to the discusssion we gave of Eq.(10). As an essential quantity in quaternionic quantum mechanics, they discussed a quantity η[3] which they did not know how to specify exactly. Later on [17], η was referred to as a conversion quantity from dimensionless probabilities in spinor space into real time. This notion comes close to the idea of an information flow or production rate as it has been addressed in a previous paper [18]. Due to Finkelstein, dynamical law can ultimately not be considered as an invariant. This issue relates to the system – specificity of the commutator K in (10) which ultimately applies to each individual event (point in spacetime).

Finally, and here we mention the main point concerning our subsequent considerations, Finkelstein states [17]: "*what we call dynamical law actually describes the interaction between exosystem and endosystem, and should be regarded as a statistical*

[3]Not to be confused with the quantity η which we shall introduce below. In our subsequent notation η will be used as an internal space coordinate.

description of the exosystem from the viewpoint of the endosystem." This quotation stresses the duality of internal and external viewpoints (endosystem and exosystem), as they apply to one and the same system.[4] For instance, consider a single photon constituting a very simple system embedded in ordinary Minkowski spacetime. Then we know that the photon internally does not age (its proper time element is $d\tau = 0$) because it moves with velocity c. However, an external observer observes the photon during a (parameter) time interval $dt = \frac{ds}{c}$, such that externally a time interval $dt > 0$ has passed by if the photon moves ($ds > 0$).

In dual quaternionic spacetimes there is room for two different notions of space as well as of time. These different notions can be regarded to correspond to an internal view (endosystem) and to an external view (exosystem). The key point to appreciate the meaning of this duality adequately is that *any system is an endosystem and an exosystem equivalently.* With respect to the example given above, proper time τ would represent internal time while t were an external parameter time. With η and q as associated spatial coordinates one can denote events from an internal point of view by:

$$\eta_1, \eta_2, \eta_3, \tau \tag{18}$$

whereas events from an external point of view are given by:

$$q_1, q_2, q_3, t \tag{19}$$

with $\underline{R} = \underline{\eta} + i\underline{q}$ and $T = t + i\tau$. At this point it becomes clear that the traditional consideration of real parts only (and the resulting neglection of imaginary parts) can lead to models representing an inappropriate mixture of endosystem and exosystem descriptions.

Note that external coordinates are those of Minkowski spacetime in its original formulation with a metric (–,–,–,+). It was also Minkowski [20] who recognized first that t and τ satisfy a mathematical transformation which reflects a change between external and internal viewpoints toward a system. If an observer is situated within the system, then he considers it from the internal viewpoint of an endosystem, with coordinates η and τ. Otherwise, an external observer uses external (exosystem–) coordinates q and t.

Essential differences between both points of view occur, if the system under consideration is a dynamical system enabling the existence of a time operator. Then the eigenvalues of this operator (also called the age of the system) provide τ, whereas an external time can be defined by an external clock (which can also be regarded as an endosystem, of course). To indicate an example, consider a system S containing distant objects moving with constant velocity relative to each other. Assume that the system, considered as an exosystem, follows dynamical laws identical to those describing the expansion of the universe. (This is implicitly the case for any system within the universe.) Then spatial and temporal scales of the endosystem S remain

[4]In this context, Lloyd's description [19] of the relation among causality and statistics is particularly relevant. Using the framework of Bayesian networks, he represents events (facts) and "relations of cause and effect" (models, laws) as nodes and links of the network. Here the duality of facts and models can be considered to manifest itself in a concrete graphical way, namely by the interchange of nodes and links. In the spirit of the quotation from Finkelstein, a change from internal to external representation (and *vice versa*) accompanies such a duality.

statistically invariant ($\frac{d\eta}{d\tau} = 0$): for an internal observer (e.g., a particle) who is exposed to the same dynamical laws, spatial and temporal distances look the same on an average. In contrast, external space and time scales for the exosystem S vary due to the same dynamical laws: an external observer sees how the system expands.

The determination of distances within the endosystem S (i.e., by an internal observer) is sketched in Fig.2. An internal observer $\mathcal{O}(\tau)$ at internal position η is shown who observes radiation from a distant light source $\mathcal{Q}(\tau)$. In internal coordinates η, τ, the line $\overline{\mathcal{O}\mathcal{Q}}$ is a vertical line because $d\tau = 0$ for photons. Since $\mathcal{O}$ does not move with respect to his own restframe, $\overline{\mathcal{O}(\tau)\mathcal{O}(\tau + d\tau)}$ is a horizontal line. While the observer ages by a time interval $d\tau$, his distance from $\mathcal{Q}$ increases by an amount dl. The transformation between dt (external time) and $d\tau$ (internal time) has first been given in Ref.[20] and can directly be obtained from Fig.2:

$$d\tau = dt\ \sqrt{1 - \frac{v^2}{c^2}} \tag{20}$$

where $v = \frac{dl}{dt}$ is the velocity of $\mathcal{Q}(\tau)$ relative to $\mathcal{O}(\tau)$.

In the context of cosmology and astrophysics of the early universe, the role of human observers is predominantly that of an internal observer, an approach which has particularly been stressed under the heading of a participatory anthropic principle [21]. Corresponding empirical consequences have been indicated recently [22, 23]. On a microscopic scale, an internal observer might be thought of as a particle within a many – particle system with its specific dynamical features. Concerning empirical predictions in this respect, we refer to the derivation of lifetime dependent correction terms to quantum electrodynamics as proposed in Ref.[24].[5]

It is important to emphasize that the cases of purely internal as well as of purely external viewpoints are limiting cases, and, as such, of approximative value only. A pure exosystem description is relevant in the hypothetical case of a system completely decoupled from its environment. Strictly speaking, such a system is unrealistic since it would exclude any interaction with an observer. Hence, no measurement would be allowed and no knowledge could be inferred about the system. On the other hand, a pure endosystem approach seems ill–defined since it would require a sufficiently separable subsystem of the entire system which could – as an exosystem – be used to check detailed predictions. As in the algebraical framework, strict duality is artificial insofar as structure and dynamics, space and time, endosystem and exosystem are inseparably interwoven.

Nevertheless, distinguishing both limiting cases might provide insight into fundamental unresolved problems. For instance, the introduction of a "two–dimensional" time, if taken seriously, offers an interesting new view toward the question whether irreversibility or reversibility is the basic feature of temporal evolution. As participants of the evolution of the universe as a whole, considering it as an endosystem, we have to use internal time, which (by its definition using semigroups) provides irreversibility. The evolution of an exosystem, on the other hand, can (probably in general) be formulated in a reversible manner, since those terms indicating irreversible endosystem

[5]Of course, for reasons mentioned above, cosmological corrections have to be expected in addition. However, their size increases with distance [22, 23] and will consequently be extremely small for laboratory systems.

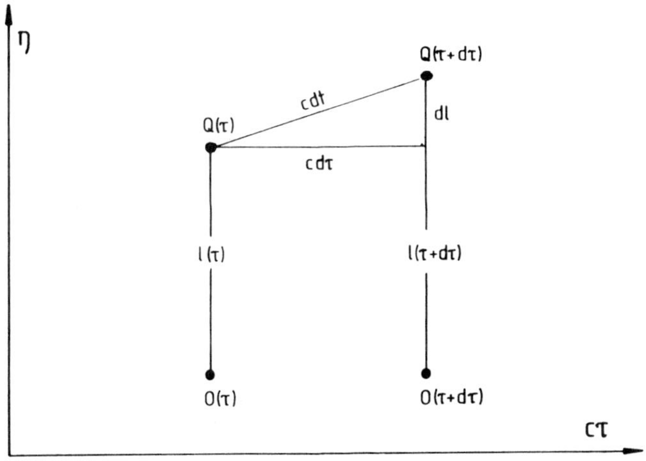

Figure 2: Determination of the velocity of a distant light source by an internal observer. For details see text.

evolution may be "reversibilized" by taking interactions with the environment into account. This view would require an integral combination of endosystem and exosystem approach, which goes beyond the dual formulations given by (18) and (19).

Here we are already in the realm of speculations (hopefully controlled speculations), so it is perhaps worthwile to add a final idea on the relation of the geometrical structure discussed to string theory. String theory represents a branch of theoretical physics on which many hopes for a unified view of the physical world are based. A string (or superstring, if the theory is to include supersymmetry) is a one–dimensional structure ascribed to internal degrees of freedom of particles. Using this model it is possible to remove those infinities and anomalies occuring if particles are considered as point–like entities in quantum field theory. The lowest–dimensional superstring models which are considered to be relevant at present need 10 dimensions, four for ordinary spacetime, and six further dimensions usually interpreted to be "curled up" into distance scales not empirically accessible. The one–dimensional string lives within these "hidden" dimensions.

The model of dual quaternionic spacetimes can be interpreted in a way resembling this approach. Namely, each event in ordinary global four–dimensional spacetime could be described by a dual, but local, four–dimensional internal spacetime. So we could imagine four–dimensional extensions of strings with internal coordinates, representing tangent (local) manifolds on a four–dimensional global manifold in an 8–dimensional vectorspace. (In recent work of Caianiello *et al.* [25] such an approach plays a central role, although it does not explicitly use the formalism of string theory.) It would be very interesting to find out whether working with four–dimensional "extended strings" could bring the necessary dimension of external spacetime down to four, if one uses the formalism of string theory.

6. Summary and Perspectives

In one of T.S. Eliot's "Four Quartets", *Burnt Norton*, one finds the line "*The detail of the pattern is movement*". This line is a beautiful poetic expression of what is here called the complementarity of structure and dynamics. In the sense of the foregoing arguments it would even be justified to complement that line by its inverse, "*The detail of movement is pattern*". Pattern and movement, structure and dynamics, space and time are to be understood as artificially separated notions of a basic unity of nature.

In an abstract theoretical manner, this unity can be demonstrated mathematically. However, it should be clear that the corresponding rational reconstruction (which will be summarized subsequently) consistently misses its dualistic counterpart: guidelines for experiencing this unity in a concrete way. This kind of experience will (also consistently) be an individually subjective affair (e.g., an aesthetic experience), it will not be further addressed here. It must not be mixed up with empirical tests for the dualistic notions referred to in Sec.5. Those tests may merely serve as indicators for the relevance of the idea of an underlying unity. This is so, because the transfer from notions and concepts established in an algebraical context onto the realm of geometry necessarily involves the unity of both in some sense. If empirical tests do reliably confirm predictions made in the framework of both dual branches, then the quest for a unity of both will turn out to be an inevitable demand.

The flow of ideas described above can be visualized in the scheme shown in Fig.3. The central point is the commutation relation of the Liouville operator $\mathcal{L}$ and an information operator $\mathcal{M}$, related to an incommensurability of energy and time which is mathematically rigorous. In contrast to the position/momentum incommensurability, it refers to energy and time as internal quantities, given by the eigenvalues of the associated superoperators. It can be demonstrated that the logic corresponding to the incommensurability of $\mathcal{L}$ and $\mathcal{M}$ gives rise to the same propositional lattice as the logic of the usual quantum mechanical incommensurability of position and momentum does. Nevertheless there is a remarkable difference between both, justifying a distinction between dynamical and structural logic. This difference simply appears in the dual formulations of the relevant lattice theoretical relations. In particular, this has been found for the non–distributivity of the lattice, which causes the corresponding logic to be non–Boolean. Both dual aspects, that of structure and that of dynamics, carry features of indeterminism with respect to measurement and prediction, respectively, with themselves. In logical terms, the duality of both aspects characterizes a tautology. This motivates the idea of their complementarity from an algebraical point of view.

The algebraical, lattice theoretical approach can be transferred to geometry if the lattice is modular. Then, the specific properties of the lattice impose restrictions on the type of number system on which the corresponding geometry is based. In case of a non–distributive lattice, a skew–symmetric geometry is required, the simplest case of which is provided by quaternionic vector spaces. The duality of those spaces has been assigned to the duality of internal and external viewpoints toward a system to be considered. (These viewpoints represent limiting cases in which the modularity of the lattice is trivially satisfied. Whether and how these limiting cases can be extended toward the general situation of a non–modular lattice accounting for the inseparability of internal and external viewpoints is unknown so far.)

With respect to them, an external description of the system as a whole is global (e.g., four–dimensional Minkowski space), while internally a local approach from the viewpoint of individual constituents of the system (events) is relevant. In this picture

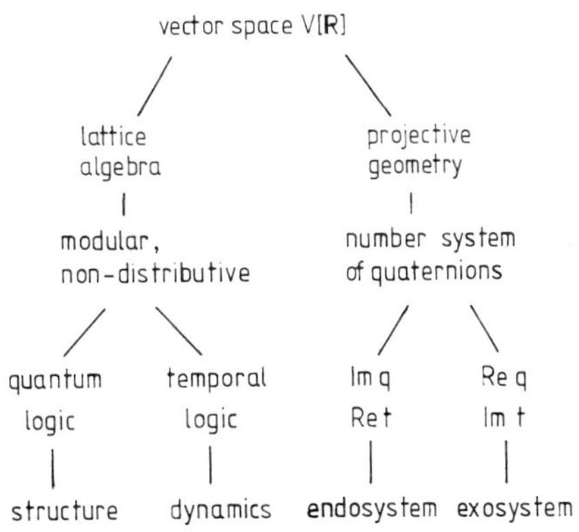

Figure 3: Illustration of the main lines of reasoning of the article.

each event has to be imagined as to be associated with an internal four–dimensional spacetime dual to ordinary Minkowski spacetime.

The geometrical counterpart of the algebraical view calls for transformations among internal and external spacetimes. Specific suggestions have already been made [22-24] and the chances for empirical results concerning them are not too bad for the near future.

The approach presented in this article uses the principle of mathematical duality to offer a model for unification on a fairly fundamental level. The proposed complementarity of structure and dynamics is built upon this notion of duality. In this respect, it is different from those complementarities expressed by non–commuting operators. How a formulation in terms of operators would look like is an open question. It might be suspected, that such a formulation could be possible on the basis of information theoretical concepts.

The way in which information theory is involved in the subjects presented above is purely temporal. Concepts like that of an information metric [26] are designed to account for uncertainty and limited knowledge in the context of temporal as well as spatial features. Information dynamics in this sense would certainly include a more general approach than given by an information operator as defined by $\mathcal{M}$. But still in this extended sense, the emergence of new types of structure and dynamics as it is observed at instabilities, bifurcations, phase transitions, etc., would not at all be covered. It may be allowed to guess (presumably this is an easy guess) that at this point, if not earlier, the context–dependence of concepts like complexity and meaning will have to be taken into account explicitly.

References

1. M. Jammer. *The Philosophy of Quantum Mechanics.* Wiley, New York, 1974.
2. R. Penrose and C.J. Isham, eds. *Quantum Concepts in Space and Time.* Clarendon, Oxford, 1986.
3. J.A. Wheeler and W.H. Zurek, eds. *Quantum Theory and Measurement.* Princeton University Press, Princeton, 1983.

4. G. Birkhoff and J. von Neumann. *Ann. Math.* **37**, 823 (1936).
5. W. Pauli. *Die allgemeinen Prinzipien der Wellenmechanik.* In *Handbuch der Physik*, Vol.24, 2nd edition. Eds. H. Geiger and K. Scheel. Springer, Berlin, 1933; Reprinted in *Handbuch der Physik* (*Encyclopedia of Physics*) Vol.5. Ed. S. Flügge. Springer, Berlin, 1958, pp.1–168.
6. I. Prigogine. *From Being to Becoming.* Freeman, San Francisco, 1980.
7. D. Finkelstein, J.M. Jauch, S. Schiminowich, and D. Speiser. *J. Math. Phys.* **3**, 297 (1962).
8. B. Misra. *Proc. Ntl. Acad. Sci. USA* **75**, 1627 (1978).
9. I. Prigogine. *Non–Equilibrium Statistical Mechanics.* Interscience, New York, 1962.
10. R. Balian, Y. Alhassid, and H. Reinhardt. *Phys. Rep.* **131**, 1 (1986).
11. R. Shaw. *Z. Naturforsch.* **36 a**, 80 (1981).
12. P. Grassberger. Information and complexity measures in dynamical systems. This volume.
13. H. Atmanspacher and H. Scheingraber. *Found. Phys.* **17**, 939 (1987).
14. B. Misra, I. Prigogine, and M. Courbage. *Physica* **98 A**, 1 (1979).
15. H. Atmanspacher. In *Parallelism, Learning, Evolution.* Eds. J. Becker, F. Mündemann, and I. Eisele. Springer, Berlin, 1990.
16. H. Atmanspacher, F.R. Krueger, and H. Scheingraber. In *Parallelism, Learning, Evolution.* Eds. J. Becker, F. Mündemann, and I. Eisele. Springer, Berlin, 1990. See also F.R. Krueger. *Physik und Evolution.* Parey, Berlin, 1984, Appendix.
17. D. Finkelstein. In *The Universal Turing Machine.* Ed. R. Herken. Oxford University Press, Oxford, 1988.
18. H. Atmanspacher. *Found. Phys.* **19**, 553 (1989).
19. S. Lloyd. Causality and information flow. This volume.
20. H. Minkowski. In *The Principle of Relativity.* Dover, New York, 1952.
21. J.A. Wheeler. In *Foundational Problems in the special sciences.* Eds. R.E. Butts and J. Hintikka. Reidel, Dordrecht, 1977, p.3.
    See also J.D. Barrow and F.J. Tipler. *The anthropic cosmological principle.* Clarendon Press, Oxford, 1986.
22. J. Charon. *Complex Relativity.* Paragon, New York, 1988.
23. H. Atmanspacher and H. Scheingraber. "Endophysical derivation of a nonlinear cosmological velocity – distance relation". Preprint 1990.
24. I. Prigogine and T. Petrosky. *Physica* **147 A**, 461 (1987).
25. E.R. Caianiello, A. Feoli, M. Gasperini, and G. Scarpetta. *Int. J. Theor. Phys.* **29**, 131 (1990).
26. E.R. Caianiello. In *Topics in the General Theory of Structures.* Eds. E.R. Caianiello and M.A. Aizerman. Reidel, Dordrecht, 1987, p.199.
    W.H. Zurek. *Nature* **341**, 119 (1989).
    J.P. Crutchfield. In *Nonlinear Structures in Physical Systems – Pattern Formation, Chaos, and Waves.* Eds. L. Lam and H.C. Morris. Springer, Berlin, 1990.

INFORMATION AND DYNAMICAL SYSTEMS

Ioannis E. Antoniou

Theoretische Naturkunde and Faculté des Sciences
Free University of Brussels
B-1050 Bruxelles, Belgium

Introduction

Operational limitations in information processing give rise to realizations of unstable dynamical systems as Markov stochastic processes. Such exact constructions are possible for both conservative and dissipative dynamical systems with a certain kind of instability.

The constructions developed so far correspond to three classes of dynamical systems:

- Conservative dynamical systems with internal time, classical or quantum;
- Conservative large Poincaré non–integrable systems, classical or quantum;
- Conservative or dissipative dynamical systems admitting finite Markov coarse – graining.

The origin of the ideas behind these constructions is the problem of irreversibility, being the focus of activity of the Brussels school directed by Professor I. Prigogine. In this view, statistical physics becomes essentially a consistent theory of processes in which we have limited information.

The contents of this communication are the following:

1. The language of dynamical systems
2. Markov processes and exact dynamical systems
3. The problem of irreversibility
4. Kolmogorov systems and internal time
5. Relativistic systems with internal time
6. Quantum systems with internal time
7. Large Poincaré non–integrable systems
8. Finite Markov coarse–graining
9. "Music" from "chaos"

Information Dynamics, Edited by H. Atmanspacher and
H. Scheingraber, Plenum Press, New York, 1991

1. The Language of Dynamical Systems

A one parameter evolution semigroup S_t in the space Y of states defines a dynamical system. The variable t signifies time and is continuous for flows and discrete for cascades. For reversible systems S_t is a group of automorphisms of Y.

The space Y is equipped with a topology in order to formulate nearness and stability, and with a measure structure, compatible with the topology, for the description of the statistical properties. The measure structure includes an algebra $\mathcal{B}$ of measurable sets and a S_t – invariant measure μ representing an equilibrium distribution.

The semigroup S_t induces the Frobenius–Perron operators V_t on the density functions $\rho(y)$ according to the formula

$$\int_{\Delta} d\mu(y) V_t \rho(y) = \int_{S_t^{-1}\Delta} d\mu(y)\rho(y)$$

for any measurable set Δ in $\mathcal{B}$.

If the evolution is reversible, we have the Koopman formula for V_t:

$$V_t\rho(y) = \rho(S_t^{-1}y) \qquad y \in Y$$

For a discussion of these ideas including recent developments, see for example the books of Sinai [1] and Lasota and Mackey [2] and references therein.

2. Markov Processes and Exact Dynamical Systems

Irreversibility marks observable manifested processes. The key feature of irreversible processes is the distinction of the direction of time or the violation of time reversal symmetry. Entropy increases in the direction of our future. Markov processes are the standard mathematical formulation of a large number of irreversible processes as the transition–reaction–kinetic processes (Boltzmann and Pauli equations), diffusion processes (Fourier heat law, Fick law, Fokker–Planck and Langevin equations for Brownian motion), and reaction–diffusion equations responsible for self–organization.

The Markov processes considered here are characterized by the following properties:

a) There exists a unique equilibrium density ρ_{eq}.

b) Equilibrium is approached monotonically in the far future,

$$|W_t\rho - \rho_{eq}| \;\to\; 0 \qquad \text{as} \quad t \to \infty$$

where ρ is any initial density evolving under the positivity preserving contraction semigroup W_t.

Uniqueness of equilibrium (a) means that the invariant measure is unique, i.e., that the system is ergodic. Monotonic approach towards equilibrium (b) means that W_t is the Frobenius–Perron operator of an exact dynamical system. Such systems have the property:

$$\mu\{S_t\Delta\} \;\to\; \mu\{Y\} = 1 \qquad \text{as} \quad t \to \infty$$

for any measurable set Δ in $\mathcal{B}$.

This property means that any set with non–zero measure eventually covers the whole available space and reflects an underlying instability, or that the system has positive Lyapunov exponents, or "chaos" according to the current fashion.

Exact dynamical system were introduced by Rokhlin [3] and they are also characterized by monotonic entropy increase [2,4], or equivalently by strictly positive Kolmogorov–Sinai entropy production [1]. Exact dynamical systems arise also as models of the asymptotic evolution of dynamical systems on strange attractors [5].

Typical examples of exact systems are the piecewise monotonic maps of the unit interval, such as the dyadic map:

$$Sy = 2y \ (\text{mod} 1)$$

and the logistic map:

$$Sy = 4y(1-y)$$

3. The Problem of Irreversibility

How do the conservative reversible dynamics give rise to entropy increasing irreversible evolutions formulated as exact Markov semigroups? Several answers have been given to this question since Boltzmann, which evoke elements imposed extraneously to dynamics, such as subjectivity, arbitrary coarse–graining, coupling with the environment, or approximations.

The approach developed in Brussels [6] is somehow the reverse: are there intrinsic mechanisms hidden in the conservative dynamical laws, responsible for the emergence of the arrow of time? The answer is yes, and the mechanism is instability.

Due to the instability dynamics is intertwined with a Markov process through non–unitary transformations Λ acting on the densities ρ according to Figure 1.

Two particular types of instability have been found which are responsible for non–unitary realizations of dynamics as Markov processes:

- The instability of Kolmogorov systems, expressed by the presence of the time asymmetric K–partition and the internal time operator acting on the density functions. The intertwining transformation Λ operationally incorporates the objective limitations to predictability arising from the unlimited refinements of the K-partition.

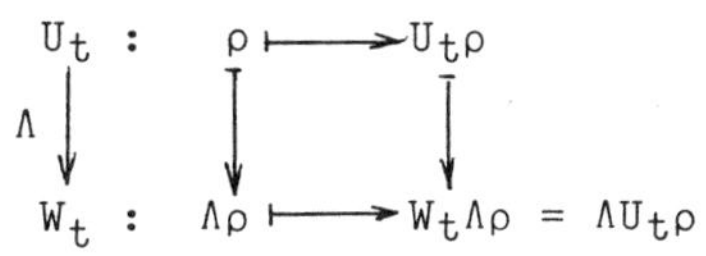

Figure 1: The intertwining of conservative dynamics with constant entropy (upper line), with irreversible Markov evolution with growing entropy (lower line).

- The instability due to the presence of continuous sets of resonances. These systems are non–integrable in the sense of Poincaré; there are no analytic and unitary diagonalizing transformations. However, suitable analytic but non–unitary diagonalizing transformations give rise to dissipative realizations of the conservative dynamics.

It has also to be emphasized that the very existence of the Λ transformation is equivalent to the existence of a Lyapunov observable M having the meaning of a microscopic entropy operator. Such observables cannot be ordinary observables, i.e., phase functions or self–adjoint operators, but represent superobservables, i.e., operators acting on the density functions or on the density operators. This was shown by Misra [7, 8] who in this way went beyond the limitations of Poincaré's recurrence theorem, namely that "a phase function will infinitely often assume its initial state". A Lyapunov observable is given by the formula:

$$M = \Lambda^{+}\Lambda$$

The function

$$h(\rho, t) = (\rho_t, M\rho_t)$$

with

$$(\rho_1, \rho_2) = \begin{cases} \int d\mu\ \rho_1\rho_2 & \text{for classical systems} \\ \mathrm{tr}\ \rho_1\rho_2 & \text{for quantum systems} \end{cases}$$

is a monotonically decreasing function.

The entropy observable M as well as the Λ transformation do not preserve the purity of states. Pure states, being phase points or wave functions lose their privileged position. This is also expressed by the non–locality of the Λ transformation: Λ–transformed states cannot be localized on phase points or on wave functions.

4. Kolmogorov Systems and Internal Time

Kolmogorov systems are highly unstable dynamical systems characterized by a stable and an unstable partition of the initial data into cells evolving asymmetrically in time. The cells of the stable or K-partition ξ are progressively refined, time going from past to future. In the far future, $t \to +\infty$, the cells are completely refined; $S_{+\infty}\xi$ is the finest point partition. In the far past, $t \to -\infty$, the partition approaches the coarsest one cell partition. Points in the same cell of ξ have a common future, but a different past. The unstable partition ξ^u has the reverse behavior, or the same behavior with respect to reversed time; i.e., from the future to the past. The trajectories of points in the same cell of ξ^u diverge from each other in the future, even though they are as close as we want in the past. The description in terms of partitions is a generalization of the stable and unstable manifolds of hyperbolic systems and arose from the study of completely non–deterministic stochastic processes. For K–systems the concept of trajectory is lost much before the Poincaré recurrence time, thus invalidating immediately the traditional argument against irreversibility.

The behavior of K–systems is illustrated in the Baker cascade on the unit square:

$$B(q, p) = \begin{cases} (2q, \frac{1}{2}p) & \text{for } q \in [0, \frac{1}{2}) \\ (2q - 1, \frac{1}{2}p + \frac{1}{2}) & \text{for } q \in [\frac{1}{2}, 1) \end{cases}$$

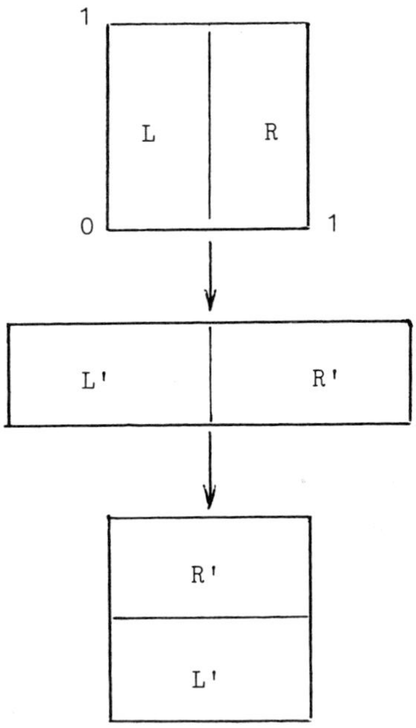

Figure 2: Baker's transformation. The first step squeezes the 1×1 square to a $2 \times 1/2$ rectangle. The second step cuts the $2 \times 1/2$ rectangle in halfs and puts the right half piece on top of the left one to produce a new 1×1 square.

As the name suggests, Baker's transformation resembles the kneading dough process as visualized in Fig.2.

The cells of the K–partition are the vertical lines of length one and contract in the future. The unstable K–partition consists of the horizontal lines which expand in the future. This behavior is shown in Fig. 3.

It is clear that it becomes increasingly more difficult to distinguish operationally points in the same K-cell as time goes on. Therefore the notion of a single trajectory ceases to be meaningful and only suitable averagings over the K-cells acquire meaning.

The averaging projections P_τ onto the cells of the partitions $S_\tau\xi$ (τ is the time parameter) are the spectral projectors of the self–adjoint internal time operator:

$$T = \int \tau dP_\tau$$

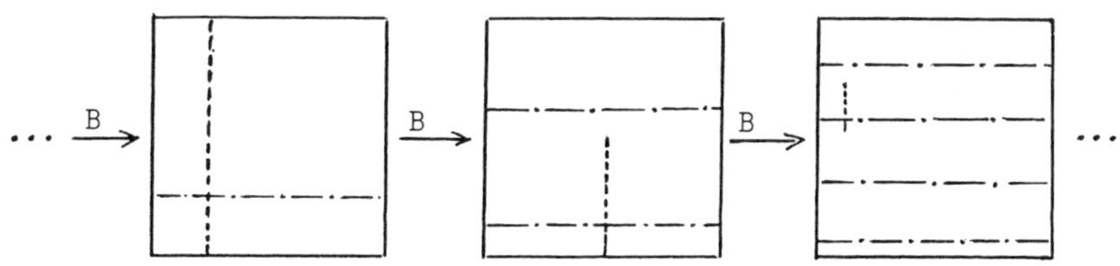

Figure 3: The contracting and expanding fibres of Baker's transformation are the cells of the K–partition and the unstable K–partition correspondingly.

The time operator is canonically conjugate to the Liouville generator L of the unitary evolution of the density functions:

$$\frac{1}{i}[L,T] = -I$$

$$U_t^+ T U_t = T + tI \qquad\qquad (U_t = e^{-iLt})$$

The expectation $(\rho, T\rho)$ is the age of the state ρ, keeping step with the external clock time t:

$$\begin{aligned}(U_t\rho, TU_t\rho) &= (\rho, U_t^+ T U_t \rho)\\ &= (\rho, T\rho) + t\end{aligned}$$

Since finite observers cannot follow *ad infinitum* the progressive refinements of the K–partition, the contributions of the higher age components acquire less and less operational significance as time goes on. Therefore the operationally attainable evolution corresponds to appropriately weighed densities

$$\tilde{\rho} = \Lambda\rho \qquad \text{with} \quad \Lambda = \Lambda(T) = \int \Lambda(\tau) dP_\tau$$

where $\Lambda(\tau)$ is a monotonically decreasing, positive, logarithmically concave function with $\Lambda(\tau) \to 0$ as $\tau \to \infty$ and $\Lambda(\tau) \to 1$ as $\tau \to -\infty$.

The non–unitary transformation Λ provides a weight over the precision characterising the progressively refined family of conditional expectation projections P_T, and converts the unitary group U_t into a Markov semigroup W_t approaching the equilibrium in the far future:

$$W_t = \Lambda U_t \Lambda^{-1}$$

An intertwining (but not invertible) transformation is also provided by any spectral projection $P = P_{t_o}$, $t_o > 0$, of the internal time operator. In this case the higher age components do not contribute to the observable evolution. The resulting Markov semigroup

$$W_t = PU_tP$$

is the projection of the group U_t onto the subspace of coarse–grained densities, $P\rho$.

In any case, the resulting Markov semigroup W_t is a best prediction for the unitary evolution U_t made by an observer who cannot follow precisely the progressive refinements of the K–partition *ad infinitum*.

The construction of the Markov semigroups from K–systems through the K–partitions is the essence of the Misra – Prigogine – Courbage theory of irreversibility [6, 9-11].

It has also to be remarked that the Markov semigroup resulting from the projection P_o is the Frobenius–Perron semigroup associated with the exact system corresponding to the K–system. It is well known [3, 1] that any K–system is the natural extension of an exact system. For example, the Baker transformation is the natural extension of the dyadic map mentioned in Sect.2.

Many physically interesting systems are known to be K–systems; the hard spheres in a box, the convex billiard, the Lorentz gas, geodesic flows in spaces with negative curvature, the infinite ideal gas and hard rods, shifts in certain Ising models, the infinite harmonic lattice, the Mixmaster cosmological model, classical relativistic fields.

5. Relativistic Systems with Internal Time

Any classical relativistic field equation, namely the scalar wave, the electromagnetic radiation, and the linear gravitational field in flat space–time, defines a Gaussian flow on the space of initial data or equivalently on the space of solutions, which is a Kolmogorov flow [12, 13].

The K–property arises from the relativistic transformation properties of the fields. An internal time operator can be constructed on the space of solutions or on the space of initial data as a function of the generators of the relativistic symmetry group. The spectral projections of the time operator allow us to test whether two initial data y_1, y_2 are in the same cell of the K–partition.

For the simple case of the wave equation

$$\partial_t^2 y = \Delta y$$

the internal time on the Hilbert space of square integrable solutions of the wave equation has the form

$$T_R = -\sum_{a=1}^{3} (N^a P^a |\vec{P}|^{-2} + |\vec{P}|^{-2} P^a N^a)$$

where P^a are the generators of space translations, and N^a are the generators of Lorentz boosts. The explicit form of the relativistic internal time is:

$$T_R = -x^o - x^a \partial_a \partial_o \Delta^{-1} - \frac{1}{2} \partial_o \Delta^{-1}$$

The presence of the inverse Laplace operator Δ^{-1} in the expression for T_R is expected, since the internal time, being a non–local operator, cannot generate field transformations implementable by point transformations in Minkowski space.

The internal time T_R is related by unitary equivalence to another time operator T_E:

$$T_R = |P^o|^{1/2} \, T_E \, |P^o|^{-1/2}$$

where P^o is the generator of the time evolution. The operator T_E is self–adjoint on the space of initial data with finite energy. The spectral projection P_o of T_E corresponds to the Lax–Phillips [14] incoming waves which vanish inside the past light cone. The spectral projections of T_R do not admit such a simple physical interpretation. However, these projections give the possibility to construct incoming subspaces algebraically, without invoking Huyghens' principle [13, 15].

Limits to predictability for the wave equation arise from the fact that future predictions at time $t > 0$ require knowledge of the field values at the points $|\vec{x}| = ct$ on the initial spacelike hypersurfaces as shown in Fig.4.

Therefore for long future prediction times, knowledge of the field on a large initial hypersurface is necessary. But the finiteness of human observations sets limits to the knowledge of the initial hypersurface and therefore to the control and prediction of the field in the far future.

The K–property of the relativistic fields implies that their evolution is asymptotically unstable. This is not in contradiction with their stability for finite times, expressed as continuous dependence of the solutions upon the initial data.

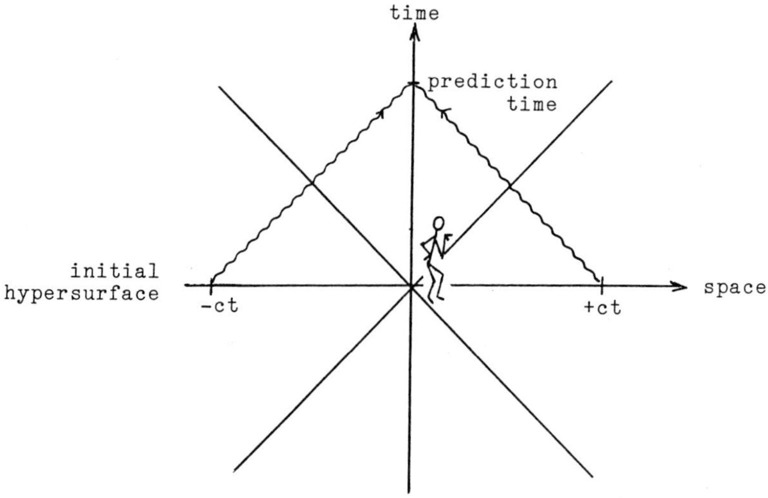

Figure 4: Relativistic limitations to predictability. Knowledge of the field values along the initial spacelike hypersurface is required to predict the future.

The relativistic fields provide concrete representations of the algebra generated by the relativistic Poincaré algebra and the internal time. The relativistic internal time algebra is an infinite dimensional Lie algebra [13, 15, 16]. The internal time commutes with the generator of rotations J_a, but does not commute with the generators P^a and N^a of space translations and Lorentz boosts, thus giving rise to the velocity V^a and to the internal position Q^a:

$$
\begin{aligned}
[P^o, T] &= -I \\
[P^a, T] &= -P^a P^o = -V^a \\
[J^a, T] &= 0 \\
[N^a, T] &= TV^a = Q^a
\end{aligned}
$$

The velocity and position generate the infinite monomials V^aV^b, $V^aV^bV^c, \ldots$, TV^aV^b, $TV^aV^bV^c, \ldots$:

$$
\begin{aligned}
[N^a, V^b] &= V^aV^b - \delta_{ab} I \\
[N^a Q^b] &= 2TV^aV^b - \delta_{ab} T \\
\ldots &= \ldots
\end{aligned}
$$

The relativistic internal time algebra allows to find the transformation properties of the internal time T and the internal position Q^a under Lorentz boosts:

$$
\begin{aligned}
(T)_b &= \frac{T\sqrt{1-\left(\frac{u^1}{c}\right)^2}}{1-\frac{u^1V^1}{c^2}} \\
(Q^1)_b &= \frac{Q^1 - TV^1}{\sqrt{1-\left(\frac{u^1}{c}\right)^2}} \frac{1}{\left(1-\frac{u^1V^1}{c^2}\right)^2}
\end{aligned}
$$

The formulae differ from Einstein's formulae and show clearly that the internal spacetime T, Q^a differs from the Minkowski spacetime of localized observations. The presence of a field endows space with new properties such as age and non–homogeneity. The field is like a landscape and allows a characterization of spacetime points in terms of the field's age.

The choice $\Lambda(T) = \exp(-kT)$ for the Λ–transformation maps the wave equation into the dissipative telegraph equation [13, 17]:

$$\partial_t^2 y = \Delta y - 2k\partial_t y - k^2 y$$

The telegraph equation describes waves propagating with exponential dissipation of the amplitude or it arises if we introduce finite propagation velocity into the heat equation [18].

6. Quantum Systems with Internal Time

Quantum K–systems are defined in an analogous way. The classical K–partition of the phase cells corresponds to a partition of the wavefunctions which becomes refined in the future. Quantum K–systems have been used in the measurement problem [19].

The time–energy uncertainty relation of quantum theory may arise from the existence of a time operator on the Liouville space of density operators [20]. It is well known since Pauli [21] that a time operator, canonically conjugate to the Hamiltonian, cannot exist on the Hilbert space of wavefunctions, because the canonical commutation relation implies that the spectrum of the Hamiltonian is not bounded from below.

The necessary and sufficient condition for the existence of a time operator is that the Liouville–von Neumann operator $L = [H, .]$ has absolutely continuous spectrum extending over the entire real line. This condition is also equivalent to the existence of an incoming subspace D with respect to the group $U_t = \exp(-iLt)$. D satisfies the conditions:

- $U_t D \supset D$ for $t > 0$.
- $\bigcup_{t \geq 0} U_t D$ is dense in the Liouville space.
- $\bigcap_{t<0} U_t D = 0$.

The subspace D is the range of the spectral projections P_o of the time operator. There have been attempts to interpret the subspace D as the subspace of decaying states of an unstable quantum system. The projected evolution $P_o U_t P_o$ describes the decay [22, 23].

7. Large Poincaré Non–Integrable Systems

In classical dynamical systems the presence of resonances destroys the analytic invariants as shown by Poincaré [24]. These are the classical Poincaré non–integrable systems [6], related to the problem of small denominators. The theorem of Kolmogorov, Arnold, and Moser [25] gives a bound on the magnitude of perturbations which "preserve" the phase portrait. However, for large systems (degrees of freedom $\to \infty$) where every point is resonant, we know nothing analogous to the KAM theorem.

Poincaré's theorem is not directly applicable to bounded quantum systems, since they have a discrete spectrum. But in the limit of large quantum systems with continuous spectrum, Poincaré's theorem acquires meaning analogous to large classical systems. Two possibilities arise [26] in the limit of large systems: either we have continuous sets of resonances, or we may neglect the role of resonances. The existence of continuous sets of resonances is equivalent to the non–vanishing of the asymptotic collision operator, which emerged in the study of the asymptotic behavior of the Liouville – von Neumann equation [27, 28].

Large Poincaré non – integrable systems are characterized by a non – vanishing asymptotic collision operator. These systems are very common in nature. The dynamical models of kinetic theory and Brownian motion, the interaction between matter and light, as well as interacting fields are typical examples.

The presence of resonances causes a breakdown of the perturbation methods which are analytic in the coupling parameter. However, a natural time ordering of the dynamical states leads to a well defined prescription for the propagators which lifts the divergence. Then a dynamical theory is obtained, which incorporates the second law of thermodynamics. The construction is based on an extension of the usual real eigenvalue problem to the complex plane. In simple cases like the Friedrichs model, the time ordering can be performed in an extended space of pure states. In general, however, we have to go to the Liouville space of densities, where the time ordering is introduced on the basis of the flow of correlations [29].

8. Finite Markov Coarse – Graining

A realization of dynamical systems as Markov chains has been recently proposed by G., J., C. Nicolis and M. Courbage [30-34] based on coarse–graining through suitable finite partitions $\xi = \{\Xi_k\}$ of the state space Y. The dynamical evolution S_t induces transitions between the cells Ξ_k such that the projected or coarse–grained process is Markovian.

The condition which guarantees the Markov property of the projected process

$$W_t = PU_tP$$

is the equality:

$$PU_tP = (PUP)_t \tag{$*$}$$

where U_t is the Frobenius–Perron operator associated with S_t (Sect.1), P is the conditional expectation projection over the cells Ξ_k of the partition ξ. The action of P on the densities ρ is:

$$P\rho = \sum_k \rho_k \delta_{\Xi_k}$$

where

$$\rho_k = \int_{\Xi_k} d\mu(y)\rho(y)$$

is the average density on the cell Ξ_k, and

$$\delta_{\Xi_k}(y) = \begin{cases} 1 & \text{for } y \in \Xi_k \\ 0 & \text{otherwise} \end{cases}$$

is the indicator of the cell Ξ_k.

In case of cascades (discrete time) the resulting Markov transition matrix is:

$$W = (W_{ik}), \qquad W_{ik} = \mu\{S^{-1}\Xi_i | \Xi_k\}$$

Partitions whose cells are not destroyed by the dynamics S^n are reasonable candidates for Markovian coarse–grainings. Every cell is transformed into a union of cells under S:

$$S\Xi_k \cap \Xi_i \neq \emptyset \;\Rightarrow\; S\Xi_k \supset \Xi_i \qquad (**)$$

Property $(**)$ guarantees that the partition ξ gives rise to exact Markov coarse–graining for any reversible cascade S^n. This can be readily seen as follows.

Every coarse–grained density $P\rho$ transforms under the Frobenius–Perron operator U into a coarse–grained density, therefore the Markovian condition $(*)$ is satisfied. The equality

$$UP = PUP$$

follows from the fact that condition $(**)$ implies $S\xi \subseteq \xi$. Indeed:

$$\begin{aligned} UP\rho(y) &= U\sum_k \rho_k \delta_{\Xi_k}(y) \\ &= \sum_k \rho_k \delta_{\Xi_k}(S^{-1}y) \\ &= \sum_k \rho_k \delta_{S\Xi_k}(y) \\ &= \sum_k \rho_k P\delta_{S\Xi_k}(y) \\ &= PUP\rho(y) \end{aligned}$$

Property $(**)$ is not sufficient to guarantee the Markov property $(*)$, if S is an endomorphism, i.e., irreversible. However, for piecewise linear maps of the unit interval, such as the dyadic map, and their equivalents, such as the logistic map, property $(**)$ guarantees [35, 36] the Markov property.

A 12–cell partition for the logistic map

$$Sy = 4y(1-y)$$

can be constructed from the inverse images of the fixed point $y_F = 0.75$ and the centre 0.5 (see Fig.5). This partition gives rise to an exact Markov chain with transition matrix:

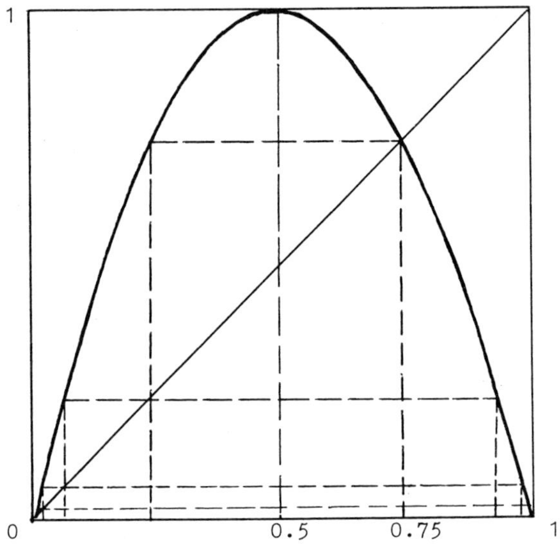

Figure 5: Construction principle of a 12–cell partition for the logistic map at $r = 4$.

$$W = (W_{ik}) = \begin{pmatrix} .513 & 0 & 0 & 0 & 0 & 0 & 0 & 0 & 0 & 0 & .483 \\ .252 & 0 & 0 & 0 & 0 & 0 & 0 & 0 & 0 & 0 & .271 \\ .235 & 0 & 0 & 0 & 0 & 0 & 0 & 0 & 0 & 0 & .246 \\ 0 & 1 & 0 & 0 & 0 & 0 & 0 & 0 & 0 & 1 & 0 \\ 0 & 0 & 1 & 0 & 0 & 0 & 0 & 0 & 1 & 0 & 0 \\ 0 & 0 & 0 & 1 & 0 & 0 & 0 & 1 & 0 & 0 & 0 \\ 0 & 0 & 0 & 0 & .259 & .249 & 0 & 0 & 0 & 0 & 0 \\ 0 & 0 & 0 & 0 & .262 & .252 & 0 & 0 & 0 & 0 & 0 \\ 0 & 0 & 0 & 0 & .113 & .136 & 0 & 0 & 0 & 0 & 0 \\ 0 & 0 & 0 & 0 & .128 & .120 & 0 & 0 & 0 & 0 & 0 \\ 0 & 0 & 0 & 0 & .238 & .243 & 0 & 0 & 0 & 0 & 0 \end{pmatrix}$$

The invariant measure

$$d\mu(y) = \frac{dy}{\pi\sqrt{y(1-y)}}$$

gives the unique equilibrium distribution of the 12–state Markov process:

$$\rho_{eq}(k) = \frac{1}{\pi}(.26, .13, .13, .26, .26, .52, .52, .26, .26, .13, .13, .26)$$

The transitions may be represented diagrammatically as shown in Fig.6.

9. "Music" from "Chaos"

It is tempting to listen to the "music" of dynamical systems considered to be "chaotic", such as the logistic map, through compatible partitions such as the one constructed in the previous section [37].

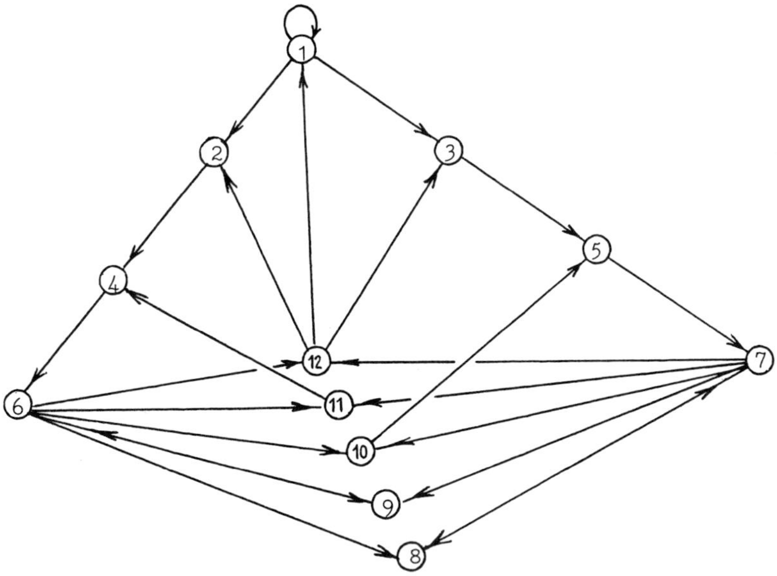

Figure 6: Diagram representing possible transitions according to the matrix (W_{ik}).

The idea is to map each cell to a tone, specify the relative duration of the tones to be analogous to the invariant measure of the corresponding cell, and then let the system go. In this way we have an acoustic representation of "chaos" and probably new means for musical synthesis. For example, we may listen to the musical piece defined by the correspondences listed in Table 1.

The investigation of the statistical properties of these musical pieces of dynamic origin allows a comparison with conventional music. For example, the mutual information for the above mentioned piece goes down to zero after a few iterations (Fig.7a) whereas Bach's partita No.1 for solo violin is rich in "innovations" (Fig.7b).

Table 1: Correspondences of cells, tones, and their duration in the acoustic representation of the logistic map. The numbers added in the tones column indicate the octave.

Cell	Tone	Relative Duration
1	Re 5	2
2	Sol 5	1
3	Sol 5	1
4	La^{+} 5	2
5	Mi 5	2
6	Re 6	4
7	Do 5	4
8	Re 5	2
9	Do^{+} 6	2
10	Mi 5	1
11	La 5	1
12	Re^{+} 4	2

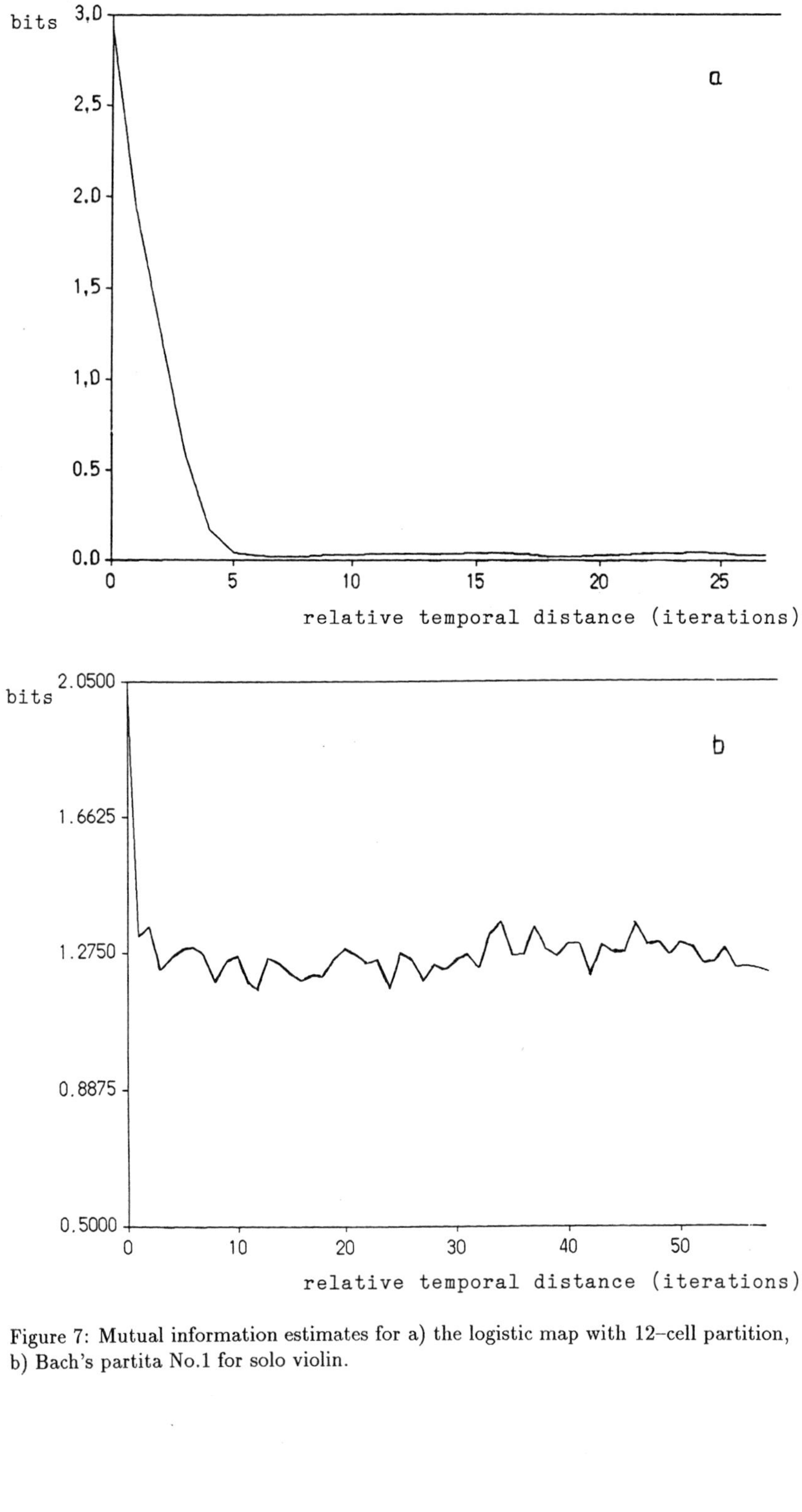

Figure 7: Mutual information estimates for a) the logistic map with 12–cell partition, b) Bach's partita No.1 for solo violin.

The mutual information is a generalization of the correlation function applicable to symbolic (not necessarily numerical) sequences, like a musical piece, or a linguistic text. Mutual information is the natural tool to detect correlations in symbolic sequences [38].

Acknowledgments

It is a pleasure to thank Professors I. Prigogine, B. Misra, G. Nicolis, J. Nicolis, M. Courbage, and T. Petrosky for several discussions which contributed to the clarification of delicate points.

The results in the last two sections were worked out in collaboration with A. Katsikas. The help of S. Pahaut, B. Saphir, and D. McKernan is also acknowledged.

The work reported here is supported by the Belgian Program on Interuniversity Attraction Poles and by the Instituts Internationaux de Physique et de Chimie fondés par E. Solvay.

References

1. Ya.B. Sinai (ed.). *Dynamical Systems*. Springer, Berlin, 1989.
2. A. Lasota and M. Mackey. *Probabilistic Properties of Deterministic Systems*. Cambridge University Press, Cambridge, 1985.
3. V. Rokhlin. *Am. Math. Soc. Trans.* **39**, *1 (1964)*.
4. K. Podgorski. In *Springer Lecture Notes in Mathematics*, Vol. 1391, p. 321. Springer, Berlin, 1989.
5. D. Ruelle. *Chaotic Evolution and Strange Attractors*. Cambridge University Press, Cambridge, 1989.
6. I. Prigogine. *From Being to Becoming*. Freeman, San Francisco, 1980.
7. B. Misra. *Proc. Ntl. Acad. Sci. USA* **75**, 1627 (1978).
8. B. Misra, I. Prigogine, and M. Courbage. *Proc. Ntl. Acad. Sci. USA* **76**, 4768 (1979).
9. B. Misra, I. Prigogine, and M. Courbage. *Physica* **98 A**, 1 (1979).
10. B. Misra and I. Prigogine. Time, probability, and dynamics. In *Long Time Prediction in Dynamics*. Eds. C.W. Horton, L.E. Reichl, and A.G. Szebehely. Wiley, New York, 1983.
11. R. Goodrich, K. Gustafson, and B. Misra. *J. Stat. Phys.* **43**, 317 (1986).
12. B. Misra. *J. Stat. Phys.* **48**, 1295 (1987).
13. I. Antoniou. Internal time and irreversibility of relativistic dynamical systems. Thesis, Free University Brussels, 1988.
14. P. Lax and R. Phillips. *Scattering Theory*. Academic Press, London, 1989.
15. I. Antoniou and B. Misra. Internal time for relativistic systems and its transformation laws. *J. Math. Phys.* To be published.
16. I. Antoniou and B. Misra. *Nucl. Phys. B (Proc. Supp.)* **6**, 240 (1989).
17. I. Antoniou and B. Misra. Non–unitary transformations of conservative to dissipative evolution. *J. Phys. A*. To be published.

18. D. Kelly. *Am. J. Phys.* **36**, 585 (1968).
19. C. Lockhart and B. Misra. *Physica* **136 A**, 47 (1986).
20. M. Courbage. *Lett. Math. Phys.* **4**, 425 (1980).
21. W. Pauli. In *Encyclopedia of Physics*, Vol. 5. Ed. S. Flügge. Springer, Berlin, 1958.
22. P. Exner. *Open Quantum Systems and Feynman Integrals.* Reidel, Dordrecht, 1985.
23. C. Lockhart. Time operators in classical and quantum systems. Thesis, University of Texas at Austin, 1981.
24. H. Poincaré. *Les Methodes Nouvelles de la Mecanique Celeste.* Reprinted by Dover, New York, 1957.
25. J. Moser. *Stable and Random Motions in Dynamical Systems.* Princeton University Press, Princeton, 1973.
26. T. Petrosky and I. Prigogine. *Physica* **147 A**, 459 (1988).
27. I. Prigogine. *Non Equilibrium Statistical Mechanics.* Wiley, New York, 1962.
28. I. Prigogine, C. George, F. Henin, and L. Rosenfeld. *Chem. Scr.* **4**, 5 (1973).
29. T. Petrosky, I. Prigogine, and S. Tasaki. Quantum theory of non–integrable systems. *Physica A.* To be published.
30. G. Nicolis and C. Nicolis. *Phys. Rev.* **A 38**, 427 (1988).
31. G. Nicolis, J. Nicolis, and C. Nicolis. *J. Stat. Phys.* **54**, 915 (1989).
32. M. Courbage and G. Nicolis. *Europhys. Lett.* **11**, 1 (1990).
33. G. Nicolis and C. Nicolis. *Physica* **163 A**, 215 (1990).
34. M. Courbage. This volume.
35. S. Grossmann and S. Thomae. *Z. Naturforsch.* **32 a**, 1353 (1977).
36. D. McKernan and G. Nicolis. Forthcoming publication.
37. I. Antoniou, A. Katsikas, S. Pahaut, and B. Saphir. Musique du Chaos. Communication at *Temps, Complexité et Oeuvres Musicales*, Les Treilles, July 1990.
38. W. Li. *J. Stat. Phys.* **60**, 823 (1990).

ON BOLTZMANN ENTROPY AND COARSE – GRAINING FOR CLASSICAL DYNAMICAL SYSTEMS

MAURICE COURBAGE

Laboratoire de Probabilités
Université Pierre et Marie Curie
CNRS
F–75252 Paris, France

Abstract

We discuss a possible construction of Markov evolutions with Boltzmann entropy, associated, through a coarse – graining, to deterministic systems which are not as strongly unstable as K – systems. Yet, this extension implies that the dynamical system should have an absolutely continuous component in its spectrum.

1. Introduction

Boltzmann entropy,

$$\int f(r,v,t)\ \log f(r,v,t)\ d^3r\ d^3v \tag{1}$$

is usually interpreted as the measure of disorder or the uncertainty of the statistical state of a system of molecules of a dilute gas evolving under the effect of binary collisions. This functional of the non – equilibrium state f has also been interpreted as expressing the microscopic analogue of the second law of thermodynamics. Yet, the concept of a measure of uncertainty was intuitive until the mathematical foundation of information theory defined the entropy of a random variable and a random process. It seems to us that the relation between these two notions of entropy can be settled owing to recent progress in dynamical system theory. This theory sheds new light also on the fundamental problem of the irreversibility of conservative dynamical systems.

Many results presented here have previously been obtained by Misra, Nicolis, Prigogine, and the author [1, 2, 3]. We discuss in some detail a possible extension to dynamical systems that are not Bernoulli systems or K – systems.

Information Dynamics, Edited by H. Atmanspacher and H. Scheingraber, Plenum Press, New York, 1991

2. Exact Master Equation and Coarse – Graining

Let us first consider an abstract classical dynamical system given by a flow S_t in phase space Γ which preserves a Liouville – like measure μ on Γ. The non – equilibrium statistical states that we shall consider here are given by probability densities $\rho_t(\omega), \omega \in \Gamma$, with respect to $d\mu$, and they evolve according to:

$$\rho_t(\omega) = \rho_o(S_{-t}\omega) \equiv (U_t\rho_o)(\omega) \tag{2}$$

U_t is the flow acting on initial densities ρ_o. It follows from the invariance of μ under the flow that the entropy

$$S(\rho_t) = -\int \rho_t(\omega)\ \log \rho_t(\omega)\ d\mu(\omega) \equiv -\Omega(\rho_t) \tag{3}$$

is constant. A well known method which leads to a monotonically decreasing entropy is to substitute for ρ_t a coarse – grained density defined as follows.

Let $\mathcal{P}$ be some partition of Γ into disjoint cells $\{P_i\}$, then the coarse – grained projection of ρ along $\mathcal{P}$ is a new density constant on each cell of $\mathcal{P}$ given by:

$$P\rho(\omega) = \frac{1}{\mu(P_i)} \int_{P_i} \rho(\omega')d\rho(\omega') \qquad \forall \omega \in P_i \tag{4}$$

If the partition $\mathcal{P}$ has an uncountable number of cells each with zero measure, then the cells are called fibers, and the coarse – grained projection is defined by the conditional expectation $E^{\mathcal{P}}(\rho)(\omega)$ which is well known in probability theory and generalizes (4). Such partitions are denoted as ξ and the corresponding coarse – graining is P_ξ.

Now we consider the coarse – grained evolution of an initial coarse – grained density given by $PU_tP\rho_o$ for $t > 0$. PU_tP is a family of contraction operators in L^2_μ preserving positivity. If this family forms a semi–group for $t > 0$ then [4] it is a Markov semi–group associated to a transition probability which satisfies the Chapman – Kolmogorov equation. This entails that the entropy functional $\Omega(PU_tP\rho)$ is monotonically decreasing. The point is to characterize the class of dynamical systems for which PU_tP is a semi–group. This condition is equivalently expressed in terms of the conditional probability $\mu(E|\mathcal{A}(\mathcal{P}))$ of a set E given the σ – algebra $\mathcal{A}(\mathcal{P})$ generated by $\mathcal{P}$; that is, the function:

$$P(t,\omega,E) = \mu\left(S_{-t}E|\mathcal{A}(\mathcal{P})\right)(\omega) \qquad (t \geq 0) \tag{5}$$

has to satisfy the Chapman – Kolmogorov equation:

$$\int_\Gamma P(t,\omega,d\omega')P(t',\omega',E) = P(t+t',\omega',E) \tag{6}$$

This result has been proven for partitions with a finite number of cells in Ref. [3]. The proof is straightforward in the general case by using the definition of conditional probability. Thus the Chapman – Kolmogorov condition (6) is a restricting condition which characterizes not only the partition but also the dynamical system.

So far we have discussed monotonicity. Now one of the characteristic features of irreversibility is the approach to equilibrium. This means that for any initial state, $\Omega(PU_tP\rho_o)$ tends to zero as $t \to \infty$. So we need that PU_tP verifies the semi–group property and the approach to equilibrium. In this case we say that P realizes a Markovian irreversible coarse – graining.

An important particular case in which a semi–group property is verified is the case in which there exists a partition ξ that becomes finer and finer under the dynamical flow. This happens in so–called K – systems which have been studied in a series of papers by Misra, Prigogine, the author, and others [1, 2, 5].

The existence of such a partition means that the σ – algebra $\mathcal{A}$ generated by ξ is invariant under S_t for $t > 0$. This immediately entails that:

$$U_{-t}P_\xi = P_\xi U_{-t} P_\xi \tag{7}$$

and provides the semi–group property,

$$\begin{aligned}(P_\xi U_{-t} P_\xi)(P_\xi U_{-t'} P_\xi) &= P_\xi U_{-t} U_{-t'} P_\xi \\ &= P_\xi U_{-(t+t')} P_\xi\end{aligned} \tag{8}$$

which is, by adjointness, equivalent to:

$$P_\xi U_t P_\xi U_{t'} P_\xi = P_\xi U_{t+t'} P_\xi$$

The strict monotonicity of the partition ξ under the flow implies that the partition has an uncountable number of cells each with zero measure.

Yet, a K – system is not the most general system admitting an irreversible Markovian coarse – graining. In fact, let us consider for simplicity, a discrete time system induced by a transformation S and a partition with a finite number of cells with positive measure. Then, the condition (6) reduces to the fact that the matrix

$$(W_t)_{ij} = \mu\left(S_{-t}P_j | P_i\right) \tag{9}$$

satisfies the Chapman – Kolmogorov equation:

$$(W_{t+t'})_{ij} = \sum_k (W_t)_{ik}(W_{t'})_{kj}$$

The irreversible approach to equilibrium is equivalent to say that the stochastic matrix W_t is irreducible and aperiodic (i.e., there exists $t > 0$ such that $(W_t)_{ij} > 0$ for any i, j).

In contrast to K – systems which have positive K-S-entropy, the class of systems of zero K-S-entropy having strongly irreversible Markovian coarse graining is not empty as can be seen from the example that will be given below. These systems necessarily have a spectral characterization: the unitary operator U_t acting on $\{1\}^\perp \subset L^2_\mu$ has an absolutely continuous component in its spectrum.

To see this, let us consider the case of a partition with two cells, P_o and P_1. Let χ_o be a normalized function taking a constant value α_r on each cell P_r and orthogonal to 1. Denote:

$$\chi_t(\omega) = U_t \chi_o(\omega), \qquad \mu(P_r) = p_r$$

Let ${}^1E$ denote the characteristic function of a measurable subset E. It is clear that ${}^1S_tP_r = p_r(1 + \alpha_r \chi_t)$. Then we have:

$$\begin{aligned}< {}^1S_tP_r, {}^1P_{r'} > &= \mu(S_tP_r \cap P_{r'}) \\ &= p_r\ (W_t)_{r,r'} \\ &=< p_r(1 + \alpha_r\chi_t),\ p_{r'}(1 + \alpha_{r'}\chi_o) > \\ &= p_r p_{r'}\,(1 + \alpha_r\alpha_{r'} < \chi_t, \chi_o >)\end{aligned}$$

and therefore:

$$< \chi_t, \chi_o >= \left[\frac{1}{p_{r'}}(W_t)_{r,r'} - 1\right] \frac{1}{\alpha_r \alpha_{r'}} \tag{10}$$

Now the matrix W_t is aperiodic and, according to a theorem on Markov chains [6], there exists λ with $|\lambda| < 1$ such that:

$$\left[(W_t)_{i,j} - p_j\right] = c_{i,j}\ \lambda^n \tag{11}$$

Let $E(x)$ be the spectral family of U, and $\overline{\mu}_t$ be the Fourier coefficients of the spectral measure associated with χ_o defined by:

$$\begin{aligned} \overline{\mu}_t &=< U_t \chi_o, \chi_o > \\ &= \int_0^{2\pi} \exp(itx)\ d < \chi_o, E(x)\chi_o > \\ &\equiv \int_0^{2\pi} \exp(itx)\ d\mu_{\chi_o}(x) \end{aligned}$$

Equations (10) and (11) imply that $\overline{\mu}_t$ tends to zero with an exponential rate. Thus the function

$$g(x) = \sum_{t=-\infty}^{\infty} \overline{\mu}_t\ \exp(-itx)$$

is the density of the measure $d\mu_{\chi_o}$ with respect to the Lebesgue measure (for any continuous function on the circle):

$$\int_0^{2\pi} \varphi(x) d\mu_{\chi_o}(x) = \int_0^{2\pi} \varphi(x) g(x) dx$$

Then, the existence of an exactly irreversible Markov coarse – graining necessarily entails some spectral instability. Yet this instability is not necessarily of the same type as arising in hyperbolic systems which are not generically Hamiltonian. The following example, introduced by Mathew and Nadkarni [7] in order to display spectral properties, turns out to be an example of a system with zero K-S-entropy having an irreversible Markov coarse–graining.

Consider first the von Neumann transformation τ on $[0,1]$ defined as follows: The interval $F_1 = [0, 1/2]$ is translated to the interval $[1/2, 1]$, the interval $F_2 = [1/2, 1/2 + (1/2)^2]$ is translated to the interval $[(1/2)^2, 1/2]$ etc. (see Figure 1 for a schematic illustration).

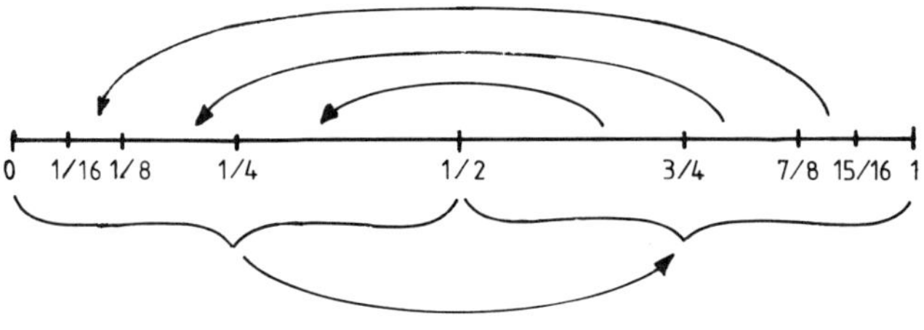

Figure 1: Schematic illustration of the von Neumann transformation on the interval $[0, 1]$.

Let F_{k1} (F_{k2}) be the first (second) half of F_k. Define the function:

$$\Phi(x) = \begin{cases} +1 & \text{if} \quad x \in F_{k1} \\ -1 & \text{if} \quad x \in F_{k2} \end{cases}$$

One can now define a new transformation T acting on a phase space $\Gamma = \{-1, 1\} \times [0, 1]$, equipped with a measure μ which is the product of the Lebesgue measure on the interval $[0, 1]$ and of the measure assigning probability 1/2 to each, –1 and +1. T is given by:

$$T(x, \sigma) = (\tau x, \Phi(x)\sigma), \qquad \sigma \in \{-1, +1\}$$

Let $\mathcal{P}$ be the partition of Γ into two cells P_o and P_1 defined by:

$$P_o = \{-1\} \times [0, 1/2[\cup \{1\} \times [1/2, 1]$$

$$P_1 = \{-1\} \times [1/2, 1] \cup \{1\} \times [0, 1/2[$$

Let $\chi(x)$ be the function

$$\chi(x) = {}^{1}F_1(x) - 1/2, \qquad x \in [0, 1]$$

To any function $f(x)$ defined onto $[0, 1]$ one associates its "odd extension" $\overline{f}(x, \sigma)$ onto Γ defined by:

$$\overline{f}(x, \sigma) = \sigma f(x) = \begin{cases} -f(x) & \text{if} \quad \sigma = -1 \\ +f(x) & \text{if} \quad \sigma = +1 \end{cases}$$

Clearly, χ is orthogonal to 1, and, for any function $\Psi(x)$, we have:

$$\begin{aligned} U\overline{\Psi}(x, \sigma) &= \Psi(\tau x, \Phi(x)\sigma) \\ &= \Phi(x)\sigma\Psi(\tau x) \\ &= \overline{\Phi(x)\Psi(\tau x)} \end{aligned}$$

If one introduces the operator V onto $L^2(0, 1)$ defined by

$$V\Phi(x) = \Phi(x)\Psi(\tau x)$$

then the above relation is written as:

$$U\overline{\Phi} = \overline{V\Phi} \tag{12}$$

Mathew and Nadkarni have proven that:

$$< V^n\chi, \chi >= 0 \qquad \forall n \neq 0$$

Then, using (12), we obtain

$$\begin{aligned} < U^n\overline{\chi}, \overline{\chi} > &=< \sigma V^n\chi, \sigma\chi > \\ &=< V^n\chi, \chi >= 0 \end{aligned}$$

and this implies that:

$$\mu\left(T^{-n}\mathcal{P}_i \cap \mathcal{P}_j\right) = 1/4 \qquad \forall n \neq 0$$

Hence, the Chapman – Kolmogorov equation is satisfied and the system is not a Bernoulli system.

This is not an isolated example of irreversible Markov coarse – graining which is not Bernoulli. If we consider the phase space $\{0,1\}^Z$ with shift, then the above example induces (by symbolic dynamics) an invariant measure μ_1 on this space satisfying the Chapman – Kolmogorov equation which, in this case, reduces to:

$$\mu_1(x_o = i, x_n = j) = \frac{1}{2^2} \tag{13}$$

This measure is different from the Bernoulli measure $\mu_0(1/2, 1/2)$. Then, owing to the convexity of the set of all invariant measures satisfying (13) (which is equivalent to the Chapman – Kolmogorov equation), any convex combination of μ_0 and μ_1 belongs to this set. Therefore, there is an uncountable number of measures belonging to this set and being different from the Bernoulli (1/2,1/2).

3. Remarks on the Relation to Information Theory

Let us come back to entropy $\Omega(PU_tP\rho)$. A simple calculation [3] shows that

$$\Omega(PU_tP\rho) = \sum \nu_t(P_i) \log \frac{\nu_t(P_i)}{\mu(P_i)} \tag{14}$$

where:

$$\nu_t(P_i) = \int_{P_i} \rho_t d\mu$$

There is an interesting relation between this entropy and information theory. In fact, the uncertainty of a random event α_i of probability p_i is equal to $(-\log p_i)$. Then, the variation of the uncertainty when the probability of the event α_i changes to q_i is equal to $\log \frac{p_i}{q_i}$, and the mean value of the random variable with respect to the distribution $\{p_i\}$ is equal to:

$$\Omega_q(p, \alpha) = \sum_i p_i \log \frac{p_i}{q_i}$$

This represents the gain or loss of information of the distribution $\{p_i\}$ with respect to $\{q_i\}$. Such a functional has been used by Kullback, Perez, and Dobrushin in the theory of transmission of information. The entropy (14) appears now as the gain of information of the non–stationary measure $\nu_t(P_i)$ with respect to equilibrium and with respect to the observation of $\mathcal{P}$ evolving under the deterministic flow.

The entropy $\Omega_q(p, \alpha)$ has some interesting properties which we list here without proof:

- $\Omega_q(p)$ is non–negative, it attains its minimum value 0 if and only if $p_i = q_i$ for any i. This entropy distinguishes between distributions.
- $\Omega_q(p) = +\infty$ if $\{p_i\}$ weighs positively some events α_i with vanishing q_i probability. In this case the gain of information is infinite and $\{p_i\}$ is not comparable to $\{q_i\}$.
- The maximum of $\Omega_q(p)$, for all p comparable to q, is attained by some distribution $\{p_i\}$ weighing only one event, that is by a deterministic distribution.

- The gain of information with respect to a composed observation $\alpha \vee \beta = \{(\alpha_i, \beta_i)\}$ is given by:

$$\Omega_q(p, \alpha \vee \beta) = \Omega_q(p, \alpha) + \Omega_q(p, \beta/\alpha)$$
$$\Omega_q(p, \beta/\alpha) = \sum p(\alpha_i)\, \Omega_q(p, \beta/\alpha_i)$$
$$\Omega_q(p, \beta/\alpha_i) = \sum_j p(\beta_j/\alpha_i) \log \frac{p(\beta_j/\alpha_i)}{q(\beta_j/\alpha_i)}$$

4. Conclusion

As a conclusion, we see that the Chapman – Kolmogorov condition implies a weaker form of irreversibility than K – systems do. This means that not all observables need to evolve to equilibrium under an irreversible Markov process but only some subspace of observables invariant under the flow U_t has to satisfy this condition. The unitary group of operators restricted to this subspace necessarily has an absolutely continuous part in its spectrum. This condition discriminates between intrinsically irreversible systems and others.

To give some insight into the meaning of the Chapman – Kolmogorov condition satisfied by the stochastic matrix W_t (9) we may again refer to the symbolic dynamics associated to the partition $\mathcal{P}$ and to Eq.(10). This condition imposes strong constraints onto the temporal correlations at times t_o and $t_o + \tau$ between the observables associated to the phase space partition $\mathcal{P}$. These correlations have to decay exponentially as $\tau \to \infty$.

Acknowledgements

I wish to thank J.P. Thouvenot for many fruitful discussions, especially concerning the example presented above, and also D. Hamdan for his contribution in studying several aspects of the problem.

References

1. B. Misra, I. Prigogine, and M. Courbage. *Physica* **98 A**, 1 (1979).
2. B. Misra and I. Prigogine. *Suppl. Prog. Theor. Phys.* **69**, 101 (1980).
3. M. Courbage and G. Nicolis. *Europhys. Lett.* **11**, 1 (1990).
4. K. Yosida. *Functional Analysis.* Springer, Berlin, 1965.
5. S. Goldstein, B. Misra, and M. Courbage. *J. Stat. Phys.* **25**, 111 (1980).
6. F.R. Gantmacher. *Théorie des Matrices*, Vol.2. Dunod, Paris, 1966.
7. J. Mathew and M.G. Nadkarni. *Bull. London Math. Soc.* **11**, 402 (1984).

ALGORITHMIC INFORMATION CONTENT, CHURCH - TURING THESIS, PHYSICAL ENTROPY, AND MAXWELL'S DEMON

Wojciech H. Zurek

Theoretical Division
Los Alamos National Laboratory
Los Alamos, NM 87545, USA

Abstract

Measurements convert alternative possibilities of its potential outcomes into the definiteness of the "record" — data describing the actual outcome. The resulting decrease of statistical entropy has been, since the inception of Maxwell's demon, regarded as a threat to the second law of thermodynamics. For, when the statistical entropy is employed as the measure of the useful work which can be extracted from the system, its decrease by the information gathering actions of the observer would lead one to believe that, at least from the observer's viewpoint, the second law can be violated.

I show that the decrease of ignorance does not necessarily lead to the lowering of disorder of the measured physical system. Measurements can only convert uncertainty (quantified by the statistical entropy) into randomness of the outcome (given by the algorithmic information content of the data). The ability to extract useful work is measured by *physical entropy*, which is equal to the sum of these two measures of disorder. So defined physical entropy is, on the average, constant in course of the measurements carried out by the observer on an equilibrium system.

1. Introduction

Algorithmic information content (also known as algorithmic randomness) of a physical entity is given by the size, in bits, of the most concise message (e.g., of the shortest program for a universal computer) which describes that entity with the requisite accuracy. Regular systems can be specified by means of concise descriptions. Therefore, algorithmic information content can be regarded as a measure of disorder.

Algorithmic randomness is defined without a recourse to probabilities. It provides an alternative to the usual ensemble measures of disorder: it quantifies randomness

Information Dynamics, Edited by H. Atmanspacher and H. Scheingraber, Plenum Press, New York, 1991

of the known features of the state of the physical system. I shall demonstrate that it is indispensable in formulating thermodynamics from the viewpoint of the *information gathering and using system* ("IGUS")—a Maxwell's demon-like entity capable of performing measurements and of modifying its strategies (for example, for extraction of useful work) on the basis of the outcomes of the measurements. Such an IGUS can be regarded as a "complex adaptive system." The aim of this paper is to review the concept of the algorithmic information content in the context of statistical mechanics and discuss its recently discovered physical applications.

2. Overview

Algorithmic randomness, an alternative measure of the information capacity of a specific physical or mathematical object, was independently introduced in the mid-60's by Solomonoff [1], Kolmogorov [2], and Chaitin [3]. It is based on an intuitively appealing idea that the information content is equal to the size, in bits, of the shortest description. Formalization of this idea will be briefly described in the next section: In its developement, it draws on the theory of algorithms and in the process makes use of the theory of computation [4, 5], establishes a firm and useful connection with Shannon's theory of information [6, 7], and benefits from its implications for coding [8].

Applications of the algorithmic measures of the information content were initially mostly mathematical in nature. More recently, Bennett, in an influential paper [9], has pointed out that the average algorithmic entropy of a thermodynamic ensemble has the same value as its statistical (ensemble) entropy and, consequently, one could attempt to build a consistent thermodynamics on an algorithmic foundation.

I have applied algorithmic randomness to the problem of measurement as seen by an observer [10-12].

Following a measurement, an observer (IGUS) is in possession of a specific record. From its intrinsic point of view, this record is quite definite. Therefore, further analysis of the measurement in terms of the ensemble language is pointless. Rather, the observer must deal with the specific measurement outcomes and with their implications for extraction of useful work. In this "Maxwell's demon" context, algorithmic randomness assumes, in part, a function analogous to the Boltzmann-Gibbs-Shannon entropy:
From the observer's point of view, the second law of thermodynamics must be formulated by taking into account both the remaining ignorance (measurements are typically far from exhaustive) and the randomness in the already available data. Thus, the *physical entropy*, the quantity which allows for the formulation of thermodynamics from the viewpoint of the observer, must consist of two contributions:

$$\begin{pmatrix}\text{physical}\\ \text{entropy}\end{pmatrix} = \begin{pmatrix}\text{remaining}\\ \text{ignorance}\end{pmatrix} + \begin{pmatrix}\text{algorithmic}\\ \text{randomness}\end{pmatrix}$$

or

$$\mathcal{S}(\rho) = H(\rho) + K(\rho) \tag{2.1}$$

where ρ is the density matrix of the system;

$$H(\rho) = -Tr\,\rho \log_2 \rho \tag{2.2}$$

is its BGS entropy, and $K(\rho)$ is the size of the shortest program capable of describing ρ:

$$K(\rho) = | p_\rho^* | \tag{2.3}$$

This recent proposal for physical entropy will be described in more detail in Sections 3 and 4.

In Section 4, I shall also discuss the importance of "compression" of the acquired data to their most concise form: thermodynamic efficiency of an IGUS-operated engine depends on its ability to find concise measurement descriptions. In turn, this efficiency can be regarded as a consequence of the IGUS's ability to "understand" or "model" the part of the Universe employed as the "engine" in terms of the regularities which can be regarded as analogs of physical laws. In this sense, "intellectual capabilities" of an IGUS are quite critical for its "success."

3. Algorithmic Randomness as a Measure of Physical Disorder

Consider a one-dimensional lattice of spins. I shall suppose that each spin can point either up (and designate this by "1") or down ("0"). A sequence of 0's and 1's with the number of digits equal to the number of spins in the system can be then regarded as a "description" of its state. Consider two such descriptions:

$$01010101010101010101 \tag{3.1}$$

and

$$10110100101100010111 \tag{3.2}$$

The first system is "regular": It can be simply and concisely described as 10 "01's." There is no equally concise description of the second spin system. To reconstruct it, one would have to have a "verbatim" description (Eq. 3.2); there is no way to "compactify" this description into a more concise message.

The concept of ***algorithmic information content*** (known also as the ***algorithmic randomness***, ***algorithmic complexity***, or ***algorithmic entropy***) captures this intuitive difference between the "regular" and "random" binary sequences.

Algorithmic information content of a binary sequence s is defined as the size of the minimal program, $s_{\mathbf{U}}^*$, which computes sequence s on the universal computer $\mathbf{U}$:

$$K_{\mathbf{U}}(s) \equiv | s_{\mathbf{U}}^* | \tag{3.3}$$

Above, vertical lines indicate the size of the binary string in bits. It is important to note that this definition of the algorithmic information content makes it explicitly subjective in the sense that it is computer dependent; hence, the subscript $\mathbf{U}$ in Eq. (3.3). However, by the very definition of the universal computer, programs executable on $\mathbf{U}$ can be also executed (and will yield the same output) on any other universal computer $\mathbf{U}'$, provided that they are preceded by a prefix $\tau_{\mathbf{UU}'}$ which depends on $\mathbf{U}$ and $\mathbf{U}'$, but not on the program. Hence, algorithmic randomness defined for two different computers will differ by at most the size of the prefix[1]:

$$| K_{\mathbf{U}}(s) - K_{\mathbf{U}'}(s) | \leq | \tau_{UU'} | \tag{3.4}$$

[1]Note that here we have used the vertical lines in two different ways: on the left-hand side they stand for the absolute value while on the right-hand they indicate the size of the binary string in bits. Only this second meaning will be employed below.

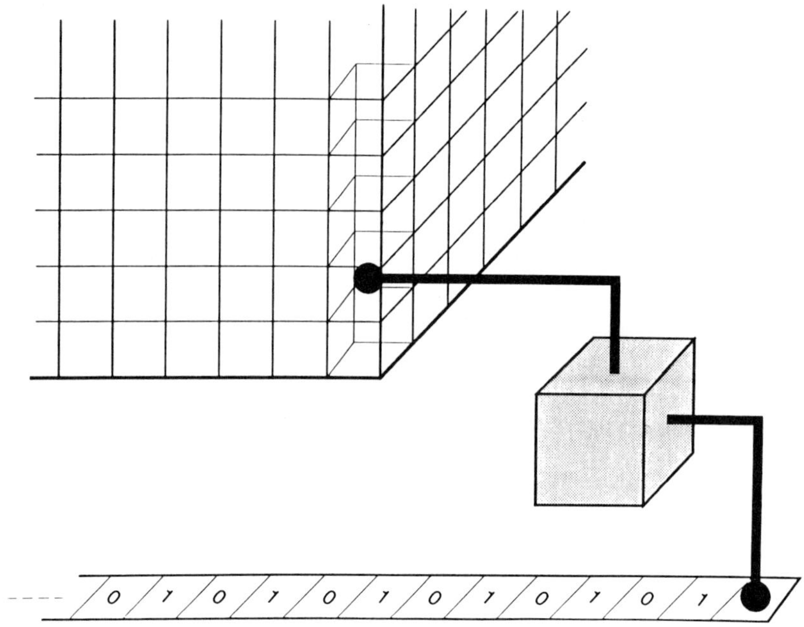

Figure 1: Turing machine **T** uses a set of instructions residing inside its "central unit" as well as the input it reads in by means of the "head" scanning the input tape to modify the content of the tape. A Universal Turing machine **U** can simulate any other Turing machine by reading on the input tape the "description" of **T**. In particular, a single-tape **U** can simulate operations of the modern computers, which can be modelled as "polycephalic" Turing machines with access to several multidimensional tapes and other kinds of hardware. Such "modern" machines (one of which is illustrated above) may be more convenient to use, but their capabilities are limited to the same range of "computable" tasks as for the original, one-tape **U**. This universality justifies the importance attached to the universal computers. In particular, it limits the subjectivity of the algorithmic information content defined by means of the minimal program, Eq. (3.3).

Such differences by a constant are usually ignored in mathematically motivated considerations, as they become relatively unimportant in the limit of very long strings. Moreover, as we shall see in Section 4, differences of algorithmic complexities (rather than their absolute value) are relevant in computations of engine efficiencies. This further reduces—but does not eliminate—the thermodynamic importance of the subjectivity.

It is worth noting that a binary string, or a physical system, which appears to be random at first sight may nevertheless have a concise description. Binary representations of π, $\sqrt{2}$, etc. are examples of such apparently random but, in fact, algorithmically regular binary strings. This difficulty in assessing a system's algorithmic complexity is related to Gödel's undecidability [3, 10, 13-14]. In spite of it, the distribution of the algorithmic randomness in physically or mathematically constructed ensembles, its value for whole classes of strings, and many of its more paradoxical (and intriguing) properties can be discussed in a rigorous manner. In particular, one can demonstrate that only exponentially few binary strings are algorithmically simple [3, 13].

The distribution of the number of strings of a fixed length as a function of their algorithmic randomness is given by:

$$\mathcal{N}(K) \simeq 2^K \tag{3.5}$$

This implies that "cryptoregular" strings (i.e., algorithmically simple but "hard to decipher" binary sequences like those associated with π, $\sqrt{2}$, etc.) are rare and a typical binary string s is algorithmically random and cannot be compressed:

$$K(s) \simeq | s | \tag{3.6}$$

In particular, a typical natural number n is algorithmically random. Hence, to the leading order:

$$K(n) \simeq \log_2 n \tag{3.7}$$

As in the "usual" information theory, one can define algorithmic randomness of sets of strings. For example, *joint algorithmic randomness* of a pair (s, t) is given by the minimal program which outputs s followed by t on the output tape. $K(s, t)$ satisfies the commuting equality

$$K(s,t) \cong K(t,s) \tag{3.8}$$

as well as the usual inequality:

$$K(s,t) \lessapprox K(s) + K(t) \tag{3.9}$$

In the equations above, I have used "approximate" ($\cong$, $\lessapprox$) rather than "exact" ($=$, $\leq$) signs to describe relationships; this is because most of the equations of the algorithmic information theory are subject to "O(1)" corrections which, in most cases, can be traced to the subjectivity in the definition of the algorithmic information content associated with the choice of the universal computer **U**.

Conditional algorithmic information content $K(s|t)$ is given by the size of the smallest program $q^*_{s|t}$ which computes string s from t (or, equivalently, both s and t from t). By analogy with Shannon's information theory, one would expect a relation:

$$K(s|t) = K(s,t) - K(t) \tag{3.10}$$

This equality does indeed hold, but only with logarithmic corrections.

It is convenient, for the reasons which will soon become apparent, to impose certain requirements on the properties of minimal programs. In particular, it is convenient to demand that they be *self-delimiting;* that is, that they must contain within them the information about the size of the input—their own size. This demand brings algorithmic information theory into a close correspondence with coding theory [8] where such codes are known as *prefix* or *prefix-free* codes and are used to guarantee the unique decodability of messages: It can be demonstrated that every uniquely decodable encoding can also be made self-delimiting without changing sizes of the words (programs) associated with different messages (outputs). This correspondence between communication theory and computation (which can be regarded as "decoding" of programs by means of universal computers) puts the powerful mathematical formalism of the coding theory at the disposal of the algorithmic information theory.

The most important consequence of the requirement of self-delimiting minimal programs is the ability to associate binary sequences with the probability that they will be generated by a randomly obtained program for the universal computer **U**: any

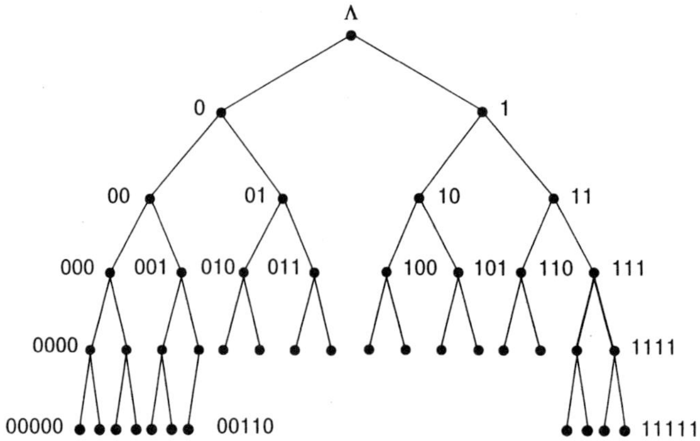

Figure 2: Lexicographic tree establishes a correspondence between binary strings and natural numbers. An algorithm which computes a certain string s from the input can be regarded as a Turing machine which "decodes" the input. The prescription for decoding is contained in the description of the computer.

random sequence of bits obtained, for instance, by coin flipping, can be used as an input to the computer **U**. To be a "legal" input program i, this bit sequence must be self-delimiting which, in practice, means that **U** will stop reading additional bits—stop asking for further flips of the coin—after a finite number of bits were supplied. Hence, the probability of generating this sequence is equal to:

$$p(i) = 2^{-|i|} \tag{3.11}$$

Probability associated with the output o will be given by the sum over all inputs which yields the given output o:

$$p(o) = \sum_{\{i|U(i)=o\}} 2^{-|i_o|} \tag{3.12}$$

Dominant contribution in this sum will come from the shortest program. Hence:

$$K(o) \simeq \log_2 p(o) \tag{3.13}$$

This connection between the probability that a given string will be generated by a random input and its algorithmic information content can be employed in proving that the average algorithmic randomness of a member of a simply described ("thermodynamic") ensemble is almost identical to its Boltzmann-Gibbs-Shannon entropy [9, 11, 16]. The relevant double inequality

$$H(\{p(s_i)\}) \leq \sum p(s_i)K(s_i) < H(\{p(s_i)\}) + K(\{s_i, p(s_i)\}) + O(1) \tag{3.14}$$

has been demonstrated by Levin [17, 18] and Chaitin [3] (see, also, Bennett [19] for a more accessible presentation and Caves [16] for a discussion in the physical context). Above, $K(s_i, \{p(s_i)\})$ is the size of the minimal description of the ensemble. Bennett

[9] has pointed out that, in the case when the BGS entropy is large compared with the size of the ensemble description

$$H(\{p(s_i)\}) = -\sum_{\{s_i\}} p(s_i) \log_2 p(s_i) \gg K(s_i, \{p(s_i)\}) \tag{3.15}$$

one could base thermodynamic formalism on the average algorithmic randomness of the ensemble.

In a recent paper I have considered an application of algorithmic randomness to the situation in which an observer attempts to extract maximum useful work from the system on the basis of partial measurements. In the next section I shall discuss this situation which forces one to consider *physical entropy* defined as the sum of the remaining ignorance $H(\{p(s_i)\})$ and of the cost of storage of the available information $K(s_i, \{p(s_i)\})$.

The last quantity which can be defined in the algorithmic context is the algorithmic information distance given by the sum of the conditional information contents:

$$\Delta(s,t) \equiv K(s|t) + K(t|s) \tag{3.16}$$

Algorithmic information distance satisfies the requirements expected of a metric [10].

In addition to the "simple" distance defined by Eq. (3.16), one can consider several related quantities. For example,

$$K(s!t!u) \equiv K(s|t,u) + K(t|u,s) + K(u|s,t) \tag{3.17}$$

is also positive, reflexive, and satisfies the obvious generalization of triangle inequality. Hence, $K(s!t!u)$ and its further generalizations involving more strings can be regarded as direct extensions of $\Delta(s,t) = K(s!t)$.

It is sometimes useful to express distance as the difference between the joint and mutual information content

$$\Delta'(s,t) = K(s,t) - K(s:t) \tag{3.18}$$

where the mutual information is given by:

$$K(s:t) = K(s,t) - (K(s) + K(t)) \tag{3.19}$$

The quantity Δ' defined by Eq. (3.18) differs from the "original" distance in Eq. (3.16) by logarithmic terms because of the similar logarithmic errors entering into Eq. (3.10). The advantage of employing Eq. (3.18) is its intuitive appeal: The distance between two binary strings is the information which they contain but do not share.

Mutual information can also be used to define algorithmic independence of two strings: s and t are independent when $K(s:t)$ is small; for example:

$$K(s:t) \ll \min(K(s), K(t))$$

Information distance can also be defined for statistical (that is, BGS) entropy. In this case, Δ and Δ' coincide. Indeed, information distance was independently discovered in the domain of Shannon's information theory by at least three authors before it was discussed (again without the benefit of knowledge of these references) by this author [10] in the algorithmic context.

4. Maxwell's Demon as An Adaptive Complex System

4.1. *Maxwell's Demon and the Church–Turing Thesis*

Smoluchowski [20] (see also Feynman [21] for an accessible up-to-date treatment) ruled out the possibility that a simple automated device (such as a trapdoor) can be used to violate the second law of thermodynamics; he has shown that the thermal fluctuations which inevitably occur whenever a trapdoor is coupled to the heat reservoir make it "jitter", which in turn makes the selection process it is supposed to implement completely unreliable. Thus, it cannot be sufficiently "selective" about its actions to create and maintain temperature or pressure differences in a closed system. Smoluchowski has, however, raised the question of whether an "intelligent being" employed as a "demon" could be more successful than the simple trapdoor. Until now, no definite answer to this question could be given. (Although Szilard [22] did write a celebrated and influential article, the title of which suggested that it would address the question of the second law and "intelligent beings." While intelligent beings never really enter into Szilard's considerations, it raises the issue of the relationship between thermodynamic entropy and information, and I shall make extensive use of its approach below.)

The reason for this state of affairs is easy to explain. After all, how can one analyze actions of an "intelligent being" within the realm of physics? In the first place, the very idea sounds "sacrilegious." More importantly, a physically sensible definition of what is an "intelligent being" was not easy to come by, especially in the times of Szilard.

In the intervening years, a compelling mathematical definition of an intelligent being became available. It is suggested by the so-called Church's thesis (or the Church-Turing thesis, or the Church-Tarski-Turing thesis, see Hofstadter [23] and Penrose [24] for an accessible discussion). In simple terms, it states that "What is human computable is universal computer computable." Thus, it equates information-processing capabilities of a human being (or of any intelligent being) with the "intellectual capacities" of a universal Turing machine. This may appear to be a gross oversimplification. Indeed, Church's thesis (CT) might be misunderstood as a claim that "all brains are created equal." For the time being, we shall only note that no counterexamples to CT have been discovered (but see Penrose [24] for a critical assessment of the extent to which computers may be able to simulate "consciousness"). Moreover, while with Penrose and others one might be concerned about the ability to simulate consciousness or the validity of CT for the "right-hemisphere" functions, it is harder to find a reason to question CT with regard to the logical and mathematical operations usually associated with the "left hemisphere" of the brain. Below, in the thermodynamic considerations, I shall need only the uncontroversial "left hemisphere" part of the thesis.

4.2. *Szilard's Engine and the Cost of Erasure*

The connection between the mathematical model of an "intelligent being" and thermodynamics goes back to the above-mentioned paper by Szilard [22]. In the analysis of the famous one-gas-particle engine (Figure 3), Szilard concluded that the second law could be indeed violated by a fairly simple "demon" unless the cost of measurements is no less than $k_B T$ per bit of acquired information. Further, essential

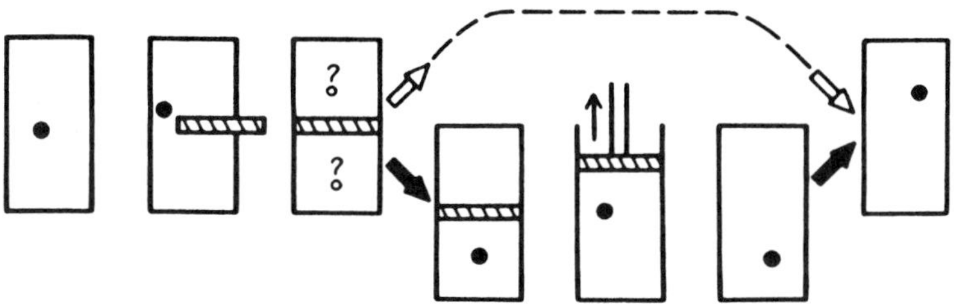

Figure 3: Szilard's engine employs one-molecule gas in contact with a heat bath at temperature T to extract $\sim T \ln 2$ of work per cycle (which is illustrated in a self-explanatory manner above). The measurement which establishes the location of the molecule is crucial. The importance of the cost of erasure for the proper accounting for the net energy gain is discussed in the text.

clarification of the situation is due to the recent work by Bennett [9, 25] who, basing his discussion on the earlier considerations of Landauer [26, 27] on the costs of information erasure, concluded that it is the "resetting" of the measuring apparatus which is thermodynamically expensive and must be responsible for restoring the validity of the second law in Szilard's engine. (Indeed, this observation was anticipated, if only in a somewhat half-hearted manner, by Szilard [22].)

Algorithmic randomness proved essential in attempts to generalize this discussion of Maxwell's demon [10-12, 16]. The validity of the original experiment about the cost of erasure was limited to the context of Szilard's engine. In that case, the outcome of the measurement can always be described by a single bit. Hence, the gain of useful work in the course of the expansion is given by:

$$\Delta W^{+} = k_B T \log_2 \left[\frac{V}{V/2} \right] = k_B T \tag{4.1}$$

Note that above we are using Boltzmann's constant k_B which differs from the usual one by a factor of $\ln 2$. This distinction reflects the difference between entropy measured in "bits" and "nats."

This gain of useful work is "paid for" by ΔW^{-} of the energy needed to restore the memory part of the "brain" of the "demon" to the "blank," "ready to measure" state:

$$\Delta W^{-} = -k_B T \tag{4.2}$$

Therefore, the net gain of work is at best nil:

$$\Delta W = \Delta W^{+} + \Delta W^{-} = 0 \tag{4.3}$$

It is, nevertheless, far from clear how to apply this "cost of erasure" argument to less idealized and more realistic situations.

One simple (although not very realistic) generalization is to consider a sequence of measurements on the Szilard's engine and to postpone the "erasure" indefinitely. This requires a demon with a significant memory size. One can then, as noted by Bennett [9, 25], use Szilard's engine to extract $k_B T$ of work per cycle as long as there is "empty" tape. This is, of course, only an apparent violation of the second law since the empty tape can be regarded as a zero-entropy (and, hence, zero-temperature) reservoir. Consequently, an ideally efficient engine can, in accord with the second law and, in particular, with the Carnot efficiency formula, attain exactly $k_B T$ of work per cycle.

The cost of erasure does not have to be paid for as long as the "memory tape" is available. However, for this very reason, the process is not truly cyclic: the demon's memory is never restored to the initial "blank" state. The gain of useful work is paid for by the "clutter" in its "brain." If the outcomes of consecutive measurements are random, getting rid of this clutter would cost $k_B T$ per bit, and all the apparent gain of work would have to be "paid back" by the final costs of erasure.

4.3. *Compressibility of Information: Second Law and Coding Theory*

Consecutive "bits" in the demon's memory may have been inscribed with 0's and 1's in a regular fashion. For example, a situation of this kind would automatically arise in the "multicylinder" Szilard engine shown in Figure 4. There the recording can be made by simply writing "0" or "1" on the squares of the tape corresponding to the empty or full cylinders. The gain of useful work extractable from the n-cylinder engine in contact with the heat bath at temperature T is:

$$\Delta W^{+} = k_B T \log_2 n \tag{4.4}$$

Each measurement results in filling up n blanks of the tape with 0's and 1's. Hence, the cost of erasure would be:

$$\Delta W^{-} = k_B T \cdot n \tag{4.5}$$

Again, one could postpone erasures indefinitely and just "dump" all of the "cluttered-up" tape into the garbage can. In the final count, however, the cost of erasure (linear in n) would outweigh the gain of useful work (which is logarithmic in n).

A more efficient multicylinder engine would be achieved by a fairly obvious but more economical encoding. Instead of making a "direct" record of the cylinders' content, one could "compress" such "verbose" messages and keep in the memory only the label of the "occupied" engine cell. It is now easy to verify that both ΔW^{+} and ΔW^{-} are logarithmic in n (providing that one insists, as one must, on the unique decodability of the label). Hence, a computerized "demon" employing this second, more economical way of encoding measurement results would be able to come much closer to the "break even point" than was the case for the operating system of the multicylinder Szilard engine considered above.

The general question raised by this example concerns the "ultimate compressibility" of a set of measurement outcomes. In particular, in the context of the example considered before, one could imagine that there is a still more efficient encoding of the alternative outcomes which results in a still more concise average size of the record. Then the demon would be able to do more than just break even, as the cost of erasure would be only partially offset by the gain of useful work. In this sense, the ability to compress records representing measurement outcomes determines the efficiency of a thermodynamic process whenever information acquisition and processing are involved. Indeed, the second law could be violated unless there is a limit on the compressibility which relates the average size $\langle \Delta K \rangle$ of the record with the decreased statistical entropy ΔH of the measured system via an inequality:

$$\Delta H \leq \langle \Delta K \rangle \tag{4.6}$$

For, unless this inequality holds, the gain of useful work which is equal to

$$\Delta W^{+} = k_B T \Delta H \tag{4.7}$$

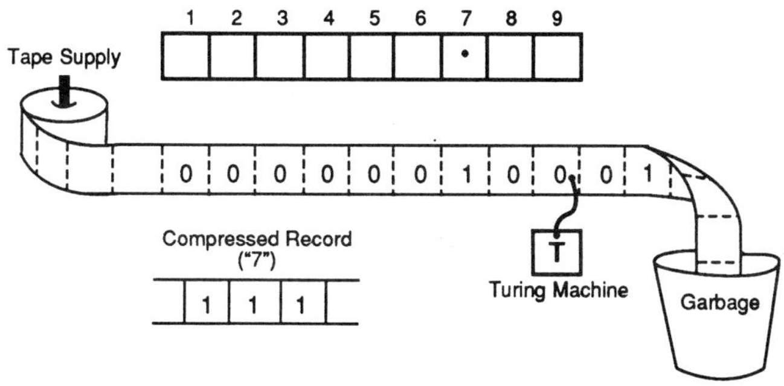

Figure 4: "Multicylinder" Szilard's engine. A single gas molecule is enclosed in a cylinder with n partitions, each of them with the same volume ΔV. The engine cycle begins with a measurement to determine which of the positions contains the molecule. Expansion following a measurement yields $\sim \ln(V/\Delta V)$ of useful work per cycle. The cost of erasure and, consequently, the efficiency of the engine depends on the manner in which the information is recorded. The most concise record (the size of which is given by the algorithmic information content) will result in the maximum efficiency. Thus, algorithmic information content inescapably enters the "demon's own" formulation of thermodynamics.

could exceed the cost of erasure which is given by:

$$\langle \Delta W^- \rangle = -k_B T \langle \Delta K \rangle \tag{4.8}$$

Hence, the net average gain of useful energy per cycle would be:

$$\langle \Delta W \rangle = k_B T (\Delta H - \langle \Delta K \rangle) \tag{4.9}$$

The second law demands that $\langle \Delta W \rangle \leq 0$, which leads to the inequality (4.6).

Fortunately, this inequality is indeed respected: It is an immediate consequence of the left-hand side of the inequality in Eq. (3.14). Indeed, it follows from the first basic result of Shannon's theory of communication (the so-called noiseless channel coding theorem; see Shannon and Weaver [7], Khintchin [6], Hamming [8], and Caves [16] for discussion): The average size of minimal "descriptions" needed to unambiguously describe measurement outcomes cannot be made smaller than the statistical entropy of the "source" of information (in our case, of the measured physical system). In this context, the second law can be regarded as a direct consequence of the Kraft inequality [8] which plays a basic role in the coding theory and thus enters physics [10, 11]: Suppose that $\{K_i\}$ are sizes (in the number of bits) of distinct symbols (programs) $\{s_i\}$ which correspond to different signals (measurement outcomes). Then one can prove that in order for the encoding to be uniquely decodable, the following inequality must be obeyed:

$$\sum_i 2^{-K_i} \leq 1 \tag{4.10}$$

The inequality (4.6) follows from Kraft inequality (4.10) since it can be immediately rewritten as

$$\log_2 \sum_i \left(\frac{p_i(s_i) 2^{-K_i}}{p(s_i)} \right) \leq 0$$

where $p(s_i)$ are the probabilities corresponding to signals (states) s_i. Now, employing convexity of the logarithm, one can write

$$\sum_i p(s_i) \log \frac{1}{p(s_i)} \leq \sum_i p(s_i) K_i \tag{4.11}$$

which establishes the desired result, Eq. (4.6).

Thus, the most efficient encoding corresponding to entropy decrease (information gain) ΔH will have to fill up, on the average, at least $\langle \Delta K \rangle \geq \Delta H$ bits in the memory of the "demon."

The second law cannot be violated because the average size of the record cannot be compressed to less than the corresponding gain of information. The success of an intelligent Maxwell's demon is ruled out as a consequence of a theorem which was proposed a century after the second law in a very different context of the theory of communication!

4.4. *From Ignorance to Disorder: The Role of Measurements*

The ability of living organisms to perform measurements and "profit" by exploiting their outcomes can be analyzed in the algorithmic terms of the above discussion. Measurements decrease ignorance about the specific state of the system, but increase the size of the record necessary to encode the acquired information. This process is illustrated in Figure 5(a) for an equilibrium ensemble: The approximate equality of the increase of the size of the record and of the decrease of ignorance is guaranteed by the coding theorem. To approach the optimal coding efficiency $\langle \Delta K \rangle = \Delta H$, the demon can assign to different outcomes symbols with length K_i given by:

$$K_i = \lceil \log_2 (1/p(s_i)) \rceil \tag{4.12}$$

Here the symbol $\lceil a \rceil$ signifies the smallest natural number that is $\geq a$. It is easy to see that with this assignment, Kraft inequality (Eq. (4.11)) is automatically satisfied. Therefore, there exists a uniquely decodable "labeling" with the word length given by Eq. (4.13). Moreover:

$$K_i < \log_2 1/p(s_i) + 1$$

Therefore:

$$\langle \Delta K \rangle < \Delta H + 1 \tag{4.13}$$

Thus, the encoding satisfying Eq. (4.12) (and known as Shannon-Fano coding [8]) is quite efficient in that it exceeds the absolute lower limit on the minimal average size of the description given by $\langle \Delta K \rangle = \Delta H$ by no more than one bit.

Indeed, Shannon-Fano coding can suggest a near-optimal "labeling" strategy for a demon dealing with an unknown ensemble: The demon could perform sufficiently many measurements to estimate probabilities $p(s_i)$ from the frequencies of the different outcomes, and then adopt short descriptions for the more frequent outcomes and longer descriptions for the rare ones in accordance with the Shannon-Fano prescription (Eq. (4.12)). For a demon dealing with an equilibrium system (Figure 5(a)), it would guarantee an (almost) optimal performance: The demon would lose no more than $k_B T$ of useful work (corresponding to the "extra" bit) per measurement. (The demon could "break even" if the probabilities $p(s_i)$ were exactly equal to the inverse

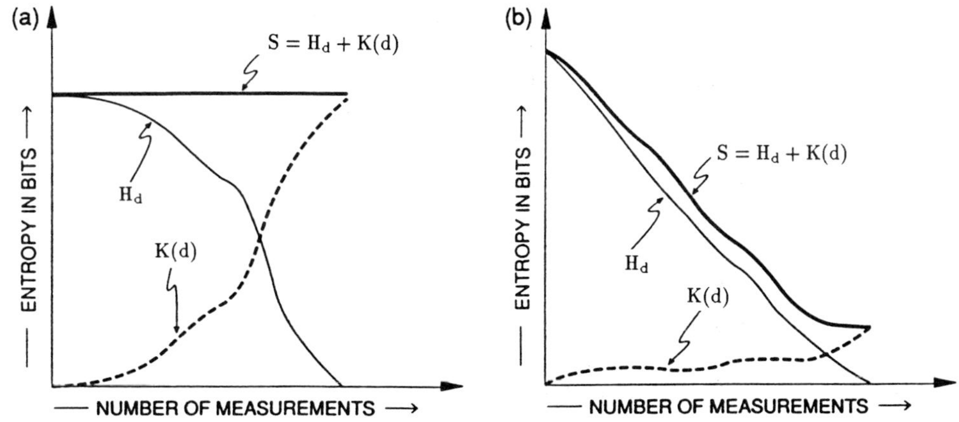

Figure 5: The effect of measurements on (i) The Shannon entropy H_d in presence of the partial information—data d; (ii) The algorithmic information content of the data $K(d)$; and (iii) The *physical entropy* $\mathcal{S}_d \equiv H_d + K(d)$, which measures the net amount of work that can be extracted from the system given the information contained in the data d. (a) When the measurements are carried out on the equilibrium ensemble, the randomness in the data increases at the rate given by the decrease of ignorance. (b) For systems far from equilibrium the increase of randomness is smaller than the decrease of ignorance, which allows the observer to extract useful work and makes measurements energetically attractive.

powers of 2.) Of course, no useful work could be extracted—Shannon's noiseless channel coding theorem guarantees the validity of the second law.

To restate the content of Figure 5(a) one can regard measurements as a way of turning statistical uncertainty into algorithmic randomness. Moreover, for equilibrium ensembles, the two quantities are, on the average, changing in such a way that their sum

$$\mathcal{S} = H + K \tag{4.14}$$

remains approximately constant. Furthermore, as it was already discussed above, the amount of useful work which can be extracted by a "demon" from a system is given by:

$$\Delta W = k_B T(\Delta H + \Delta K) = k_B T \Delta \mathcal{S} \tag{4.15}$$

Therefore, from the point of view of the observer (the demon) evaluating its ability to extract *net* useful work from the physical system, the hybrid quantity $\mathcal{S}$ plays the role of entropy. I shall therefore refer to $\mathcal{S}$ as *physical entropy*.

It is important to emphasize that the actual record r_i corresponding, in the demon's memory, to the state s_i will rarely be minimal; the inequality

$$\mid r_i \mid > K(r_i) \tag{4.16}$$

will usually be satisfied. This is because the minimal programs cannot be found in a systematic manner. Moreover, there is usually no incentive to look for the truly minimal record. A reasonably short record, often with built-in redundancies to allow for error-correction, can be far more useful. Consequently, the actual gain of useful work will be given by:

$$\Delta W \simeq k_B T(\Delta H + \Delta \mid r \mid) \tag{4.17}$$

One might be tempted even to use the quantity $H + \mid r \mid$ as a measure of entropy. As M. Gell-Mann emphasizes, this would be incorrect: $H + \mid r \mid$ can be lowered by *computation* at no thermodynamic expense. Hence, if $H + \mid r \mid$ was regarded as entropy, the second law could be violated by reversible computation. Such violation is impossible with $\mathcal{S}$ where the record part of $\mathcal{S}$ is already as short as it can be. The second law is safe when formulated in terms of the physical entropy $\mathcal{S}$.

Indeed, physical entropy $\mathcal{S}$ has the great advantage of removing the "illusion" that entropy decreases in the course of measurement. Of course, the ignorance (measured by H) does indeed decrease, but only at the expense of the increase of the minimal record size K. Hence, in measurements performed on equilibrium ensembles $\mathcal{S} \equiv H + K$ is, on the average, constant (Figure 5(a)).

By contrast, measurements performed on far-from-equilibrium systems can result in a decrease of ignorance which is much larger than the resulting increase in the record size (Figure 5(b)). Fortunately, the Universe that we inhabit is precisely such a nonequilibrium environment: It pays to measure. Moreover, it is not really necessary to look for truly minimal programs: the possible gain of useful work outweighs even substantial inefficiencies in the record size optimization (as well as in other more "traditional" thermodynamic inefficiencies). Nevertheless, the ability to recognize in the measurement outcome the opportunity for such a gain of useful work is essential for the well-being of IGUS's.

Acknowledgments

I would like to thank Charles Bennett, Carl Caves, Stirling Colgate, Doyne Farmer, Murray Gell-Mann, James Hartle, Rolf Landauer, Seth Lloyd, Bill Unruh, and John Wheeler for stimulating and enjoyable discussions on the subject of this paper. The warm hospitality of the Aspen Center for Physics, the Institute for Theoretical Physics in Santa Barbara, and the Santa Fe Institute is very much appreciated.

References

1. R.J. Solomonoff. *Info. & Control* **7**, 1–22 (1964).
2. A.N. Kolmogorov. *Information Transmission* **1**, 3–11 (1965).
3. G.J. Chaitin. *J. ACM* **13**, 547–569 (1966).
4. M. Davis. *Computability and Unsolvability.* Dover, New York, 1973.
5. H. Rogers. *Theory of Recursive Functions and Effective Computability.* McGraw-Hill, New York, 1967.
6. A.I. Khintchin. *Information Theory.* Dover, New York, 1957.
7. C.E. Shannon and W. Weaver. *The Mathematical Theory of Communication.* Univ. of Illinois Press, Urbana, 1949.
8. R.W. Hamming. *Coding and Information Theory.* Prentice–Hall, Englewood Cliffs, 1987.
9. C.H. Bennett. *Int. J. Theor. Phys.* **21**, 305–340 (1982).
10. W.H. Zurek. *Nature* **341**, 119–124 (1989).
11. W.H. Zurek. *Phys. Rev. A* **40**, 4731–4751 (1989).
12. W.H. Zurek. In *Proceedings of the International Symposium on Quantum Mechanics.* Ed. Y. Murayama. Physical Society of Japan, Tokyo, 1990.

13. G.J. Chaitin. *Sci. Am.* **232(5)**, 47–52 (1975).
14. K. Gödel. *Monatshefte Math. Phys.* **38**, 173–198 (1931).
15. A.M. Turing. *Proc. Lond. Math. Soc.* **42**, 230–265 (1936).
16. C.M. Caves. In *Complexity, Entropy, and the Physics of Information.* Ed. W.H. Zurek. Addison–Wesley, Reading MA, 1990.
17. L.A. Levin. *Dokl. Akad Nauk SSSR* **227** (1976).
18. L.A. Levin. *Sov. Math Doklady* **17**, 522–526 (1976).
19. C.H. Bennett. In *The Universal Turing Machine – A Half-Century Survey.* Ed. R. Herken. Oxford University Press, Oxford, 1988.
20. M. Smoluchowski. In *Vorträge über die kinetische Theorie der Materie und der Elektrizität.* Teubner, Leipzig, 1914.
21. R.P. Feynman, R.B. Leighton, and M. Sands. *Feynman Lectures on Physics,* Vol. 1, Sect. 46. Addison–Wesley, Reading MA, 1964.
22. L. Szilard. *Z. Phys.* **53**, 840–856 (1929).
23. D. Hofstadter. *Gödel, Escher, Bach.* Random House, New York, 1979.
24. R. Penrose. *The Emperor's New Mind.* Oxford University Press, Oxford, 1989.
25. C.H. Bennett. *Sci. Am.* **255(11)**, 108–117 (1987).
26. R. Landauer. *IBM J. Res. Dev.* **3**, 113–131 (1961).
27. R. Landauer. In *Signal Processing.* Ed. S. Haykin. Prentice-Hall, New York, 1989, pp. 18–47.

THE ENTROPY CREATION NECESSARY TO PERFORM A MEASUREMENT AND TO RESET IT

THÉO MULLER

Centre de Recherche Nucléaire
Université Louis Pasteur
F-67037 Strasbourg, France

Abstract

For a measurement with only two possible outcomes 0 and 1, or **a** and **b**, the entropy creation necessary to perform the reset of the measuring apparatus is not automatically $\Delta S \geq k \ln 2$ in all cases. The necessity to be in agreement with determinism imposes only the condition that at least one of the resets $0 \to 0$ or $1 \to 0$, or else $\mathbf{a} \to 0$ or $\mathbf{b} \to 0$, is subject to an entropy creation $\Delta S \geq k \ln(1 + n\epsilon)$, whereas the other one may be performed in a conservative way. Examples are given to illustrate this result, which contradicts the proclaimed view that every reset creates an entropy $\Delta S \geq k \ln 2$.

1. Historical Introduction

In 1929, Leo Szilard [1] proposed a thought experiment, here given with some minor modifications: Take a box of volume V in contact with a thermostat at temperature T, and with *one* molecule M inside. Without doing work, put a separation P through the middle of the box dividing V into two equal parts $V/2$. One of them contains the molecule M. An automatic measuring device arranges things in such a way that P, now used as a piston, is always moved (by a suitable gear, either a "forward" gear or a "reverse" gear, coupled to a flywheel) in such a way as to provide an isothermal expansion from $V/2$ to V on the molecule M (never a compression; so in half of the cases the gear has to be set "reverse" by the measuring apparatus). If the measuring process and the reset of the measuring apparatus would not themselves create an entropy $\Delta S \geq k \ln 2$ (k = Boltzmann's constant), then the second law of thermodynamics would be violated, because the isothermal expansion produces mechanical energy $E = kT \ln 2$. At the end of the expansion, P can be taken out automatically, reinserted in the middle of the box, and the whole process can be repeated all over again. This would mean that an unlimited amount of mechanical energy could be generated from a single heat source, every isothermal expansion of one molecule by a factor of two permitting an entropy reduction of $\Delta S \leq k \ln 2$. Szilard concluded that in order to make it impossible to produce a *systematic* entropy decrease, *every* measurement and its reset together have to create at least an entropy $\Delta S = k \ln 2$.

2. General Argument

In order to examine the necessity of entropy creation as required by the laws of nature, all parts of the measuring apparatus and of the reset device studied here are supposed to work in a conservative way, except when required by the laws and specified explicitly.

It is currently admitted that only the reset of a measurement, but not the measurement itself or its memorisation requires entropy creation. The current view [2, 3] is that for every reset of a measurement with (for simplicity) only two possible outcomes 0 and 1 the entropy created for every reset is $\Delta S \geq k \ln 2$. As soon as the measurement is recorded or used, this prevents the reset from being conservative. The states of the measuring apparatus appear then necessarily to be random states for an automatic reset device AM in charge of this reset. An automatic reset device AM of suitable design is always able to reset one of the states in a conservative way, but then the reset of the other state has to create entropy because of the uniqueness of the solutions of the equations of physics. This is seen immediately by reversing mentally the two now supposedly conservative (and therefore reversible) motions of the reset device AM ($1 \to 0$ and $0 \to 0$). In this case the reset state, now considered as the initial state, could lead in a conservative way to two different final states, and this is prevented by the uniqueness of the solutions of the laws. (The same reasoning applies if the measuring apparatus has three states: the reset state **c** and two states **a** or **b** resulting from the measurement.) Consequently, one of the resets (not both), either $0 \to 0$ or $1 \to 0$ (the choice is free and depends on the design of AM) has to create an entropy $\Delta S \geq k \ln(1 + n\epsilon)$, whereas the other reset ($1 \to 0$ or $0 \to 0$) may be conservative, $\Delta S = 0$, $n = 1, 2, \ldots$. For a well designed reset, n may be as low as 1; ϵ corresponds to the smallest possible phase space cell. In what follows the measurement is performed in a conservative way. For simplicity, the measuring apparatus may take only the two states 0 and 1, where 0 or 1 means that an event P has not or has occured during the measuring time Δt_1. State 0 is also the state of the measuring apparatus between two measurements, when it has been reset during the time interval Δt_2, but 0 characterizes the reset state in this case. The time interval $\Delta t = \Delta t_1 + \Delta t_2$ is the duration of a measuring cycle and repeats at regular intervals as determined by the working cycle of the automatic reset device AM. The possible states of the measuring apparatus are set by P without any relation to AM. Moreover, they are random for the reset device AM which resets all of them to the same final state 0.

A similar type of reasoning may be applied to measurement. In this case a property of the event P is translated into a property of the measuring apparatus. The measuring apparatus is always in reset state 0 when the measurement is beginning, and it is always in state 1 when P has been detected. It remains in state 0 if P has not been detected during the interval Δt_1. Both sequences, $0 \to 1$ and $0 \to 0$, are deterministic. They may *therefore be performed in a conservative way*, and $\Delta S = 0$. The event P is merely transcribed into a more convenient form by the measurement. This motivates the view that a measurement may be performed, at least in principle, in a conservative way. On the other hand, the state of the measuring apparatus is random for the reset device AM (it may be 0 or 1 with some probability), but when it is reset, then it is always in the same final state. This process is therefore not deterministic and *the reset requires entropy creation.*

3. Example: Superconducting Coil

For simplicity the measuring apparatus, here a superconducting coil B, may take two states 0 and 1. They are represented by the quantized flux F_Q and the corresponding supercurrent i_B which are given by $F_Q = i_B = 0$ or $F_Q = i_B = 1$ in appropriate units. The transitions $0 \to 1$ or $0 \to 0$ are performed in a conservative way [4, 5, 6] when B is traversed or is not traversed by the primary event P during the measuring interval Δt_1. The primary event is either an incoming magnetic monopole M, or one pole M_D of an extended magnetic dipole, the other pole of which is sufficiently far away so that its effect may be neglected. M_D has the same strength as M, and only poles of one polarity are considered. Here, measurement is nothing else than the transcription of the primary event P (the transient passage of M through B) into a form which is easier to be used (the permanent state F_Q, i_B) by the measuring instrument (the superconducting coil B). There is now a permanent record (1 or 0) of the primary event P, which itself has disappeared. Because the primary event is no longer available, no reversible reset by inversion of the motion is possible. *Apparently* one could conclude that the corresponding state can *only* be reset with entropy creation, since the permanent record has to be destroyed. This is not so, as will be discussed below. (A reset with $\Delta S = 0$ is possible, for instance, if M is reflected back through B by a preexisting static magnetic field of suitable strength and polarity which M encounters after having traversed B. In this case the measurement is always reset with $\Delta S = 0$ but it can never be used. As soon as $F_Q = 1$ triggers some permanent change, the measurement can be used, but then the reversible reset is no longer possible because the field in B to be reset is no longer the same and $\Delta S \neq 0$.)

In order to reset the measuring instrument B considered in the foregoing example to 0, the coil B may be opened and reclosed by AM before the start of each measuring cycle. When B is in state 1, then the reset destroys $F_Q = i_B = 1$ with entropy creation. If no event has been detected, then B is in state 0 and the action of AM (opening and reclosing B) may be conservative. So one of the resets has been performed in a conservative way whereas the reset of the other state necessarily requires entropy creation. The entropy creation necessary in order to reset the measurement is conditioned by the intervention of AM on random states, and determinism allows one of both resets ($1 \to 0$ or $0 \to 0$, the choice is free) to be performed in a conservative way. The possibility of this choice will be shown now.

Consider the case already discussed, where M has disappeared to infinity while producing $F_Q = i_B = 1$ in B which is therefore in state 1. In order to reset the measuring apparatus B from 1 to 0, the automatic reset device AM pushes the pole M_D (of a dipole of the same strength as M) through B in such a way as to cancel $+F_Q$ and $+i_B$ by inducing $-F_Q$ and $-i_B$ (*). Then AM opens the coil B (*), retracts M_D (*), and closes the coil B (*). All those operations denoted by (*) (called trajectory of AM) may be performed in a conservative way and are supposed to be done so. State 1 of B has therefore been reset to 0 by AM in a conservative way, $\Delta S = 0$, and the general argument leads to the conclusion that the reset of state 0 to 0 by AM, performing of course the *same* trajectory, will necessarily be dissipative. In this case no monopole M has traversed B during Δt_1 (B is in state 0) which is followed by Δt_2 where AM performs the *same* trajectory as before to reset B (already at 0) to 0. As before, M_D is pushed through B inducing $-F_Q$ and $-i_B$ (*). Then AM opens B, this time creating entropy by transforming $-F_Q$ and $-i_B$ into heat and electromagnetic radiation, $\Delta S \geq k \ln(1 + n\epsilon)$. Then AM retracts M_D (*) and closes B (*) which is then in state 0 (after the heat has been absorbed by a thermostat). State 0 of B has therefore been reset, $0 \to 0$, by AM with entropy creation, whereas the reset $1 \to 0$ has been conservative.

It is just as easy to reset state 0 of B to 0 in a conservative way by a suitable design of AM, but then the reset $1 \to 0$ will necessarily require entropy creation. To do this, it is sufficient that AM performs the *same* trajectory as before with a pole M_D of strength zero, $M_D = 0$ (therefore M_D *is inexistent*). Then state 1 ($F_Q = i_B = 1$) is *not* cancelled by M_D, but it is subsequently destroyed with entropy creation by opening of B. Then M_D is retracted as before, and B is closed. B is then in state 0, ready for the next measuring cycle. On the other hand, if B is in state 0 ($F_Q = i_B = 0$) after Δt_1, then the *same* reset trajectory with $M_D = 0$ is performed by AM. State 0 of B has been reset in a conservative way, whereas now the reset $1 \to 0$ creates entropy. Zurek has argued that for a large coil B the phase of the wavefunction is not the same at both ends of the opened coil. So when the coil is reclosed it may contain $0, \pm 1, \pm 2, \ldots$ flux quanta F_Q. Their energy in this case may be $E \ll kT$, but then the measurement does not mean anything. For a small coil, the phase is always the same at both ends of the opened coil. When this coil is closed it will always be in state $F_Q = 0$. But then on the other hand opening the coil with $F_Q = 1$ is an irreversible process with $\Delta E > kT$ [7]. However, the conclusion from the general argument remains valid, namely, that one reset may be performed with $\Delta S = 0$.

4. Almost Conservative Reset: A Potential Model

If the measurements are expressed by different positions of a pointer (e.g., from an ammeter) the pointer of the measuring instrument comes always to the 0 state by viscous damping, replacing here AM, whatever the initial value has been. The produced heat is proportional to the deviation of the pointer and is absorbed by the thermostat. If the pointer is in state 0, no entropy is created by the reset. If the knowledge about the system is sufficiently detailed, then information is still conserved. To quote von Neumann [8], "for a classical observer who knows all coordinates and momenta, the entropy is zero". Information could never be destroyed in this case.

In another model for nearly conservative reset, both positions (0 or 1) may be represented by the equilibrium positions of a pointlike particle P with mass m and charge $+e_R$ free to move only along the $+x$, $-x$–segment. Three additional charges $+e$ of equal sign and value are located at $+x$, at O (the origin of the $\{x, y\}$ system), and at $-x$. They are much larger than the elementary charge and fixed, e.g., on electrets. Then P may take two equilibrium positions, called states 0 and 1, near $+x/2$ and $-x/2$ in the two potential wells formed by the three additional charges. As everything else, P has some thermal agitation (m, e, and x are chosen such that P remains always in that state where it has been put: "no" tunneling and "no" transition $0 \leftrightarrow 1$ due to thermal agitation is allowed). If the state of P is unknown, it may nevertheless be driven to state 0 in a finite time with a minimum of entropy creation $\Delta S = k \ln(1 + h\nu/kT)$. In order to produce this reset, the charge $+e_{AM}$ located at O and attached to the reset device AM is retracted along the y–axis (perpendicular to the x–axis at O) so that P shifts to O. Doing so, the double wells W_+ and W_- with minima at approximately $\pm x/2$ are deformed adiabatically such that the potential wall W between the two wells disappears and they finally form a single well with minimum at O. While this is done, P may tunnel through the lowered potential wall W, but this does not matter since the setup is symmetric with respect to the y–axis. Then AM moves the charge e_{AM} attached to it from the y–axis parallel to the x–axis to $(-x, y)$, from there around $(-x, 0)$ (this avoids infinite repulsion energy), and then along the x–axis to O where it stops, ready for the next reset. While this is done, P (starting from 0 or 1) is pushed to 0 in any case. The net work done by AM is zero since AM describes a closed path, and since finally everything is energetically the same ($0 \rightarrow 0$ or $1 \rightarrow 0$) no matter whether starting from 0 or from the symmetric position 1 which has the same energy as does 0. Since the velocity of AM along its trajectory is finite, P overshoots the "reset" position (state 0) and oscillates at least with one quantum of energy $h\nu$ around the state 0, and with a phase remembering, in principle, the original state (0 or 1). This oscillation (which may be of very low energy compared with kT by a suitable choice of m, e, and x, such that kT corresponds to a highly excited state of the oscillator) has to be suppressed by the exchange of energy with thermal agitation. This means that the whole process is performed in a conservative way, except for the small entropy creation just mentioned.

The charge e_{AM} and its electret support, which have to be displaced by AM, are at "rest" at O at the beginning of the reset cycle. (Everything is in thermal agitation, but this will only be mentioned if necessary for the discussion.) They have to be accelerated, remain at constant velocity, and have to be decelerated by AM (for instance through a cam system) in order to follow the described path. Therefore the different parts of AM are stressed and vibrations appear. In principle, the energy $E = \sum h\nu_i$, $i = 1, \ldots, n$ of these vibrations of AM can be made very small compared to kT if the accelerations are very small, and if the proper frequencies are very low. In principle, the accelerated charge also produces synchrotron radiation, but again (if at all) of very low frequency and of very low energy $h\nu_s$. The charge e_R, i.e., the particle P which has to be reset, is free to move along x in the respective potential well centered roughly around $\pm x/2$. For a deep well and a quantized oscillator there are many discrete levels (they would be equally spaced for a harmonic oscillator with infinite walls). The wells considered here are of finite depth, and change in shape adiabatically. So the level spacing changes also during the reset, but, as mentioned above, kT is an energy far below the continuum caused by the finite depth of the wells. Nevertheless, an equal spacing is a good approximation for the quantized states of e_R, and conditions (e, m, and $\pm x$) can always be chosen in such a way that the destruction of the oscillation quanta caused by overshooting corresponds to an entropy creation far below k. This means that all the vibrational energies induced by the various accelerations necessary for the transport, including that of e_R in the changing wells, may be far below kT since they depend only on the accelerations to be freely chosen by the design of AM. This means that the reset of random bits may be performed in an almost conservative way, and that the entropy creation, $\Delta S = k \ln(1 + \sum h\nu_i/kT +$ negligible higher order terms), $\sum h\nu_i \ll kT$, is far below the usually considered entropy creation $\Delta S \geq k \ln 2$. The work done on P is (to a good approximation) very similar to a conservative expansion and recompression to a similar volume of a gas molecule.

5. Conclusion

For a measurement with two outcomes (0 and 1), the general argument given in Sec.2 requires an entropy creation for their reset at least in one of the cases which may be chosen by the design of the reset apparatus (Sec.3). The amount of entropy creation may be very small compared with $\Delta S = k \ln 2$ usually considered as a lower bound. The reason is that the general argument does not fix the smallest amount of entropy to be created (Sec.4).

References

1. L. Szilard. *Z. Phys.* **53**, 840 (1929).
2. C. Bennett. *Int. J. Theor. Phys.* **21**, 905 (1982).
   J.A. Wheeler. *IBM J. Res. Dev.* **32**, 4 (1988).
   J.A. Wheeler and W.H. Zurek, eds. *Quantum Theory and Measurement.* Princeton University Press, Princeton, 1983, p.783, p.539.
   W.H. Zurek, ed. *Complexity, Entropy, and the Physics of Information.* Addison–Wesley, Reading, 1990.
3. R. Landauer. *Found. Phys.* **19**, 729 (1989).
   C. Bennett, *Sci. Am.* **255**, 108 (1987).
4. P.H. Eberhard, R.R. Ross, and L.W. Alvarez. *Phys. Rev.* **D 4**, 3260 (1971).
5. B. Cabrera. *Phys. Rev. Lett.* **48**, 1378 (1982).
6. L.J. Tassie. *Nuovo Cim.* **38**, 1935 (1965).
7. W.H. Zurek. Private communication.
8. J. v. Neumann. *Mathematical Foundations of Quantum Theory.* Princeton University Press, Princeton, 1955, p. 400.

INFORMATION THEORETICAL ASPECTS OF QUANTUM MECHANICAL ENTROPY

ALFRED WEHRL

Institut für Theoretische Physik
Universität Wien
A-1090 Wien, Austria

Abstract

Properties of von Neumann's entropy of a density matrix ρ, i.e., $S(\rho) = -k\mathrm{Tr}\rho \ln \rho$, are considered within the framework of quantum information theory. Although the von Neumann entropy shares many properties with the Boltzmann, Gibbs, Shannon, Baron-Jauch entropies, in some respects there exist considerable differences (i.e., the lack of monotonicity). Besides the von Neumann entropy one can also think of other measures (or concepts) of quantum information, some of them are discussed. The – rather tricky – problem of dynamical entropy for quantum systems is treated, too.

In quantum mechanics, a (generally mixed) state is characterized by a density matrix ρ – i.e., a positive operator with trace (Tr) 1 – in some Hilbert space $\mathcal{H}$, the special structure of $\mathcal{H}$ depends on the system under consideration (e.g., one-particle systems, many-particle-systems with Bose-Einstein or Fermi-Dirac statistics, spin systems). Its entropy is given by von Neumann's formula [1]:

$$S(\rho) = -k \ \mathrm{Tr} \ \rho \ln \rho = -k \sum p_i \ln p_i \tag{1}$$

p_i are the eigenvalues of the (compact) operator ρ; we understand $0 \cdot \ln 0 \equiv 0$. From now on we choose units such that Boltzmann's constant $k = 1$.

Although at first sight the expression in formula (1) may look just the same as Shannon's [2], it turns out that there are considerable differences. Before coming to them, let me give the information-theoretical interpretation of the von Neumann entropy:

Let φ_i be the normalized eigenvectors corresponding to the p_i. They are pairwise orthogonal since ρ, as a positive operator, is Hermitian. Taking into account the possibility that some p_i are $= 0$ we can assume that the φ_i are an orthonormal basis for $\mathcal{H}$. The projection onto φ_i – $|\varphi_i)(\varphi_i|$ – is a pure state, i.e., cannot be further decomposed. p_i is the probability of finding the system in this pure state. If one performs N measurements, we will find – at least for large N – the system $p_1 \cdot N$

Information Dynamics, Edited by H. Atmanspacher and
H. Scheingraber, Plenum Press, New York, 1991

times in the state $|\varphi_1)(\varphi_1|$, $p_2 \cdot N$ times the state $|\varphi_2)(\varphi_2|$, and so on. Notice that the density matrix does not presuppose anything about the order of the p_i. Thus in N measurements there are $N!/(p_1N)!(p_2N)!\ldots$ possibilities to "realize" ρ. For $N \to \infty$, by virtue of Stirling's formula, $1/N$ times the logarithm of the above number converges to $S(\rho)$. (This is virtually the same argument as that of Shannon, the difference comes in if one wants to compare density matrices that do not commute [3].)

Let me now list some simple properties of quantum-mechanical entropy:

(i) $S(\rho) \geq 0$, possibly $+\infty$ (which is in fact the most frequent case in some sense, we will come to this point later on). Whereas this is true for both the von Neumann and the Shannon entropy (simply because $0 \leq p_i \leq 1$, thus $-p_i \ln p_i \geq 0$), it is *not* true for the Boltzmann and the Gibbs entropy, they need not even be defined. $S(\rho) = 0$ if and only if ρ is pure since $-p_i \ln p_i > 0$ for $0 < p_i < 1$, hence p_i is either 0 or 1, but because of $\mathrm{Tr}\ \rho = \sum p_i = 1$ exactly *one* p_i can be $= 1$ [4, 5].

(ii) One can easily verify that the range of $S(\rho)$ is $[0, \infty]$.

(iii) If U is unitary, $S(U^*\rho U) = S(\rho)$ since the eigenvalues remain unchanged under unitary transformations.

(iv) Expansibility: let $\mathcal{H}_1 := \mathcal{H} \oplus \mathcal{H}_2$, $\rho_1 := \rho \oplus 0$, then trivially $S(\rho_1) = S(\rho)$ (the second Hilbert space does not provide any information) [6].

It is very important to note that in infinite dimensions entropy is *not* continuous with respect to the trace norm topology, but only lower semicontinuous (see later on).

Another simple property is additivity. Let $\mathcal{H}_1$, ρ_1 or $\mathcal{H}_2$, ρ_2, respectively, be two systems, then the information about the composed system $\mathcal{H}_1 \otimes \mathcal{H}_2$, $\rho_1 \otimes \rho_2$ equals the sum of the information about its constituents. (This is not a specific quantum-mechanical feature.) This means $S(\rho_1 \otimes \rho_2) = S(\rho_1) + S(\rho_2)$. Namely, let $p_1, p_2, \ldots$, or $q_1, q_2, \ldots$, resp., be the eigenvalues of ρ_1, or ρ_2, resp. Then the eigenvalues of $\rho_1 \otimes \rho_2$ are $p_i q_k$, thus $S(\rho_1 \otimes \rho_2) = -\sum_{ik} p_i q_k \ln(p_i q_k) = -\sum p_i \ln p_i - \sum q_k \ln q_k$ (because $\sum p_i = \sum q_k = 1) = S(\rho_1) + S(\rho_2)$.

The following properties are also not specifically quantum mechanical but their proof demands more sophisticated techniques.

Concavity: Let ρ be a convex combination of two density matrices ρ_1 and ρ_2, i.e., $\rho = \lambda\rho_1 + (1-\lambda)\rho_2$ $(0 \leq \lambda \leq 1)$, then:

$$S(\rho) \geq \lambda S(\rho_1) + (1-\lambda) S(\rho_2) \tag{2}$$

Why is this considered to be an important property? Entropy is a measure of the lack of information, or of our ignorance of a system. If two states (ensembles) are fitted together in an incoherent way (described in mathematical terms in the above manner), information is lost (one cannot reconstruct ρ_1 or ρ_2 from ρ), therefore entropy must increase.

To make the mathematical proof more transparent, we prove a more general theorem: let f be a concave function. First we note for every unit vector φ that $f(\varphi, A\varphi)$ (A Hermitian, for simplicity with purely discrete spectrum $\alpha_1, \alpha_2, \ldots$, the corresponding normalized eigenvectors being $\psi_1, \psi_2, \ldots) \geq (\varphi, f(A)\varphi)$. This implies:

$$f((\varphi, A\varphi)) = f(\sum |(\varphi, \psi_i)|^2 \alpha_i) \geq \sum |(\varphi, \psi_i)|^2 f(\alpha_i) = (\varphi, f(A)\varphi)$$

Now [7]:

$$\begin{aligned}\text{Tr } f(\rho) &= \sum f((\varphi_i, \rho\varphi_i)) \\ &\geq \sum [\lambda f(\varphi_i, \rho_1\varphi_i) + (1-\lambda) f(\varphi_i, \rho_2\varphi_i)] \\ &\geq \sum [\lambda(\varphi_i, f(\rho_1)\varphi_i) + (1-\lambda)(\varphi_i, f(\rho_2)\varphi_i)] \\ &= \lambda \text{ Tr } f(\rho_1) + (1-\lambda) \text{ Tr } f(\rho_2)\end{aligned} \tag{3}$$

(Remark: quantities of the type Tr $f(\rho)$, f being concave, are sometimes called quasi-entropies. In some respects they share properties with entropy but not in all, for instance additivity only holds for $f(\rho) = const \cdot \rho \ln \rho$. Cf. also the theorem of Aczel, Forte, and Ng, to be discussed later [8].)

There is also an upper bound for ρ as defined in (2) [9]:

$$S(\rho) \leq \lambda S(\rho_1) + (1-\lambda) S(\rho_2) - \lambda \ln \lambda - (1-\lambda) \ln(1-\lambda)$$

(This means that the loss of information cannot exceed $-\lambda \ln \lambda - (1-\lambda)\ln(1-\lambda)$.)

Actually this is a subadditivity inequality. Extending the definition of S to all positive compact operators (simply by $S(A) := -$ Tr $A \ln A$), it reads $S(\sigma_1 + \sigma_2) \leq S(\sigma_1) + S(\sigma_2)$, where $\sigma_1 = \lambda\rho_1$, $\sigma_2 = (1-\lambda)\rho_2$. (Note that $S(\sigma_1) = \lambda S(\rho_1) - \lambda \ln \lambda$ etc.) The function $s(x) := -x \ln x$ is subadditive, $s(x+y) \leq s(x) + s(y)$, as every concave function (defined on $[0, \tau]$, τ being finite or infinite), vanishing at the origin.

The key for the proof is the following lemma: let $A = \sum \alpha_i |\psi_i)(\psi_i|$ be the decomposition into eigenprojections such that $\alpha_1 \geq \alpha_2 \geq \ldots$. Suppose that there is another decomposition of A into one-dimensional projections (not necessarily being pair-wise orthogonal) $A = \sum \beta_i P_i$. Then $\alpha_1 \geq \beta_1$, $\alpha_1 + \alpha_2 \geq \beta_1 + \beta_2$, ..., $\alpha_1 + \alpha_2 + \ldots + \alpha_n \geq \beta_1 + \beta_2 + \ldots + \beta_n$, ... This is a consequence of Ky Fan's theorem: Let Q be a projection onto a subspace of dimension $\leq n$. Then Tr $QA \leq \alpha_1 + \alpha_2 + \ldots + \alpha_n$. Now insert for Q $P_1 \vee P_2 \vee \ldots \vee P_n$, then Tr $QA = \beta_1 + \beta_2 + \ldots + \beta_n$.

Now a classical theorem of analysis states – provided that the total sum of the α's equals to that of the β's – that $\sum f(\alpha_i) \leq \sum f(\beta_i)$ for a concave f. Let $\sigma_1 = \sum \gamma_i^{(1)} P_i^{(1)}$ and $\sigma_2 = \sum \gamma_i^{(2)} P_i^{(2)}$ be the decompositions into eigenprojections. Then $\sigma = [\sum \gamma_i^{(1)} P_i^{(1)} + \sum \gamma_i^{(2)} P_i^{(2)}]$, hence Tr $f(\sigma_1 + \sigma_2) \leq \sum f(\gamma_i^{(1)}) + \sum f(\gamma_i^{(2)}) =$ Tr $f(\sigma_1) +$ Tr $f(\sigma_2)$; in particular, $S(\sigma_1 + \sigma_2) \leq S(\sigma_1) + S(\sigma_2)$ [5].

Before coming to the notion of subadditivity, we should make a first acquaintance with *quantum-mechanical relative entropy*. This is the quantum-mechanical version of Renyi's information gain [10]. However, whereas Renyi's information gain has a clear information-theoretical interpretation, for the quantum-mechanical relative entropy the interpretation is not so clear. Nevertheless, this concept becomes more and more important.

The relative entropy refers to two density matrices, ρ and σ, and is defined by $S(\sigma|\rho) :=$ Tr $\rho(\ln \rho - \ln \sigma)$ [11, 12]. (Some authors prefer the notation $S(\rho|\sigma)$.)

The first non-trivial thing we want to discuss is that $S(\sigma|\rho) \geq 0$. This is the consequence of an inequality attributed to O. Klein. It generalizes a well-known fact concerning concave functions: "they always stay below the tangent". For (Hermitian) operators A, B and a concave function f is Tr $[f(B) - f(A)] \leq$ Tr $[(B-A)f'(A)]$. Inserting $f(x) = -x \ln x$, we obtain Tr $A(\ln A - \ln B) \geq$ Tr $(A - B)$ and Tr $\rho(\ln \rho - \ln \sigma) \geq$ Tr $(\rho - \sigma) = 0$.

As an application let us consider the maximum entropy principle (which in the classical case is known as Jaynes' principle [13] and may be traced back as far as to Boltzmann – equal *a priori* probability. Philosophers would say that in a sense these are versions of Laplace's principle of insufficient reason). The Gibbs state σ_β : $+Z^{-1}e^{-\beta H}$, $Z =$ Tr $e^{-\beta H}$ = partition function. Choose β in such a manner that Tr $\sigma_\beta H$ (expectation value of the Hamiltonian) is some prescribed energy E. Now if ρ is some density matrix such that Tr $\rho H \leq E$, then:

$$\text{Tr } \rho \ln \sigma_\beta = -\beta \text{ Tr } \rho H - \ln \text{Tr} e^{-\beta H}$$

$$\text{Tr } \sigma_\beta \ln \sigma_\beta = -\beta \text{ Tr } \sigma_\beta H - \ln \text{Tr} e^{-\beta H}$$

By assumption, $-$ Tr $\sigma_\beta \ln \sigma_\beta \geq -$ Tr $\rho \ln \sigma_\beta$.

The positivity of the relative entropy yields $S(\rho) = -$ Tr $\rho \ln \rho \leq -$ Tr $\rho \ln \sigma_\beta$, the last term is $\leq -$ Tr $\sigma_\beta \ln \sigma_\beta = S(\sigma_\beta)$. (In the classical case one could arrive at the same result by using Lagrangian multipliers, too.) This means that, if we do not have any more knowledge of the system than that the expectation value of the Hamiltonian does not exceed E, then the state with least information is the Gibbs state. Thus this state is the most probable one.

Now let us turn to subadditivity. We consider, as in the case of additivity, a composed system with Hilbert space $\mathcal{H} = \mathcal{H}_1 \otimes \mathcal{H}_2$ and a density matrix ρ in $\mathcal{H}$, which, however, now need not factorize. It gives rise to reduced density matrices (partial traces, conditional expectations) ρ_1 in $\mathcal{H}_1$ and ρ_2 in $\mathcal{H}_2$. One can introduce them as follows: let $T \in \mathcal{B}(\mathcal{H})$ be of the form $T_1 \otimes 1$, then Tr $\rho T =$ Tr $\rho_1 T_1$. Similarly ρ_2 is introduced. This means that ρ_1 (ρ_2) contains all the information about the first (second) system that can be deduced from ρ.

Subadditivity means that $S(\rho) \leq S(\rho_1) + S(\rho_2)$ $(= S(\rho_1 \otimes \rho_2))$.

Proof: $S(\rho_1 \otimes \rho_2 | \rho) =$ Tr $\rho[\ln \rho - \ln(\rho_1 \otimes \rho_2)] =$ Tr $\rho(\ln \rho - \ln \rho_1 \otimes 1 - 1 \otimes \ln \rho_2) =$ Tr $\rho \ln \rho -$ Tr $\rho_1 \ln \rho_1 -$ Tr $\rho_2 \ln \rho_2 \geq 0$. Thus decomposing the system causes a loss of information (which is to be expected because one cannot reconstruct ρ from ρ_1 and ρ_2).

(Remark: subadditivity is a stronger property than concavity.)

There are severe differences to the classical case. For instance it can happen (and is actually the usual case) that ρ is pure, i.e., $S(\rho) = 0$, but ρ_1 and ρ_2 are not (in that case $S(\rho_1) = S(\rho_2)$). Generally, the entropy of a subsystem can be larger than the entropy of the total system. (This may lead to speculations about the famous heat death of the universe.) Thus entropy is not monotonic ($S(\rho_1) \not\leq S(\rho)$), whereas it is in the classical discrete case. (However, in the classical continuous case it is not monotonic, too.) I do not have a good explanation for this phenomenon. – The best thing one can show is that $S(\rho_1) \leq S(\rho) + S(\rho_2)$ (and the same for 1 and 2 interchanged). Sketch of the proof: (i) ρ pure $\Rightarrow$ ρ_1 and ρ_2 have the same positive spectrum which is quite easy to show: let φ_i, ψ_k be orthonormal bases for $\mathcal{H}_1$ or $\mathcal{H}_2$, resp. Let ω be the vector $\sum_{ik} c_{ik} \varphi_i \otimes \psi_k$, C be the infinite matrix with entries c_{ik}. Calculate $|\omega)(\omega|$ and reduce. The eigenvalues of ρ_1 equal those of CC^*, the eigenvalues of ρ_2 equal those of C^*C. The positive spectra of CC^* and C^*C coincide. (ii) Let $\rho = \sum p_i |\varphi_i)(\varphi_i|$. "Double" the space. ρ_1 is the partial trace of the pure state $|\chi)(\chi|$, $\chi := \sum \sqrt{p_i}(\varphi_i \otimes \varphi_i) \in \mathcal{H} \otimes \mathcal{H}$. Collecting all together, we arrive at the "triangle inequality" $|S(\rho_1) - S(\rho_2)| \leq S(\rho) \leq S(\rho_1) + S(\rho_2)$ [4].

Subadditivity and additivity (plus unitary invariance) characterize entropy. *Theorem of Aczel, Forte, and Ng* (quantum-mechanical version): the only functionals that meet these requirements are linear combinations of entropy and the Hartley entropy S_0 [8].

Let me say a few words about the Hartley entropy [14]. $S_0(\rho)$ is the logarithm of the number of eigenvalues of ρ that are $\neq 0$. (Of course, $S_0(\rho) \geq S(\rho)$.) It is basic for the notion of *topological entropy*. In contradistinction, however, it is not even *insensitive*, by which is meant a very mild form of continuity: $S(\rho)$ is a continuous function of *finitely* many eigenvalues p_i, provided that the rest of them is kept fixed.

The proof of the Aczel-Forte-Ng theorem is rather involved. It consists of a series of steps, none of them is difficult (except for the last, number-theoretical one), but they are arranged in a very tricky manner. There is an interesting interplay between recursive and inductive arguments. I can present a *very short* outline only.

Let $\phi(\rho)$ be a function meeting the forementioned conditions. It is sufficient to consider the finite-dimensional case only. The "information function" $I_n(p_1, \ldots, p_n) = \phi(\rho)$ depends on the eigenvalues p_i of ρ only and is symmetric. Consider another space and define $I_m(q_1, \ldots, q_m)$ in a similar manner. By additivity, $I_{nm}(p_1q_1, \ldots, p_kq_j, \ldots, p_nq_m) = I_n(p_1, \ldots, p_n) + I_m(q_1, \ldots, q_m)$. Exploiting subadditivity and after some work one arrives at the relation $I_n(p_1, \ldots, p_n) = aS(p_1, \ldots, p_n) + A_n$ (a = some constant, independent of n). A_n is a number-theoretic function: $A_{nm} = A_n + A_m$. By symmetry, additivity, and subadditivity, one finds $\liminf(A_{n+1} - A_n) \geq 0$. Katai's theorem [15] (which actually is not that hard to prove) now tells us that $A_n = b \cdot \ln n$. A simple application of insensitivity (which is guaranteed by continuity, proven in the foregoing steps) yields A_n = number of p's that are $\neq 0$, so that we arrive at our desired result.

Let us turn to continuity properties of entropy. We have already mentioned that (in infinite dimensions) entropy is lower semi-continuous only: if (for density matrices) Tr $|\rho_n - \rho| \to 0$, then $S(\rho) \leq \liminf S(\rho_n)$. This means *cum grano salis* that the limiting ρ contains more information than the approximating ρ_n, which is not so surprising. Once more, the proof relies on Ky Fan's inequality. $\rho_n = (\rho_n - \rho) + \rho$, so that for the eigenvalues $p_i(\rho_n)$, $p_i(\rho)$, arranged in decreasing order, $p_1(\rho_n) + \ldots + p_n(\rho_n) \leq \text{Tr } |\rho_n - \rho| + p_1(\rho) + \ldots + p_k(\rho)$. Similarly, $p_1(\rho) + \ldots + p_k(\rho) \leq \text{Tr } |\rho_n - \rho| + p_1(\rho_n) + \ldots + p_k(\rho_n)$, that is $|(p_1(\rho) - p_1(\rho_n)) + \ldots + (p_k(\rho) - p_k(\rho_n)| \to 0$ as $n \to \infty$, consequently, $p_k(\rho_n) \to p_k(\rho)$ for all k. For finite sums:

$$-\sum_{i=1}^{k} p_i(\rho) \ln p_i(\rho) = \lim_n \left[-\sum_{n=1}^{k} p_i(\rho_n) \ln p_i(\rho_n) \right]$$

Passing to the limit $n \to \infty$ [16, 17]:

$$S(\rho) = \sup_k \sum_{i=1}^{k} (-p_i(\rho) \ln p_i(\rho)) \leq \liminf_n \sup_k \sum_{i=1}^{k} (-p_i(\rho_n) \ln p_i(\rho_n))$$

Lower semi-continuity may be regarded as something "academic" but it has very strange consequences that deserve closer inspection.

First we observe that, given ρ, in every neighbourhood $\{\sigma : \text{Tr } |\rho - \sigma| < \varepsilon, \varepsilon > 0$, but otherwise arbitrary$\}$ there exists a density matrix with infinite entropy. Let $p_1, p_2, \ldots$ be the eigenvalues of ρ, and choose simply a density matrix ρ' with the same eigenvectors and eigenvalues $p'_1, p'_2, \ldots$, where $p'_i = p_i$ for $i \leq \ell$ and, for instance,

p_i' is proportional to $(i(\ln i)^2)^{-1}$ for $i > \ell$ and ℓ is chosen large enough. Notice that for an observable (Hermitian operator) A, Tr $|\rho - \rho'| < \varepsilon$ implies for the expectation values $|\text{Tr } \rho A - \text{ Tr } \rho' A| \leq \|A\|\varepsilon$, so even if ρ and ρ' are hardly distinguishable by experiment, their entropies can be totally different.

The situation is even worse: entropy is infinite almost everywhere, namely, the set of density matrices with finite entropy is of first category only. (Still we understand that the dimension of the Hilbert space is infinite.) This is the case because, due to lower semi-continuity, the sets $\{\rho : S(\rho) \leq n,\ n \text{ integer}\}$ are closed, in view of the above statement they are nowhere dense, hence their union is of first category and $\{\rho : S(\rho) = \infty\}$ is of second category, because, as a complete metric space, the set of all density matrices is of second category.

Thus one might question whether entropy is a sensible concept at all. One way out of this enigma is (at least in physics) the following: let H be a reasonable Hamiltonian such that the partition function Tr $e^{-\beta H}$ exists, σ_β be the Gibbs state. We know already that, if Tr $\sigma_\beta H =$ (or $\leq$) E, E finite, and Tr $\rho H \leq E$, then $S(\rho) \leq S(\sigma_\beta)$, so it is finite.

On the other hand, given ρ and $\varepsilon > 0$, but arbitrarily small, there exists always a ρ' with finite entropy such that Tr $|\rho - \rho'| < \varepsilon$. (Simply take finite-rank approximations.) So, anyway, entropy turns out to be a very sensitive concept.

Besides entropy, there exist very many other measures of information, which are useful to a certain extent. (Needless to say that there are concepts that are of no use at all.) Let me single out a few of them.

We have already met the *quasi-entropies* $S_f(\rho) :=$ Tr $f(\rho)$, f concave, $f(0) = 0$, $f'(0) > 0$ (usually $= \infty$). With entropy they have the following features in common: (i) they are unitarily invariant, (ii) concave, (iii) subadditive, (iv) $\lambda \to S_f(\lambda\rho_1 + (1-\lambda)\rho_2)$ is continuous on $[0,1]$, (v) if $S_f(\rho) < \infty$, then $\lambda \to S_f(\lambda\rho)$ is continuous on $[0,1]$, (vi) if $f'(0) = \infty$, then in every neighbourhood of a given ρ there exists a density matrix ρ' with $S_f(\rho') = \infty$, the range of $S_f(\rho')$ in $\{\rho' :\ \text{Tr } |\rho - \rho'| < \varepsilon\}$ is $(\alpha, \infty]$, where $\alpha < S_f(\rho')$, (vii) S_f is lower semi-continuous in trace norm.

There are several models (linear or non-linear) of a time evolution $\rho \to \rho_t$, for which $S_f(\rho_t)$ is increasing in time.

α-entropies (Renyi-entropies) $S_\alpha(\rho) := (1-\alpha)^{-1} \ln$ Tr ρ^α. They are additive, but not subadditive except for $\alpha = 0$ or $= 1$. S_0 is the Hartley entropy, S_1 the usual entropy, $S_\infty(\rho) = -\ln \|\rho\|$. For fixed ρ, $\alpha \to S_\alpha(\rho)$ is decreasing, continuous in $[1, \infty]$. It may happen that even if $S_1(\rho)$ is finite, $S_\alpha(\rho)$ suddenly jumps to infinity as α becomes < 1.

For $\alpha > 1$, $S_\alpha(\rho)$ is neither concave nor even subadditive.

In the last years, α-entropies became more and more interesting [18].

Daroczy entropies: $f^{-1}($ Tr $\rho f(-\ln \rho))$, f being an increasing convex or concave function [3].

Measures of non-commutativity. Obviously there cannot exist a classical analog, since they measure how "skew", i.e., non-commutative, two operators are relative to each other.

The Wigner-Yanase entropy $S(\rho, K) := \frac{1}{2}$ Tr $[\rho^{1/2}, K]^2$ (ρ being a density matrix, K a fixed observable) is concave in ρ but definitely not unitarily invariant. Dyson's generalization is $S_p(\rho, K) := \frac{1}{2}\text{Tr}([\rho^p, K]\ [\rho^{1-p}, K])$ $(0 < p < 1)$, the degenerate case $p \to 0$ is $\frac{1}{2}$ Tr $([\rho, K]\,[\ln\rho, K])$. The question arises whether S_p is concave in ρ too. This was an open question for a long time [19]. The conjecture is commonly attributed to Dyson [20] but apparently has not been formulated by himself. Due to Tr $(AB\ldots C) =$ Tr $(CAB\ldots)$, $S_p(\rho, K) = -$ Tr $\rho K^2 +$ Tr $\rho^{1-p}K\rho^p K$. Noting that the first term is linear, one has to show that the second one is concave (Lieb). Besides the first proof, there are now four more proofs; in my opinion the most elegant one is due to Ando [21] and I will give a brief outline. (Cf. also Ref. [22].)

(i) Reduce the problem to showing that $A^p \otimes A^{1-p}$ (or more generally $A^p \otimes A^q$, $0 \le p, q,\ p+q \le 1$) is concave in A.

(ii) Consider $\mathcal{H} \oplus \mathcal{H}$. For A, B there are *maximal* operator X such that

$$\begin{pmatrix} A & X \\ X & B \end{pmatrix}$$

is positive. X is called the "geometric mean" $A \# B$.

(iii) If $\phi(A)$, $\psi(A)$ are concave in A, then so is $\phi(A) \# \psi(A)$. In particular, $A \to A^p$ $(0 \le p \le 1)$ is concave.

(iv) $A \to A^p \otimes A^q$ $(0 \le p, q,\ p+q \le 1)$ is concave. (Let Ω be the set of all p, q, such that $A^p \otimes A^q$ is concave. Because of (iii), with $(p_1, q_1), (p_2, q_2) \in \Omega$, also $(\frac{1}{2}(p_1+p_2), \frac{1}{2}(q_1+q_2)) \in \Omega$. $(0,0), (1,0), (0,1) \in \Omega \Rightarrow$ assertion.)

Why I have spent so much time to this concavity property (called "Lieb concavity") is that it is the key to the proof of a very important property of quantum-mechanical entropy: *strong subadditivity* [23, 24]. In the classical case this is fairly trivial but in the quantum case it is not.

It refers to three Hilbert spaces $\mathcal{H}_1$, $\mathcal{H}_2$, $\mathcal{H}_3$. Let ρ be a density matrix in $\mathcal{H}_1 \otimes \mathcal{H}_2 \otimes \mathcal{H}_3$, for the sake of clearness we will write ρ_{123} instead of ρ. ρ_{12} shall denote the reduced density matrix with respect to $\mathcal{H}_3$, ρ_1 the reduced density matrix with respect to $\mathcal{H}_2 \otimes \mathcal{H}_3$, and so on. Strong subadditivity states that $S(\rho_{123}) + S(\rho_2) \le S(\rho_{12}) + S(\rho_{23})$.

The starting point is the observation that as a consequence of Lieb concavity the relative entropy is jointly convex in ρ and σ. We have shown above (retranslating Ando's result) that $A \to \mathrm{Tr} A^p K A^q K$ (K Hilbert-Schmidt) is concave in A. But replacing

$$A \text{ by } \begin{pmatrix} A & \\ & B \end{pmatrix} \qquad \text{and} \qquad K \text{ by } \begin{pmatrix} & K \\ K & \end{pmatrix}$$

one sees that Tr $A^p K B^q K$ is jointly concave in A and B. Let us assume, for simplicity, that the Hilbert space is finite-dimensional. (The technical details in passing to the infinite-dimensional case are not so important.) Then Tr $A^p B^{1-p}$ is jointly concave in A and B

$$\frac{d}{dp} \mathrm{Tr}\ A^p B^{1-p}\Big|_{p=0} = \mathrm{Tr}\ B(\ln A - \ln B)$$

thus $S(\sigma|\rho) =$ Tr $\rho(\ln\rho - \ln\sigma)$ is jointly convex in ρ and σ. (Reverse sign than above.) Hence the "conditional entropy" $S(\rho_{12}) - S(\rho_1)$ is concave in ρ_{12}, because

$$S(\rho_{12}) - S(\rho_1) = -S(\rho_{12}|\rho_1 \otimes \frac{1}{d_2}) + \ln d_2$$

(d_2 = dimension of $\mathcal{H}_2$).

Thus we arrive at the inequality $S_1 + S_2 \leq S_{13} + S_{23}$ (with the abbreviations $S_1 = S(\rho_1)$, $S_{13} = S(\rho_{13})$, etc.), which looks somewhat like monotonicity, but actually is something different. Consider $\Delta := (S_{13} - S_1) + (S_{23} - S_2)$. Since $\rho_{123} \to \rho_{13}$ is linear, $S_{13} - S_1$ is concave in ρ_{123}. Similarly, $S_{23} - S_2$ is concave in ρ_{123}. For pure states, $S_{13} = S_2$, $S_{23} = S_1$ (a fact, that we have mentioned already), thus $\Delta = 0$. For mixed states, due to concavity, $\Delta > 0$.

Now choose a fourth space, $\mathcal{H}_4$. Given ρ_{123}, there exists a pure ρ_{1234} such that ρ_{123} is the reduction of ρ_{1234} with respect to $\mathcal{H}_4$. So, $S_{123} + S_2 - S_{12} - S_{23} = S_4 + S_2 - S_{12} - S_{14} \leq 0$, which proves strong subadditivity.

Let me just mention two applications of strong subadditvity, for translationally invariant systems: (i) it guarantees the existence of a mean entropy, (ii) for linear systems, entropy *is* monotonic in the length.

I now want to come back to quantum-mechanical aspects of information theory: the *order structure of states* [16, 18, 25-27]. Suppose that two density matrices are unitarily equivalent, $\rho' = U^*\rho U$. Then they have the same amount of information (whatever is meant by this), since the spectra coincide. What can be said about the spectrum of a convex combination of them? Denote by $e_n(\rho)$ the sum of the n biggest eigenvalues of ρ. This is a concave function of ρ: let $\rho = \lambda\rho_1 + (1-\lambda)\rho_2$. $e_n(\rho) = \max \text{ Tr } P\rho$, where P is an n-dimensional projection. Ref. [28] gives:

$$\text{Tr } P\rho = \lambda \text{ Tr } P\rho_1 + (1-\lambda) \text{ Tr } P\rho_2 \leq \lambda e_n(\rho_1) + (1-\lambda) e_n(\rho_2)$$
$$\Rightarrow e_n(\rho) \leq \lambda e_n(\rho_1) + (1-\lambda) e_n(\rho_2).$$

Now let $\sigma = \sum \lambda_\ell U_\ell^* \rho U_\ell$ (finite sum, $\lambda_\ell \geq 0$, $\sum \lambda_i = 1$), then $e_n(\sigma) \leq \sum \lambda_\ell e_n(\rho) = e_n(\rho)$. One says that σ is more mixed (or more chaotic) than ρ and writes $\sigma \succ \rho$. This notion is also kept even if we do not know whether σ has the above structure. We come back to this problem in a minute.

It is a very strong property since not only the entropy of σ is larger than the one of ρ, but also – according to a theorem already quoted – Tr $f(\sigma) \geq$ Tr $f(\rho)$ for all concave functions f. The interpretation of this could be that passing from ρ to σ, $\sigma \succ \rho$, information is lost in the worst possible way. (One knows examples of time evolution $\rho \to \rho_t$, where $\rho_{t'} \succ \rho_t$ if $t' \geq t$, even non-linear ones. Also for the Gibbs state one has $\sigma_{\beta_1} \succ \sigma_{\beta_2}$ if $\beta_1 \leq \beta_2$ – heating a system produces an enormous loss of information.)

We now come back to the question just posed. Let us say that $\rho' \sim \rho$ if $e_n(\rho') = e_n(\rho)$ for all n, i.e., if the positive spectra coincide. (The kernels need not have the same dimension.) The following assertions are equivalent:

(i) $\sigma \succ \rho$,
(ii) σ is in the weak closure of the convex hull of $\{\rho' : \rho' \sim \rho\}$,
(iii) σ is in the weak closure of the convex hull of $\{U^*\rho U : U \text{ unitary}\}$.

We are not going to prove this neither do we want to discuss the finer details but will rather turn to something entirely different on a conceptional level (although there is an intimate connection with "usual" entropy).

Whereas entropy is a function of a *state*, the quantities we will be concerned with now are functions of the *dynamics*. At this juncture we can present a short outline only.

In the classical case, the best known quantity is the **Kolmogorov-Sinai entropy** [29, 30]. It refers to a phase space (= probability space) Ω with measure μ and a one-parameter group of automorphisms (i.e., measure preserving transformations). One usually considers discrete groups, so they are generated by a single transformation T. (How to proceed in the continuous case seems to be still unknown.)

The Kolmogorov-Sinai entropy is constructed as follows: take a finite partition (Ω_i) of Ω (which means that $\Omega_i \cap \Omega_k = \emptyset$ if $i \neq k$, $\bigcup \Omega_i = \Omega$). The entropy of such a partition (we call it ω) is:

$$S(\omega) := -\sum \mu(\Omega_i) \ln \mu(\Omega_i)$$

Now consider two partitions $\omega_1 = (\Omega_i^{(1)})$, $\omega_2 = (\Omega_k^{(2)})$. They generate a partition $\omega_1 \vee \omega_2$ consisting of all intersections $\Omega_i^{(1)} \cap \Omega_k^{(2)}$. Let, for the moment, $p_i := \mu(\Omega_i^{(1)})$, $q_k := \mu(\Omega_k^{(2)})$, $r_{ik} := \mu(\Omega_i^{(1)} \cap \Omega_k^{(2)})$. Then $S(\omega_1 \vee \omega_2) = -\sum r_{ik} \ln r_{ik}$, hence by (classical) subadditivity $S(\omega_1 \vee \omega_2) \leq S(\omega_1) + S(\omega_2)$. (Clearly, by assumption, $S(T\omega) = S(\omega)$.) Subadditivity guarantees (by a classical theorem) the existence of

$$\lim \frac{1}{n}(S(\omega \vee T\omega \vee \ldots \vee T^{n-1}\omega)) = \inf \frac{1}{n}(S(\omega \vee T\omega \vee \ldots \vee T^{n-1}\omega)) =: s(\omega, T)$$

(the argument of S being defined in an obvious way).

Consider the (classical) conditional entropy of two partitions:

$$\begin{aligned} S(\omega_1, \omega_2) &:= -\sum q_k \left(\frac{r_{ik}}{q_k}\right) \ln \left(\frac{r_{ik}}{q_k}\right) \\ &= -\sum r_{ik} \ln r_{ik} + \sum q_k \ln q_k = S(\omega_1 \vee \omega_2) - S(\omega_2) \end{aligned}$$

(The term "conditional" stems from probability theory; $\frac{r_{ik}}{q_k} = \frac{\mu(\Omega_i^{(1)} \cap \Omega_k^{(2)})}{\mu(\Omega_k^{(2)})}$ is called a "conditional expectation".) The difference $\Delta_n := S(\omega \vee \ldots \vee T^n\omega) - S(\omega \vee \ldots \vee T^{n-1}\omega) = S(T^n\omega, \omega \vee \ldots \vee T^{n-1}\omega)$. By strong subadditivity, $S(\alpha, \beta) \geq S(\alpha, \beta \vee \gamma)$ for any partitions α, β, γ, hence the limit of Δ_n exists and, since $S(\omega \vee \ldots \vee T^{n-1}\omega) = S(\omega) + \Delta_1 + \ldots + \Delta_{n-1}$:

$$\lim \frac{1}{n}(S(\omega \vee \ldots \vee T^{n-1}\omega)) = \lim S(T^n\omega, \omega \vee \ldots \vee T^{n-1}\omega) = \lim \Delta_n = s(\omega, T)$$

(Actually this construction is more suited for generalizations.) The Kolmogorov-Sinai entropy is then $s(T) = \sup s(\omega, T)$, the supremum being taken over all partitions.

It has various properties that have been studied extensively, for instance (i) it is an invariant of the dynamical system, even a conjugacy invariant, which means: two transformations T_1 and T_2 are conjugate, if there exists a measure-algebra isomorphism ϕ such that $\phi \circ T_2^{-1} = T_1^{-1} \circ \phi$ [31], (ii) if a partition α exists, which is a "generator" in the following sense: the σ-algebra generated by the sets $T^m A$ ($m = 0, \pm 1, \pm 2, \ldots$; $A \in \alpha$) is just Ω, then $s(T) = s(\alpha, T)$. Needless to say that there are many more interesting properties.

Two ways of looking at it are: (i) one can interpret it as a mean entropy, or (ii) one can think of it as a measure of the speed (or strength) of mixing, i.e., how fast the system becomes "chaotic" (keep in mind, that this word has different meanings).

As far as quantum mechanics is concerned, one might be tempted to proceed in an analogous manner, replacing the Ω_i by pairwise orthogonal projections P_i with $\sum P_i = 1$, the measure μ by a density matrix ρ and define the entropy as $S(\omega) := -\sum \text{Tr}\,(\rho P_i) \ln \text{Tr}(\rho P_i)$ (ω being the partition (P_i)). But this does not work. Let T be an automorphism, i.e., a unitary mapping, $\varphi \to U\varphi$, of the Hilbert space, or likewise $P_i \to U^* P_i U =: TP_i$. Unfortunately, $\omega \vee T\omega$ does not make very much sense because the algebra generated by $\{P_i, TP_k\}$ can be tremendously large (even in very simple cases $= \mathcal{B}(\mathcal{H})$).

There were several attempts to construct a sensible quantum analog of the Kolmogorov-Sinai entropy but the success was poor. The problem was brought to a solution quite recently only.

The first usable result was obtained by Connes and Størmer [32] when they studied the conjugacy problem for the hyperfinite II_1-factor. Later on, the final answer was given by Connes, Narnhofer, and Thirring [33].

The first step to do is to "translate" the classical theory into an algebraic formalism. This amounts simply to replace the measure algebra by a commutative von Neumann algebra [34]. (In that case, the forementioned difficulties do not occur.)

In order to pass to the non-commutative case, it is wise to consider – instead of the entropy just mentioned before – a functional $H(\mathcal{A}_1 \vee \ldots \vee \mathcal{A}_n)$, where the $\mathcal{A}$'s are finite-dimensional von Neumann algebras containing 1 (we denote the set of all these algebras by $\mathcal{F}$), in addition H should be symmetric in its arguments. The following requirements are more or less self-evident:

(i) $A_i \subset B_i \in \mathcal{F} \Rightarrow H(\mathcal{A}_1, \mathcal{A}_2, \ldots, \mathcal{A}_n) \leq H(\mathcal{B}_1, \mathcal{B}_2, \ldots, \mathcal{B}_n)$,

(ii) $H(\mathcal{A}_1, \ldots, \mathcal{A}_k, \mathcal{A}_{k+1}, \ldots, \mathcal{A}_n) \leq H(\mathcal{A}_1, \ldots, \mathcal{A}_k) + H(\mathcal{A}_{k+1}, \ldots, \mathcal{A}_n)$,

(iii) $\mathcal{A}_1, \ldots, \mathcal{A}_k \subset \mathcal{A} \in \mathcal{F} \Rightarrow H(\mathcal{A}_1, \ldots, \mathcal{A}_k, \mathcal{A}_{k+1}, \ldots, \mathcal{A}_n) \leq H(\mathcal{A}, \mathcal{A}_{k+1}, \ldots, \mathcal{A}_n)$,

(iv) for minimal projections (P_i) (should they exist) with $\sum P_i = 1$ is $H(\mathcal{A}) = \sum s(\tau(P_i))$ (τ being the faithful normal trace).

(Conditions (i) and (ii) refer to monotonicity, or subadditivity, resp.)

Now, following Connes and Størmer, one defines the entropy of a set $\{\mathcal{A}_1, \ldots, \mathcal{A}_n\} \subset \mathcal{F}$ by

$$H(\mathcal{A}_1, \ldots, \mathcal{A}_n) := \sup_{x \in S_n} \left(\sum_{I_n} s(T(x_{I_n})) - \sum_{k=1}^{n} \sum_{i_k} \tau \circ s(E_{\mathcal{A}_k}(x_i^k)) \right)$$

(I_n = multi-index, $S_n = \{x \equiv (x_{I_n}),\ x_{I_n} = 0$ for all but finitely many indices$\}$, E = conditional expectation). The conditional entropy with respect to $\mathcal{A}$, or $\mathcal{B}$, $\in \mathcal{F}$ is:

$$H(\mathcal{A}|\mathcal{B}) := \sup \sum_{i \in I} (T \circ s(E_{\mathcal{B}}(x_i)) - T \circ s(E_{\mathcal{A}}(x_i)))$$

Once one has accomplished all necessary proofs, the Connes-Størmer entropy is defined in the obvious way: $h(\mathcal{A}, T) := \lim \frac{1}{n} H(\mathcal{A}, T\mathcal{A}, \ldots, T^{n-1}\mathcal{A})$, and finally $h(T) := \sup_{\mathcal{A} \in \mathcal{F}} h(\mathcal{A}, T)$. It is – among other things – a conjugacy invariant: two automorphisms T_1, T_2 of a von Neumann algebra are called "conjugate", if there exists a *-automorphism σ such that $T_2 = \sigma^{-1} \circ T_1 \circ \sigma$. (Remark: in the hyperfinite case there is a variety of further results.)

The *ultima ratio* has been found out by Connes, Narnhofer, and Thirring. Let $\mathcal{A}$ be a C^*-algebra, ω be a state, γ_k be completely positive maps $\mathcal{A}_k \to \mathcal{A}$ ($\mathcal{A}_k$ finite-dimensional). In order to find a quantum-mechanical analog of the Kolmogorov-Sinai entropy one should not too much deviate from the classical situation. Now an *Abelian model* consists of a mapping P: $\mathcal{A} \to \mathcal{B}$ (with subalgebras $\mathcal{B}_k$ and a state μ such that $\mu \circ P = \omega$, the original state). $\mathcal{B}$ is supposed to be finite-dimensional and Abelian. $Q^k_{i_k}$ shall be the (existing) minimal projections in $\mathcal{B}_k$, the conditional expectations will be denoted be E_k ($\mathcal{B} \to \mathcal{B}_k$, $\mu_{|\mathcal{B}_k} \circ E_k = \mu$), the completely positive mappings $\rho_k : \mathcal{A}_k \to \mathcal{B}_k$ are defined by $\rho_k := E_k \circ P \circ \gamma_k$. Then the entropy of $\gamma_1, \ldots, \gamma_n$ is

$$H_\omega(\gamma_1, \ldots, \gamma_n) :=$$

$$\sup_{\mu \circ P = \omega} \left[S(\mu_{|\mathcal{B}}) - \sum_{k=1}^{n} S(\mu_{|\mathcal{B}_k}) + \sum_{k=1}^{n} \sum_{i_k} \mu_{|\mathcal{B}_k}(Q^k_{i_k}) S(\mu_{\mathcal{B}_k} \circ \rho_k|_{Q^k_{i_k}} \circ \rho_k) \right]$$

the supremum being taken over all Abelian models. (In the case of an Abelian von Neumann algebra one is led to the usual situation.) Remark: Complete positivity is convenient, but after all not really needed. Schwartz positivity would do the same.

The Connes - Narnhofer - Thirring entropy is now defined as follows: let T be a *-automorphism of a C^*-algebra, ω be an invariant state ($\omega = \omega \circ T$), further let $\mathcal{P}$ be the family of all completely positive mappings of finite-dimensional C^*-algebras in $\mathcal{A}$.

(i) $h_\omega(T, \gamma) := \lim_{n\to\infty} H_\omega(\gamma, T \circ \gamma, \ldots, T^{n-1} \circ \gamma)$. (It is easy to see that this limit exists – due to subadditivity.)

(ii) $h_\omega(T) := \sup_{\gamma \in \mathcal{P}} h_\omega(T, \gamma)$. (Again this is a conjugacy invariant, which follows immediately from the construction.)

Of course, in the majority of cases the dynamical entropy alone cannot decide whether two transformations are isomorphic (or conjugate), only different values of the entropy are simply saying that this cannot be the case.

There are many features in common between quantum entropy, or quantum information theory, resp., and their classical analogs, but definitely not every classical statement has a quantum-mechanical counterpart.

Acknowledgement

I wish to thank T. Hudetz for valuable support.

References

1. J. von Neumann. *Gött. Nachr.* 273 (1927).
2. C. Shannon and W. Weaver. *The Mathematical Theory of Communication.* University of Illinois, Urbana, 1949.
3. A. Wehrl. *Rev. Mod. Phys.* **50**, 221 (1978).
4. E.H. Lieb. *Bull. Am. Math. Soc.* **81**, 1 (1975).
5. A. Wehrl. *Found. Phys.* **9**, 939 (1979).
6. W. Ochs. *Rep. Math. Phys.* **8**, 109 (1975).
   W. Ochs. *Rep. Math. Phys.* **9**, 135 (1976).
7. D. Ruelle. *Statistical Mechanics.* Benjamin, New York, 1969.
8. J. Aczel, B. Forte, and C.T. Ng. *Adv. Appl. Prob.* **6**, 131 (1974).

9. D.W. Robinson and D. Ruelle. *Commun. Math. Phys.* **5**, 288 (1967).
10. A. Renyi. *Wahrscheinlichkeitsrechnung.* Deutscher Verlag der Wissenschaften, Berlin, 1966.
11. H. Umegaki. *Kodai Math. Sem. Rep.* **14**, 59 (1962).
12. G. Lindblad. *Commun. Math. Phys.* **33**, 305 (1973).
13. E.T. Jaynes. *Phys. Rev.* **106**, 620 (1957).
14. R.V. Hartley. *Bell Syst. Techn. J.* **7**, 535 (1928).
15. S. Katai. *Ann. Univ. Sci. Budapest, Eötvös Sect. Math.* **12**, 81 (1967).
16. A. Wehrl. *Rep. Math. Phys.* **6**, 15 (1974).
    A. Wehrl. *Rep. Math. Phys.* **20**, 401 (1984).
17. A. Wehrl. In *Proc. CIME Summer School on Statistical Mechanics* (Bressanone, 1976).
18. W. Thirring. *Quantenmechanik großer Systeme.* Springer, Wien - New York, 1980.
19. E.H. Lieb. *Adv. Math.* **11**, 267 (1973).
20. F.J. Dyson. *J. Math. Phys.* **8**, 1538 (1967).
21. T. Ando. *Topics on Operator Inequalities.* Sapporo, Hokkaido, Japan, 1979.
22. A. Uhlmann. *Commun. Math. Phys.* **54**, 21 (1976).
23. E.H. Lieb and M.B. Ruskai. *Phys. Rev. Lett.* **30**, 434 (1973).
24. E.H. Lieb and M.B. Ruskai. *J. Math. Phys.* **14**, 1938 (1973).
25. A. Uhlmann. *Wiss. Z. Karl-Marx-Univ. Leipzig* **20**, 638 (1971).
    A. Uhlmann. *Wiss. Z. Karl-Marx-Univ. Leipzig* **21**, 427 (1972).
    A. Uhlmann. *Wiss. Z. Karl-Marx-Univ. Leipzig* **22**, 139 (1973).
26. P. Alberti and A. Uhlmann. *Dissipative Motion in State Spaces.* Teubner, Leipzig, 1981.
27. P. Alberti and A. Uhlmann. *Stochasticity and Partial Order.* Deutscher Verlag der Wissenschaften, Berlin, 1981.
28. Fan Ky. *Proc. Ntl. Acad. Sci. USA* **35**, 652 (1959).
29. A.N. Kolmogorov. *Dokl. Akad. Nauk. SSSR* **124**, 754 (1959).
30. Ya.G. Sinai. *Dokl. Akad. Nauk. SSSR* **25**, 899 (1961).
    Ya.G. Sinai. *Usp. Math. Nauk.* **20**, 232 (1965).
31. P. Walters. *An Introduction to Ergodic Theory.* Graduate Texts in Mathematics **79**. Springer, New York, 1982.
32. A. Connes and E. Størmer. *Acta Math.* **134**, 289 (1975).
33. A. Connes, H. Narnhofer, and W. Thirring. *Commun. Math. Phys.* **112**, 691 (1987).
34. T. Hudetz. Diplomarbeit, Wien (1990).

DYNAMICAL ENTROPY FOR INFINITE QUANTUM SYSTEMS

THOMAS HUDETZ

Institut für Theoretische Physik
Universität Wien
A-1090 Wien, Austria

Abstract

We review the recent physical applications of the so–called Connes–Narnhofer–Thirring entropy, which is the successful *quantum mechanical* generalization of the classical Kolmogorov–Sinai entropy and, by its very conception, is a dynamical entropy for *infinite* quantum systems. We thus comparingly review also the physical applications of the *classical* dynamical entropy for *infinite* classical systems.

1. Introduction

The concept of *dynamical* (or Kolmogorov–Sinai) *entropy* plays a central rôle in classical "information dynamics" (cf. Atmanspacher and Scheingraber [1]), and thus presumably also in some of the other contributions collected in this proceedings volume. In particular, the two final pages (before the conclusion) of the included review by Wehrl [2] could even be read as a *mathematical* introduction to the subject of this present contribution: Here, we want to review the recent *physical* applications of the so–called Connes–Narnhofer–Thirring (CNT) entropy (mathematically defined in full generality at the end of Wehrl's text), which is the successful *quantum mechanical* generalization of the classical Kolmogorov–Sinai (KS) entropy.

As briefly indicated by Wehrl [2], here the "generalization" has to be understood in a mathematical sense: Classically, (bounded) *observables* of a physical system are represented by (bounded and possibly continuous, but in any case "measurable") real–valued functions on the phase (or "configuration"–, see section 2) space of the system, endowed with a natural algebraic structure given by point–wise addition and (commutative) multiplication of these functions; and the classical (measure theoretic) KS theory can be equivalently "translated" into this (function–) algebraic language (cf. Hudetz [3, 4, 5]). The (bounded) observables of a *quantum* system, however, are represented by self–adjoint (and bounded, but not necessarily *all* of the self–adjoint bounded) operators on a suitable Hilbert space, with the natural algebraic structure of operator addition and (*non*–commutative) multiplication; and in this sense *classical* systems are just “quantum” systems with a *commutative* (or Abelian) algebra of observables (technically speaking, an Abelian von Neumann– or general C^*–algebra; cf. Thirring [6, 7]).

From this far–sighted point of view, the quantum mechanical generalization of the classical dynamical entropy amounts to extending the algebraic formulation of the KS entropy theory from commutative to *non*–Abelian operator algebras; and this generalization was performed *successfully* (in a “physical” sense which should become clear from the following), partly by Connes and Størmer [8] and “completely” by Connes, Narnhofer and Thirring [9], as also cited by Wehrl [2]. The success of this algebraic approach to quantum dynamical entropy, within the general framework of *algebraic* quantum mechanics, was made possible by the fact that the latter applies not only to microscopic quantum systems (with finitely many degrees of freedom, such as single atoms or small molecules; cf. Thirring [6]) but also to macroscopic systems (i.e., “very large” many–particle systems which can be idealized as *infinite*, with infinitely many degrees of freedom; cf. Thirring [7]). An easily readable exposition of this generalized (algebraic) quantum mechanical framework can be found in part I of the book by Sewell [10], where an introductory statement explicitly says that “the quantum theory of infinite systems (...) contains physically relevant structures, that do not occur in the quantum theory of finite systems”; and this fully applies also to the physically *relevant* concept of quantum dynamical entropy, as developed by Connes, Narnhofer and Thirring [9]. We thus claim that, by its very conception, this generalized KS entropy is a dynamical entropy for *infinite* quantum systems; and our aim in this review paper is to convince the reader thereof, by “physical” arguments and comparing with the physical applications of the *classical* dynamical entropy to classical statistical mechanics of *infinite* systems (in the same sense as above).

The comparative reading of section 2 (classical dynamical entropy) and section 3 (quantum dynamical entropy, both for infinite systems respectively) is enabled by a simultaneous subdivision of the two sections; and the reader is advised to leaf back and forth but to keep the historical succession of classical before quantum theory. The review list of references, as far as they concern the development and applications of the CNT entropy, is tentatively complete; but again the historical order is preferable to the alphabetic one (which in turn would be preferable here to the order of citation as chosen by the editors): Distinguishing *two* historical “phases”, the “mathematical phase” was set about by Connes and Størmer [8] with their definition (cited by Wehrl [2]) of the dynamical entropy with respect to a normalized trace on a von Neumann algebra (which can be reinterpreted “physically” as a quantum state at *infinite* temperature; cf. Thirring [7]), temporarily rejected again by Connes and Størmer [11] themselves. This definition has been applied “mathematically” by Shukhov [12], Besson [13], Pimsner and Popa [14], Quasthoff [15], and Choda [16]. But already in 1985, a “phase transition” to the “mathematical physics” (and *finite* temperature) phase manifested itself, announced by Connes [17] with distant correlation to Narnhofer and Thirring [18], and fully developed by Connes, Narnhofer and Thirring [9] together. Immediately after the transition, in Narnhofer and Thirring [19] there occurred a certain “theory–dynamical” discontinuity with respect to the mathematical phase (cf. Størmer and Voiculescu [20]); and since then the “mathematical physics” phase has coexisted in Narnhofer and Thirring [21, 22], Benatti and Narnhofer [23, 24], Narnhofer [25, 26, 27] and Hudetz [28, 3, 4, 5]. But only recently, the work of Gaspard [29] seems to announce a possible second “phase transition” to a really “physical” phase (see (2.7) resp. (3.7) below).

2. Classical Dynamical Entropy

(2.1) The KS entropy was introduced by Kolmogorov and refined by Sinai as a conjugacy invariant of measure preserving transformations on probability spaces (cf. Wehrl [2] and the references there). In physical applications, it refers to a transformation (e.g., unit time evolution) $T : \Omega \to \Omega$ of the phase (or "configuration"–, see (2.4) below) space Ω of a classical system and assigns a characteristic real number $h_\mu(T) \geq 0$, still depending on the invariant probability measure μ on Ω (e.g., an "equilibrium state" for the time evolution); and we emphasize this dependence by our notation in contrast to the notation $s(T)$ used by Wehrl [2]. As it is well–known, this entropy $h_\mu(T)$ has been applied successfully to (spatially) *finite* classical systems with *finitely* many degrees of freedom.

(2.2) But $h_\mu(T) = 0$ *identically* (i.e., for all possible T) for spatially finite classical systems with a phase space Ω which has a *finite* "grain–size" (mathematically, for an *atomic* measure space Ω), with any "physically reasonable" measure μ. This is so because then the *total* entropy $S(\mu)$ of the partition of Ω into its "atoms" is *finite*, and the entropy of *any* partition of Ω (as defined, e.g., in Wehrl, [2]) is bounded above by $S(\mu) < \infty$. From the definition of $h_\mu(T)$ as in Wehrl [2] it then follows that $h_\mu(T) \leq \lim_{n\to\infty} S(\mu)/n = 0$.

(2.3) Put differently: There is *no intrinsic* (minimal) "scale of size" in Ω for $h_\mu(T) > 0$ (cf. also the remarks on classical vs. quantum chaos by Lindblad [30]). But careful: Ω need not be the *phase* (or even position) space of the physical system; and in particular for classical systems with *infinitely* many degrees of freedom and *infinite* spatial extension (i.e., in classical statistical mechanics) we can have $h_\mu(T) > 0$ *in spite* of an intrinsic (minimal) length scale in physical *position* space!

(2.4) In fact, the KS entropy was first applied to classical statistical mechanics (of infinite systems) by Robinson and Ruelle [31] for the *classical spin models*: For simplicity, we first consider a *one*–dimensional "classical spin" *chain*, where $\Omega_0 = \{s_0, \ldots, s_N\}$ is a set of $(N+1)$ different "spin values", and Ω is given by the spin *configuration space* $\Omega = \times_{k\in\mathbf{Z}}\Omega_k$ with $\Omega_k \equiv \Omega_0\ \forall k \in \mathbf{Z}$ (i.e., the space of two–sided sequences $x = (x_k)_{k\in\mathbf{Z}}$ with elements $x_k \in \Omega_0$, and with the measurable sets generated by "cylinder sets", as is well–known). Let $T : \Omega \to \Omega$ be induced by the *lattice shift* along the chain, i.e., for $x = (x_k)_{k\in\mathbf{Z}}$ let $Tx = (x_{k+1})_{k\in\mathbf{Z}}$. Then, for a T–invariant probability measure ("state") μ on Ω: $h_\mu(T) = s(\mu)$, where $s(\mu) \leq \log(N+1)$ denotes the *entropy density* of the state μ on the chain (in the sense of classical statistical mechanics; cf. Ruelle [32]).

(2.5) Robinson and Ruelle [31] proved this equality already for n–dimensional (cubic) spin lattice systems (physically $n = 2, 3$), implicitly using the definition and properties of the *joint* dynamical entropy $h_\mu(S_1, \ldots, S_n)$ of n *commuting* transformations $(S_1, \ldots, S_n) = \vec{S}$ (symbolic notation) of Ω ("exactified" e.g. by Conze, [33]); where for the lattice translations $S_1, \ldots, S_n$ on $\Omega = \times_{\mathbf{k}\in\mathbf{Z}^n}\Omega_{\mathbf{k}}$ with $\Omega_{\mathbf{k}} \equiv \Omega_0\ \forall \mathbf{k} \in \mathbf{Z}^n$ and with $\vec{S}$–invariant state μ again: $h_\mu(\vec{S}) = s(\mu)$, with the entropy density of μ on the latticc (cf. Ruelle, [32]).

(2.6) But conversely: For infinite classical systems of infinitely many particles moving in a *continuous* physical position space ($\mathbf{Z}^n$ in (2.5) replaced by $\mathbf{R}^n$), "typically" both $h_\mu(\vec{S}) = \infty$ for the unit space translations $\vec{S} = (S_1, \ldots, S_n)$ in the coordinate directions, and $h_\mu(T) = \infty$ for the unit time evolution, *independently* of the particles' interaction (mutual or with an external field)! Goldstein [34] thus considered the so–called *space–time entropy* $h_\mu(\vec{S},T)$ (defined as in Conze [33]) of such systems with translationally invariant equilibrium state μ for the time evolution T (generated by a translationally invariant interaction or external field): He showed that $h_\mu(\vec{S},T) = 0$ for the infinite classical *ideal gas*, whereas $h_\mu(\vec{S},T) > 0$ for the infinite *periodic Lorentz gas* (where $\vec{S}$ is given by the periodicity translations, acting on the space Ω of particle "configurations" in the one–particle phase space X), for a Maxwellian momentum distribution as well (inducing a probability measure μ on Ω given by a "Poisson" distribution of *independent* particles in X; cf. Cornfeld, Fomin and Sinai [35]).

(2.7) Only recently, Gaspard [29] announced an explicit *estimate* for the space resp. time entropy of the (here *one*–dimensional) infinite *ideal gas*: For the unit space translation S and for a partition ξ of the configuration space Ω (as in (2.6) above), induced by a fine–graining of the one–particle phase space X with position (resp. velocity) grain size Δx (resp. Δv), he gets "asymptotically" (as $\Delta x \Delta v \to 0$) for the entropy of S with respect to ξ (cf. Wehrl, [2]):

$$h_\mu(S,\xi) \cong \rho \int_{-\infty}^{\infty} dv\, f(v) \log \frac{e}{\rho\, f(v) \Delta x \Delta v}$$

where ρ (resp. $f(v)$) denotes the constant spatial particle number density (resp. the velocity distribution density with respect to the measure dv). A detailed and rigorous derivation of this "heuristic" formula will be part of a forthcoming joint publication by Gaspard and Hudetz. Similarly, then, we get for the unit time evolution T of the infinite ideal gas:

$$h_\mu(T,\xi) \cong \rho \int_{-\infty}^{\infty} dv\, |v|\, f(v) \log \frac{e}{\rho\, f(v) \Delta x \Delta v}$$

For a *Maxwellian* velocity distribution $f(v)$ at inverse temperature β (times Boltzmann's constant), inducing the equilibrium state μ_β of the ideal gas, one then gets:

$$h_{\mu_\beta}(S,\xi) \cong \rho \log \frac{(2\pi m)^{1/2}}{\rho \beta^{1/2} \Delta x \Delta p} + \frac{3\rho}{2}$$

where $\Delta p \equiv m \Delta v$ with particle mass m, resp.

$$h_{\mu_\beta}(T,\xi) \cong \frac{2\rho}{(2\pi m\beta)^{1/2}} \log \frac{(2\pi m)^{1/2}}{\rho \beta^{1/2} \Delta x \Delta p} + \frac{4\rho}{(2\pi m \beta)^{1/2}}$$

Note that $h_{\mu_\beta}(S) \geq \lim_{\Delta x \Delta p \to 0} h_{\mu_\beta}(S,\xi) = \infty$, resp. $h_{\mu_\beta}(T) \geq \lim_{\Delta x \Delta p \to 0} h_{\mu_\beta}(T,\xi) = \infty$, as generally stated in (2.5) above. But remember that in this case $h_{\mu_\beta}(S,T) = 0$, as shown by Goldstein [34].

(2.8) On the other hand, Goldstein [34] quite generally showed for infinite, spatially *continuous* and *periodic* systems of mutually non–interacting particles (e.g., the infinite periodic Lorentz gas, as already cited in (2.6) above) the following: Let T resp. $\vec{S}$ be the unit time evolution resp. the periodicity translations (acting on the configuration space Ω) with invariant state μ on Ω. For a minimal periodicity volume Λ, we denote by T_Λ the restriction of T to Λ by periodic boundary conditions and ignoring all particles not in Λ. Then, under some rather mild technical assumptions: $h_\mu(\vec{S},T) = h_\mu(T_\Lambda)/|\Lambda|$, where $|\Lambda|$ denotes the volume of Λ. Goldstein [34] further conjectured that for general infinite (periodic) systems of *interacting* particles: $h_\mu(\vec{S},T) = \lim_{\Lambda\to\infty} h_\mu(T_\Lambda)/|\Lambda|$, where $\Lambda \to \infty$ in the sense that the smallest sides of the increasing (periodicity) volumes Λ approach ∞.

In fact, Sinai and Chernov [36] showed this for the infinite system of hard spheres with elastic collisions ("infinite billiard") with unit time evolution T and unit coordinate translations $\vec{S}$ and with (microcanonical) invariant state μ (with *reflecting* boundary conditions for T_Λ and with "suitable" volumes $\Lambda \to \infty$). And Chernov [37] considered the infinite system of infinitely many particles interacting via a pair –, finite range and hard core potential, with unit time evolution T and unit space translations $\vec{S}$, and with (grand canonical) equilibrium state $\mu_{\beta,\rho}$ at inverse temperature β and for (sufficiently small) particle density ρ. He showed that $h_{\mu_{\beta,\rho}}(\vec{S},T)$ is *finite* and has an upper bound $\rho \cdot c(\beta)$.

3. Quantum Dynamical Entropy

(3.1) As already mentioned in the introduction, the CNT entropy was developed as a conjugacy invariant of automorphisms acting on a C^*–algebra with an invariant "state" (cf. Wehrl [2]). In physical applications, it refers to a transformation (e.g., unit time evolution) θ of a quantum system, acting (in the *Heisenberg representation*; cf. Sewell [10]) on the (bounded) observables $A \to \theta(A)$ (which are self–adjoint bounded operators on a suitable Hilbert space $\mathcal{H}$); and it assigns a characteristic real number $h_\omega(\theta) \geq 0$, still depending on the invariant quantum state ω of the system (which, e.g. for *finite* systems, is given by a density matrix ρ_ω on $\mathcal{H}$ with $\theta(\rho_\omega) = \rho_\omega$, where $\omega(A) = \mathrm{Tr}(\rho_\omega A)$ with the trace Tr in $\mathcal{H}$). Many of the properties of this entropy $h_\omega(\theta)$ directly generalize (by construction) the properties of the classical KS entropy.

(3.2) But $h_\omega(\theta) = 0$ *identically* (i.e., for all possible θ) for quantum systems with *finitely* many particles (or degrees of freedom), with any "physically reasonable" state ω. This is so because then the *total* entropy $S(\omega)$ of the system is *finite*; e.g., $S(\omega) = -\mathrm{Tr}\rho_\omega \log \rho_\omega < \infty$ for states ω with finite *energy*, cf. Wehrl [2], Thirring [7]. The H_ω–functional, as cited and used in definition (i) at the end of Wehrl's text, is always bounded above by $S(\omega)$; and from this definition (i) and (ii) of $h_\omega(\theta)$ it then follows that $h_\omega(\theta) \leq \lim_{n\to\infty} S(\omega)/n = 0$; cf. Narnhofer [26].

(3.3) Roughly speaking, the "quantum mechanical phase space" has grain size h^N (with Planck's constant h and the number N of degrees of freedom of the system; cf. Lindblad [30]); and we can get $h_\omega(\theta) > 0$ only for $N \to \infty$, either in *finite* spatial volume V (i.e., in quantum *field* theory) or together with the "thermodynamic limit" $V \to \infty$ (and $N/V \to \rho$, i.e., in quantum *statistical* mechanics). Put differently, a "quantization" of a *finite* (but continuous) *chaotic* classical system (here in the sense that $h_\mu(T) > 0$), if it is performed "chaos–preserving" (e.g., in the sense that $h_\mu(T) = h_\omega(\theta)$ with obvious notations), has to lead to an *infinite* (macroscopic or "mesoscopic", cf., e.g., Dittrich and Graham, [38]) quantum system, but *not* necessarily continuous in space.

(3.4) For the well–known classical toy–model of the "Baker's transformation", this "quantization" is performed explicitly (e.g.) in Narnhofer, Pflug and Thirring [39]; leading first (by conjugacy and equivalent algebraic "translation", cf. the introduction) to a one–dimensional classical spin model (2.4), and then to a *quantum* spin chain (cf. also Benatti, Narnhofer and Sewell [40] for a "quantization" of Arnold's "cat map"). In fact, the CNT entropy was first applied to quantum statistical mechanics by

Connes [17] for these *one*–dimensional *quantum spin models*, where the Hilbert spaces $\mathcal{H}_0 = \mathbf{C}^{(N+1)}$, $\mathcal{H}_k \equiv \mathcal{H}_0\ \forall k \in \mathbf{Z}$, and $\mathcal{H} = \bigotimes_{k\in\mathbf{Z}} \mathcal{H}_k$ mean a "spin $N/2$–particle" at each lattice site, with θ given by the lattice shift $\theta : \mathcal{B}(\mathcal{H}_k) \to \mathcal{B}(\mathcal{H}_{k+1})\ \forall k \in \mathbf{Z}$. Then, for a θ–invariant quantum state ω of the spin chain: $h_\omega(\theta) \leq s(\omega)$ at first, where $s(\omega) \leq \log(N+1)$ denotes the *entropy density* of the state ω (in the sense of quantum statistical mechanics, cf. Bratteli and Robinson [41]).

This general inequality was sharpened to $h_\omega(\theta) = s(\omega)$ (as also classically, cf. (2.4)) by Narnhofer and Thirring [21] at least for a θ–invariant *short range* interaction between near–by spins, with an *equilibrium* state ω for the generated time evolution. And also for a one–dimensional *Fermi lattice gas* with "quasi–free" state ω invariant with respect to the lattice shift θ (cf., e.g., Narnhofer, Pflug and Thirring [39]) $h_\omega(\theta) = s(\omega)$, as shown by Connes, Narnhofer and Thirring [9].

(3.5) The definition and properties of the *joint* quantum dynamical entropy $h_\omega(\sigma_1, \ldots, \sigma_n)$ of n *commuting* quantum transformations $(\sigma_1, \ldots, \sigma_n) = \vec{\sigma}$ (symbolic notation) were extended from (2.5) by Hudetz [28] (following the classical route of Conze, [33]); and the author generalized the quantum *inequality* of (3.4) to n–dimensional (cubic) quantum spin lattice systems with the lattice translations $\sigma_1, \ldots, \sigma_n$ and $\vec{\sigma}$–invariant state ω: $h_\omega(\vec{\sigma}) \leq s(\omega)$, with the quantum entropy density of ω on the lattice (cf. Bratteli and Robinson [41]). Again, $h_\omega(\vec{\sigma}) = s(\omega)$ for an n–dimensional Fermi lattice gas with $\vec{\sigma}$–invariant "quasi–free" state ω (cf. Hudetz [3]).

(3.6) Narnhofer and Thirring [19] performed the *continuum limit* for the one–dimensional Fermi lattice gas as in (3.4) above, leading to the one–dimensional "ideal" Fermi gas with translationally invariant, quasi–free equilibrium state ω, which is determined by the (two–point function) number density distribution in momentum space: $n_\omega(k) = (1+\exp[\beta(\epsilon(k)-\mu)])^{-1}$, where β (times Boltzmann's constant) is the inverse temperature, k is momentum ("wave number" times Planck's reduced constant $\hbar$), $\epsilon(k)$ is the one–particle energy–momentum spectrum, and μ is a "chemical potential" of the grand canonical equilibrium state ω (cf. Bratteli and Robinson [41]). With the "fermionic" entropy functional $s_F(n) = -n\log n - (1-n)\log(1-n)$ of $n \in [0,1]$, Narnhofer and Thirring [19] expressed the entropy of the unit space translation σ as:

$$h_\omega(\sigma) = \int_{-\infty}^{\infty} \frac{dk}{2\pi\hbar}\, s_F(n_\omega(k))$$

As mentioned in the introduction, the derivation of this (physically *correct*) formula involved a certain "physical" discontinuity with respect to the "mathematical" phase before, which is still present by now (cf. Størmer and Voiculescu [20]) but will presumably disappear in the future. And for the one–particle implemented ("quasi–free") unit time evolution τ given by $\epsilon(k)$ as one–particle Hamiltonian in momentum representation, the *dynamical* entropy of the "ideal" Fermi gas was deduced (by conjugacy invariance) from $h_\omega(\sigma)$:

$$h_\omega(\tau) = \int_{-\infty}^{\infty} \frac{dk}{2\pi\hbar} \left|\frac{d\epsilon(k)}{dk}\right| s_F(n_\omega(k))$$

(3.7) Gaspard [29] performed for both $h_\omega(\sigma)$ and $h_\omega(\tau)$ the *high temperature – dilute gas* limit, where the mean *particle density* $\rho = (2\pi\hbar)^{-1} \int n_\omega(k)dk$ tends to zero such that the "fugacity" is approximately $\exp(\beta\mu) \approx 2\pi\hbar\rho(\beta/2\pi m)^{1/2}$, cf. Bratteli and Robinson [41]. Then, for the "classical" free one–particle time evolution given by $\epsilon(k) = k^2/2m$, he gets approximately (to first order in ρ):

$$h_\omega(\sigma) \approx \rho \log \frac{(2\pi m)^{1/2}}{2\pi\hbar\rho\beta^{1/2}} + \frac{3\rho}{2}$$

$$h_\omega(\tau) \approx \frac{2\rho}{(2\pi m\beta)^{1/2}} \log \frac{(2\pi m)^{1/2}}{2\pi\hbar\rho\beta^{1/2}} + \frac{4\rho}{(2\pi m\beta)^{1/2}}$$

Note that, with these approximations, exact "correspondence" to (2.7) is established by $\Delta x \Delta p = 2\pi\hbar \equiv h$, illustrating the general "heuristic" statement in (3.3) above. But in contradistinction to the classical situation (2.7), $h_\omega(\sigma) < \infty$ resp. $h_\omega(\tau) < \infty$ cannot be exceeded any more in the quantum case. More generally, it can be shown for the n–dimensional continuous Fermi gas with the unit space translations $(\sigma_1, \ldots, \sigma_n) = \vec{\sigma}$ in the coordinate directions and with *any* (not necessarily quasi–free) $\vec{\sigma}$–invariant state ω with *finite* particle density ρ, that $h_\omega(\vec{\sigma}) < \infty$ (using the results of (3.5) above; cf. Hudetz [3, 28]).

(3.8) But whereas for the Fermi gas in $n = 1$ dimension (as in (3.6) above) also $h_\omega(\tau) < \infty$ for any physically reasonable quasi–energy $\epsilon(k)$, we get $h_\omega(\tau) = \infty$ for $n > 1$, as also classically in (2.6), at least with one–particle implemented unit time evolution τ and quasi–free equilibrium state ω. This (at first sight surprising) result follows (see below) from the *scaling* of the multi–entropy $h_\omega(\theta_1, \ldots, \theta_n)$ as cited in (3.5), which implies the following (also surprising) property: If $h_\omega(\theta_{i(1)}, \ldots, \theta_{i(m)}) < \infty$ for $0 < m < n$ (with $i(j) \neq i(k)$), then $h_\omega(\theta_1, \ldots, \theta_n) = 0$ (Hudetz [28]).

From this we first have to draw the following "no–go" conclusion (Hudetz [3, 28]): For an n–dimensional continuous Fermi gas with *any* translationally invariant (mutual) interaction such that the (unit) time evolution τ exists, with the unit coordinate translations $(\sigma_1, \ldots, \sigma_n) = \vec{\sigma}$ and $\vec{\sigma}$–invariant equilibrium state ω (with finite particle density), because of $h_\omega(\vec{\sigma}) < \infty$ from (3.7) together with the scaling property, the space–time entropy *vanishes*: $h_\omega(\vec{\sigma}, \tau) = 0$, even if $h_\omega(\tau) = \infty$ for $n > 1$ as announced above. For these infinite quantum systems of *interacting* particles in $n > 1$ dimensions (physically $n = 3$), the space–time entropy thus can *not* distinguish between different time evolutions τ (as it *does* in the classical case (2.8)): $h_\omega(\tau) \equiv \infty$ but $h_\omega(\tau, \vec{\sigma}) \equiv 0$!

But the above scaling property can be "circumvented" by omitting one of the coordinate translations $\vec{\sigma}$ (without loss of generality, σ_1) such that the "time–space" entropy $h_\omega(\tau, \sigma_2, \ldots, \sigma_n)$ can become *positive* and non–trivial: Again for the one–particle implemented (*non*–interacting) unit time evolution τ of the "ideal" Fermi gas (for quasi–energy $\epsilon(\vec{k})$) with quasi–free equilibrium state ω determined by a density–momentum distribution $n_\omega(\vec{k})$ as for $n = 1$ in (3.6), the following expression was derived by Hudetz [3] as for $n = 1$ by Narnhofer and Thirring [19]:

$$h_\omega(\tau, \sigma_2, \ldots, \sigma_n) = \int_{\mathbf{R}^n} \frac{d^n k}{(2\pi\hbar)^n} \left| \frac{\partial \epsilon(\vec{k})}{\partial k_1} \right| s_F(n_\omega(\vec{k}))$$

This "time–space" entropy is *finite* (for reasonable $\epsilon(\vec{k})$) and of course *positive*, from which by the scaling property it follows that $h_\omega(\tau) = \infty$ as announced above.

Acknowledgements

I thank the organizers of "Information Dynamics" for the invitation to this conference, and P. Gaspard for the kind permission to cite his work prior to publication. Financial support by "Fonds zur Förderung der wissenschaftlichen Forschung in Österreich" (Project P7101–PHY), and a travel grant from the Austrian ministery of science, is gratefully acknowledged.

References

1. H. Atmanspacher and H. Scheingraber. *Found. Phys.* **17**, 939 (1987).
2. A. Wehrl. Information theoretical aspects of quantum mechanical entropy. This volume.
3. T. Hudetz. Dynamical entropy: From classical to quantum theory. Thesis (in German, unpublished), Univ. Vienna (1989).
4. T. Hudetz. In *Selected Topics in QFT and Mathematical Physics.* Eds. J. Niederle and J. Fischer. World Scientific, Singapore, 1990, p. 110.
5. T. Hudetz. In *Nonlinear Dynamics and Quantum Dynamical Systems.* Eds. G.A. Leonov, V. Reitmann and W. Timmermann. *Mathematical Research* 59, p. 27. Akademie–Verlag, Berlin, 1990.
6. W. Thirring. *Quantum Mechanics of Atoms and Molecules.* Springer, New York, 1981.
7. W. Thirring. *Quantum Mechanics of Large Systems.* Springer, New York, 1982.
8. A. Connes and E. Størmer. *Acta Math.* **134**, 289 (1975).
9. A. Connes, H. Narnhofer, and W. Thirring. *Commun. Math. Phys.* **112**, 691 (1987).
10. G.L. Sewell. *Quantum Theory of Collective Phenomena.* Oxford University Press, Oxford, 1986.
11. A. Connes and E. Størmer. In *Operator Algebras and Group Representations. Monographs and Studies in Mathematics* 17, p. 113. Pitman, 1984.
12. A.G. Shukhov. *Funct. Anal. Appl.* **15**, 154 (1981).
13. O. Besson. *Ergod. Th. & Dynam. Sys.* **1**, 419 (1981).
    O. Besson. In *Quantum Probability and Applications II.* Eds. L. Accardi and W. von Waldenfels. *Lecture Notes in Mathematics* **1136**, p. 81. Springer, Berlin, 1985.
14. M. Pimsner and S. Popa. *Ann. Scient. Éc. Norm. Sup.*, 4e *série*, **19**, 57 (1986).
15. U. Quasthoff. *Math. Nachr.* **131**, 101 (1987).
    U. Quasthoff. On automorphisms of factors related to measure space transformations. Univ. Leipzig preprint. To appear in *Rep. Math. Phys.* (1990).
16. M. Choda. Entropy for *–endomorphisms and relative entropy for subalgebras. Univ. Osaka preprint. To appear in *J. Operator Theory* (1990).
17. A. Connes. *C. R. Acad. Sci. Paris* **301 I**, 1 (1985).
18. H. Narnhofer and W. Thirring. *Fizika* **17**, 257 (1985).
19. H. Narnhofer and W. Thirring. *Lett. Math. Phys.* **14**, 89 (1987).

20. E. Størmer and D. Voiculescu. Entropy of Bogoliubov automorphisms of the canonical anticommutation relations. Univ. Berkeley preprint. To appear (1990).

21. H. Narnhofer and W. Thirring. *Lett. Math. Phys.* **15**, 261 (1988).

22. H. Narnhofer and W. Thirring. *Commun. Math. Phys.* **125**, 565 (1989).

23. F. Benatti and H. Narnhofer. *J. Stat. Phys.* **53**, 1273 (1988).

24. F. Benatti and H. Narnhofer. Strong asymptotic Abelianess of entropic K–systems. Univ. Vienna preprint UWThPh–1989–34. To appear in *Comm. Math. Phys.* (1990).

25. H. Narnhofer. *Rep. Math. Phys.* **25**, 345 (1988).

26. H. Narnhofer. In *IXth International Congress on Mathematical Physics*. Eds. B. Simon, A. Truman, I.M. Davies. Adam Hilger, Bristol and New York, 1989, p. 64.

27. H. Narnhofer. Dynamical entropy, quantum K–systems, and clustering. Univ. Vienna preprint UWThPh–1989–3. To appear in *Quantum Probability and Applications V* (1990).

28. T. Hudetz. *Lett. Math. Phys.* **16**, 151 (1988).

29. P. Gaspard. Bound, quasibound and resonant quantum states. Univ. Brussels preprint. To appear in the proceedings of *Adriatico Research Conference and Miniworkshop on Quantum Chaos.* World Scientific, Singapore, 1990.

30. G. Lindblad. In *Fundamental Aspects of Quantum Theory*. Eds. V. Gorini and A. Frigerio. Plenum Press, New York, 1986, p. 199.

31. D.W. Robinson and D. Ruelle. *Commun. Math. Phys.* **5**, 288 (1967).

32. D. Ruelle. *Thermodynamic Formalism. Enc. Math. Appl.* **5**. Addison–Wesley, Reading, 1978.

33. J.P. Conze. *Z. Wahrscheinlichkeitstheorie verw. Geb.* **25**, 11 (1972).

34. S. Goldstein. *Commun. Math. Phys.* **39**, 303 (1975).
S. Goldstein, J.L. Lebowitz, and M. Aizenmann. In *Dynamical Systems, Theory and Applications*. Ed. J. Moser. *Lecture Notes in Physics* **38**. Springer, Berlin, 1975, p. 112.

35. I.P. Cornfeld, S.V. Fomin and Ya.G. Sinai. *Ergodic Theory. Grundlehren der mathematischen Wissenschaften* **245**. Springer, New York, 1982.

36. N.I. Chernov and Ya.G. Sinai. *Trudy Sem. I.G. Petrovsky* **8**, 218 (1982).

37. N.I. Chernov. In *Mathematical Problems of Statistical Mechanics and Dynamics*. Ed. R.L. Dobrushin. Reidel, Dordrecht, 1986, p. 125.

38. T. Dittrich and R. Graham. *Naturwissenschaften* **76**, 401 (1989).

39. H. Narnhofer, A. Pflug, and W. Thirring. In *Symmetry in Nature*. Volume in honour of L.A. Radicati di Bronzolo. Scuola Normale Superiore, Pisa, 1989, p. 597.

40. F. Benatti, H. Narnhofer, and G.L. Sewell. A non–commutative version of the Arnold cat map. Preprint. To appear in *Lett. Math. Phys.* (1990).

41. O. Bratteli and D.W. Robinson. *Operator Algebras and Quantum Statistical Mechanics II.* Springer, New York, 1981.

QUANTUM CHAOS IN OPEN SYSTEMS

Thomas Dittrich

Institut für Physik
Universität Augsburg
D-8900 Augsburg, FRG

Robert Graham

Fachbereich Physik
Universität GHS Essen
D-4300 Essen 1, FRG

Abstract

Quantum systems exhibiting chaotic dynamics in their classical limit provide a stimulating testing ground for an information theoretical interpretation of dynamical properties: In isolated quantum systems, the positive information production characteristic of classical chaotic systems is replaced by quasiperiodic behavior. We show that the quantal suppression of chaos is resolved, in turn, if an information exchange with the environment is taken into account. As a specific example, we study the quantum kicked rotor under the influence of dissipation and continuous observation.

A Proviso

The present contribution is not concerned with any general aspect of the rôle of information theory in physics. Rather, it represents a case study of complex behavior in systems combining microscopic with macroscopic features. It is challenging to understand the results obtained in this field in terms of information theoretical concepts, but just a few preliminary and qualitative ideas will be presented how to proceed in this direction.

1. Introduction

It is well known that basic properties of dynamical systems can be described and interpreted in terms of information content, information flow, information produc-

Information Dynamics, Edited by H. Atmanspacher and
H. Scheingraber, Plenum Press, New York, 1991

tion etc. Chaotic systems, in particular, can be characterized in an elegant way as continuously conveying information from fine scales to coarse scales [1].

For classical systems, these ideas have been worked out in great detail by now. The investigation of complex dynamics in microscopic systems which have to be described in the framework of quantum mechanics, on the other hand, has gained thrust only in the last decade. Presently this field, called "quantum chaos" for short, is very active (a collection of reviews is listed under Ref. [2]). It has turned out that the wave nature of matter on small scales enforces far-reaching modifications of the notions adopted from the classical theory of chaotic dynamics. It is the purpose of this contribution to discuss these subtleties of the quantum-classical correspondence in a specific case which is reasonably well understood, the so-called "kicked rotor". Classically, this model is represented by the time-dependent Hamiltonian [3,4]

$$H(p,q;t) = \frac{p^2}{2} - \frac{K^2}{4\pi}\cos(2\pi q)\sum_{n=-\infty}^{\infty}(t-n) \tag{1}$$

defined on a cylindrical phase space, $-\infty < p < \infty$, $0 \le q < 1$. It corresponds, e.g., to a particle confined to a ring and subjected to a spatially uniform force field in the plane of the ring, pulsed periodically in time. The periodic driving suggests to describe the dynamics more economically in a stroboscopic way, by a map which propagates the system from one time section, say immediately after a "kick", to a time section immediately after the subsequent one [3,4]:

$$\begin{aligned} p' &= p - \frac{K}{2\pi}\sin(2\pi q') \\ q' &= (q+p) \bmod 1 \end{aligned} \tag{2}$$

It is referred to as the "standard map".

Quantum mechanically, the action and angle variables in the Hamiltonian (1) become operators with the commutation relation $[p,q] = -i\hbar$. A time step of the quantum mechanical state

$$\psi' = U\psi \tag{3a}$$

analogous to the classical map (2), is generated by the unitary operator [5]:

$$U = \exp\left(\frac{i}{\hbar}\frac{K^2}{4\pi^2}\cos(2\pi q)\right)\exp\left(-\frac{i}{\hbar}\frac{p^2}{2}\right) \tag{3b}$$

By the dimensionless parameter $\hbar$ we denote the ratio of a characteristic action of the kicked rotor to Planck's constant, so that the classical limit of this system amounts to $\hbar \to 0$.

Due to the discrete time-translation symmetry of the Hamiltonian (1), Floquet's theorem (the one-dimensional version of Bloch's theorem) applies. As a consequence, the unimodular eigenvalues $\exp(i\omega_\alpha)$ of the operator U are good quantum numbers for this system. The eigenphases ω_α, defined on the cyclically closed interval $[0, 2\pi)$, are referred to as "quasi-energies".

Both in the classical and in the quantal case, numerical iterations of the corresponding dynamical mappings (2) and (3), respectively, provide a practicable way of investigating the long-time behavior of the kicked rotor.

2. Quantum Suppression of Chaos

In the classical case, the kicked rotor passes through all stages from completely regular to almost completely chaotic motion if the parameter K is raised from $K = 0$ to large values [3, 4]. Figure 1 shows the phase space structure generated by the classical map (2) for various values of K. In particular, there is a threshold $K_c \approx 0.9716\ldots$ such that for $K < K_c$, regular motion prevails and chaotic regions in phase space are isolated from each other by invariant manifolds encircling the phase-space cylinder. For $K > K_c$, these chaotic regions become connected and form a single domain along the cylinder. For $K \gg K_c$, islands of regular motion in this "chaotic sea" progressively diminish and the dynamics, despite being deterministic, allows for an approximate description in terms of diffusion in the action variable p [5]:

$$\langle (p_n - p_0)^2 \rangle = Dn \tag{4}$$

In this regime, energy increases linearly and without bound.

By a slight generalization, the kicked rotor can be modified so as to serve as a model also for a dissipative chaotic dynamics: A dissipative version of the map (2), which, of course, cannot be derived from any underlying Hamiltonian, is the Zaslavsky map [6]:

$$\begin{aligned} p' &= \lambda p - \frac{K}{2\pi} \sin(2\pi q') \\ q' &= (q + p) \bmod 1 \end{aligned} \tag{5}$$

It takes friction proportional to the action p of the rotor into account, controlled by the additional parameter λ which is identical to the Jacobian determinant of the map (5). For $0 \leq \lambda < 1$, dissipation will eventually balance the diffusive energy increase (cf. Eq.(4)) characteristic of the conservative chaotic dynamics, and force the system into a steady state. Above some threshold $K_s(\lambda)$ in the nonlinearity parameter K, this steady state takes the form of a strange attractor [7], a typical example of which is presented in Fig.2.

In order to set the stage for a comparison of the classical and quantal long-time behavior of the kicked rotor, we briefly consider an analogous issue occurring in numerical simulations of classical dynamical systems: If a chaotic dynamics is implemented on a digital computer, care has to be taken of possible artefacts caused by the finite precision of the machine used. Specifically, round-off errors alter both the time-evolution and the phase-space structure of chaotic systems in a profound way [8, 9]:

(i) The number of accessible different states becomes finite. As a consequence, any deterministic time-evolution must eventually enter into a limit cycle of finite period, such that all instability and unpredictability disappears.

(ii) All phase-space structures are coarse-grained. In particular, fractal structures characteristic both of conservative and dissipative chaotic systems acquire a finite smallest scale and thus lose their self-similarity.

It turns out that the deviations of a quantal dynamics from its chaotic classical counterpart are surprisingly reminiscent of these effects of discretization in computing.

In a seminal work, Casati *et al.* [5] have performed a systematic numerical comparison of the time-evolution of the energy for the quantal and the classical kicked rotor. A typical result, for a value of the parameter K well in the classically chaotic

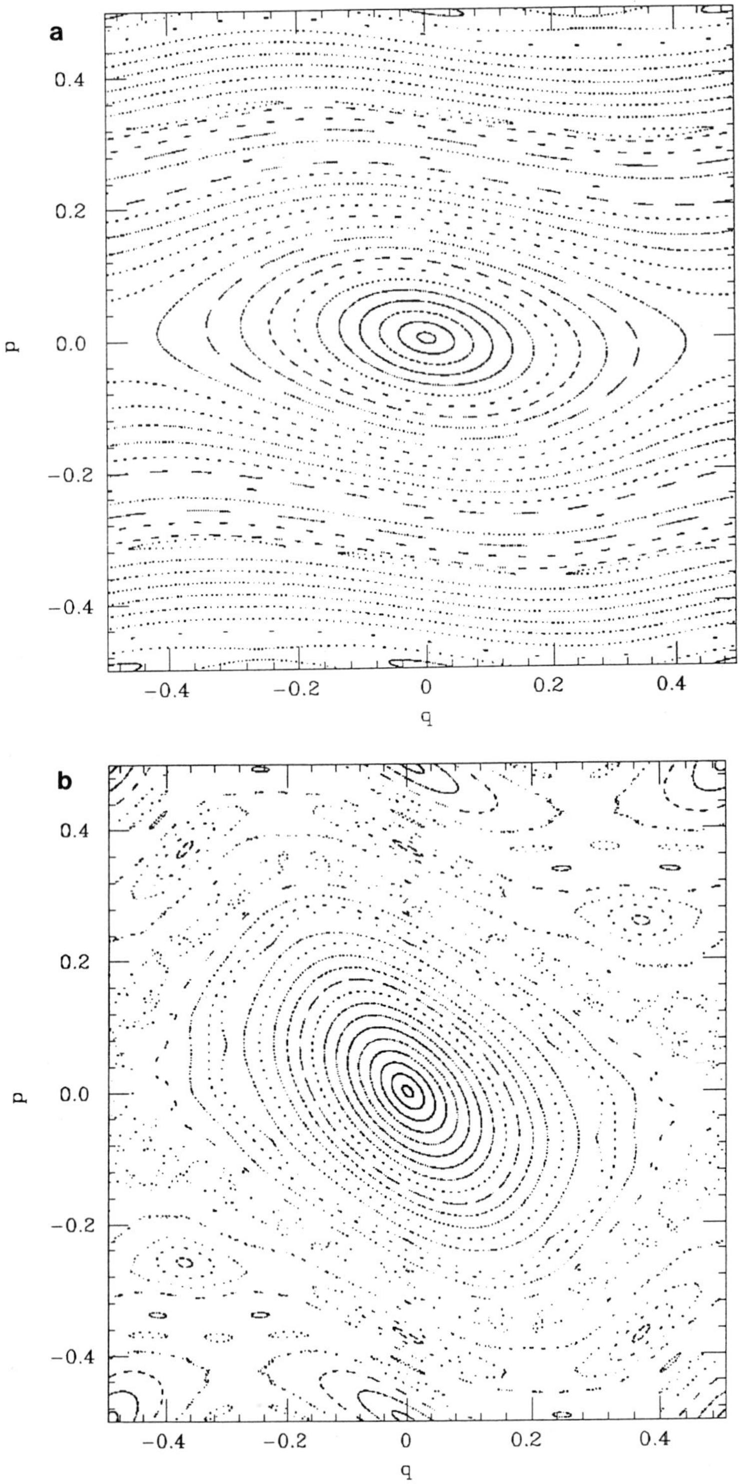

Figure 1: Phase-space portraits of the classical standard map for four different values of the nonlinearity parameter: (a) $K = 0.3$ (phase space is mostly regular), (b) $K = 0.9716$ (at the critical value K_c, see text).

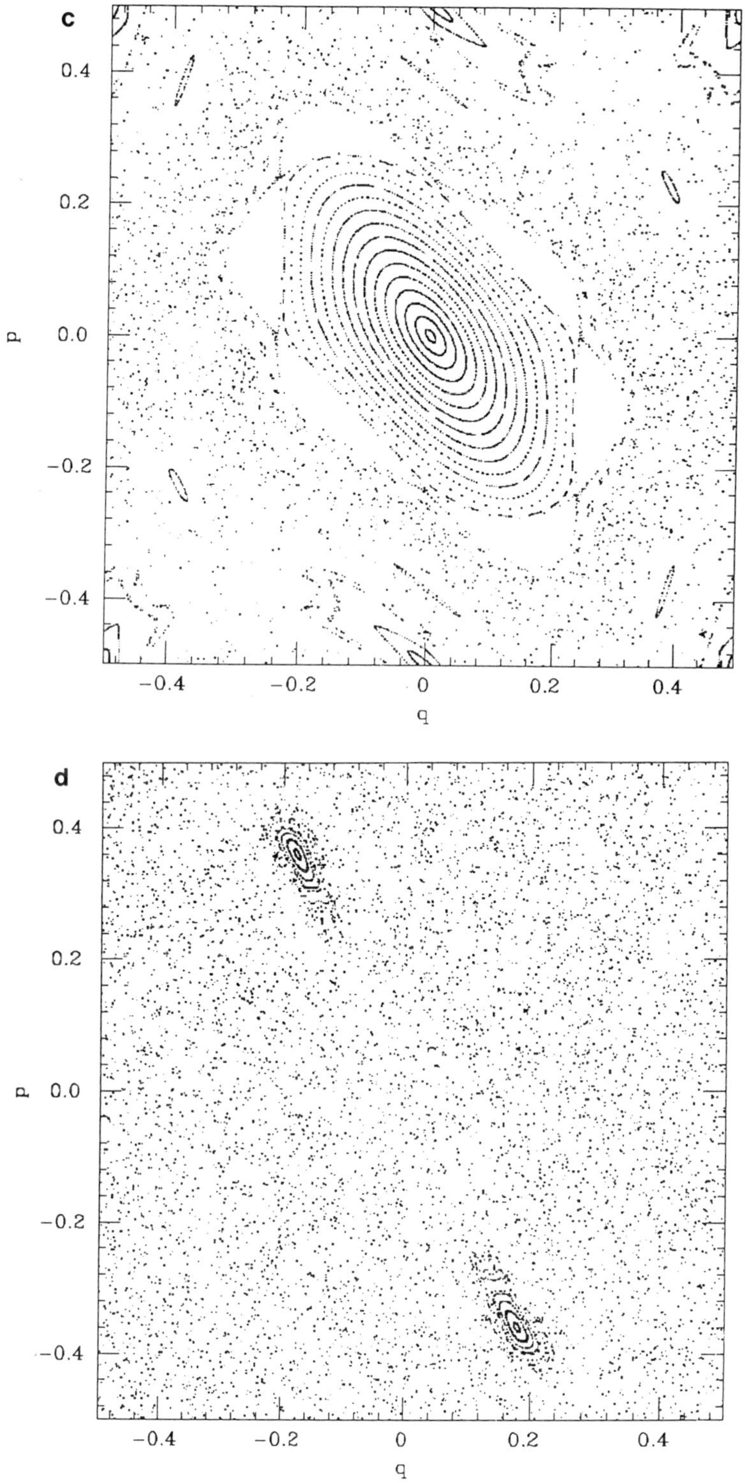

Figure 1: Phase-space portraits of the classical standard map for four different values of the nonlinearity parameter: (c) $K = 1.5$ (the main resonance is surrounded by a chain of secondary islands), (d) $K = 5.0$ (the main resonance has bifurcated, phase space is mostly chaotic).

regime, is shown in Fig.3: Initially, both systems behave in a roughly identical manner, as should be expected from the correspondence principle. However, from some characteristic time on, which depends both on K and $\hbar$, the quantum system ceases to absorb energy and enters into a dynamical regime where the energy fluctuates around a constant mean. Further investigations [10] revealed that the quantal time-evolution is quasiperiodic, i.e., it corresponds to a *discrete* quasienergy spectrum. A discrete spectrum is always associated with normalizable eigenstates of effectively finite extension and does not allow for an unbounded energy growth.

The reason for this surprising outcome was later identified as a phenomenon well known from the theory of electronic states in disordered media [11]. For waves scattered by a random potential, destructive interference always dominates on long spatial scales and leads to exponential localization. In the solid state context, this coherence effect is known as "Anderson localization". Of course, in the quantum kicked rotor, being a completely deterministic system, there is no genuine randomness. It could be shown, however, that the phase factors $\exp(-ip^2/2\hbar)$ occurring in the operator (3b) become sufficiently irregular (provided the parameter $2\pi\hbar$ is a typical irrational number [12-14]) to act in a way similar to random scattering: Quasienergy eigenstates become exponentially localized on the p-axis. Consequently, any initial state that is itself localized in p decomposes only into an effectively finite number of quasienergy states. The corresponding time-evolution must start to deviate from its classical coun-

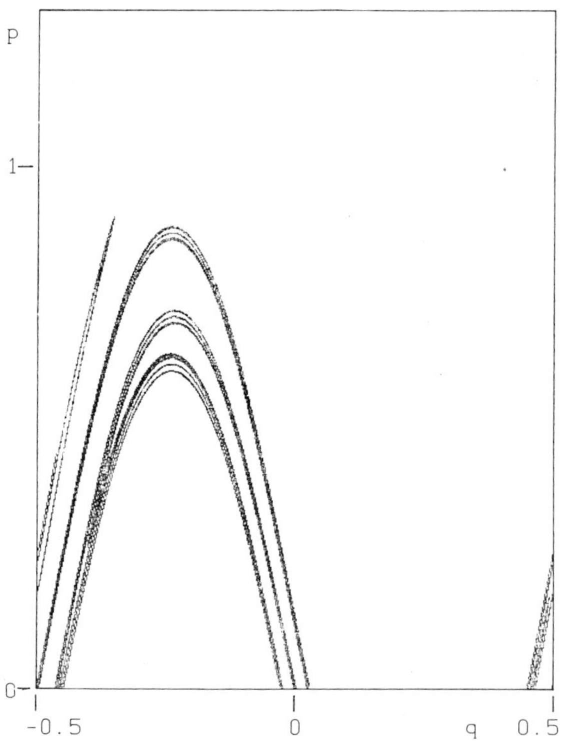

Figure 2: Strange attractor for the dissipative standard map at the parameter values $K = 5$, $\lambda = 0.3$. Because of the symmetry of the attractor, only the upper half-plane $p \geq 0$ is shown.

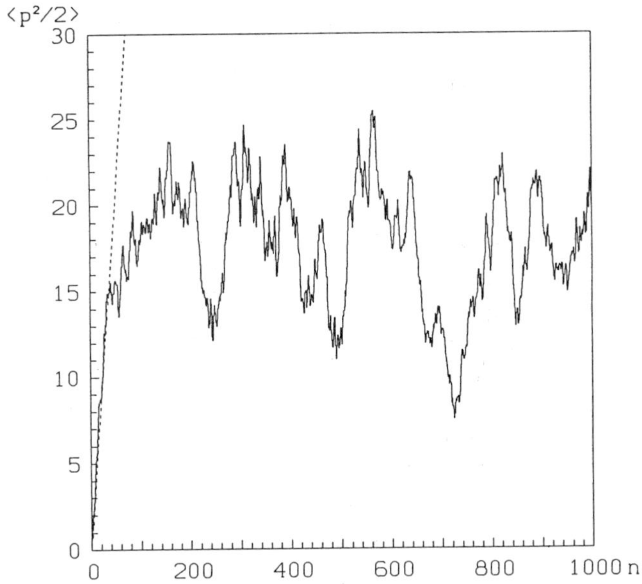

Figure 3: Time-evolution of the energy for the quantum kicked rotor (solid line) and its classical counterpart (dashed line). The parameter values are $K = 10$ and, for the quantized version, $2\pi\hbar = 0.3/(\sqrt{5}-1)$.

terpart as soon as the energy-time uncertainty relation allows the finite separation between the discrete quasienergies to be resolved. This determines the cross-over time and completes the argument [15].

3. Quantum Chaos in Open Systems

The results obtained from studies of the quantum kicked rotor and other systems show that these microscopic, few-degrees-of-freedom systems are not capable of supporting chaotic dynamics and behave more stable and predictable than their classical counterparts. This is due to subtle interference phenomena which reveal the wave nature of microscopic physics in a much more dramatic way than in the case of classically regular systems. As a consequence, there must be a particularly intricate quantum-to-classical transition bridging the simpler microscopic dynamics and the richness of complex behavior in macroscopic systems. This transition combines two important aspects: The increase in characteristic values of action corresponds to a relative decrease in the typical wavelength of quantum interference patterns and leads to a divergence of the time of deviation from classical dynamics. The increase in the number of degrees of freedom involved in a dynamics, on the other hand, allows for macroscopic, irreversible behavior through the destruction of coherence.

Using again the kicked rotor as a working example, the present authors have studied this transition in two prototypical cases where both the semiclassical, i.e., short-wavelength limit and the disruption of coherence by ambient degrees of freedom are central issues: Quantum chaos in dissipative systems [16, 17] and in systems which are continuously observed [18, 19].

The simplest way to investigate the influence of a macroscopic environment on a quantum dynamics consists in simulating it summarily as random noise. It has indeed been shown that localization in the quantum kicked rotor is destroyed in this way [20]. A more detailed description of the disruption of coherence in open systems, however, is obtained by coupling the system at issue to a heat bath [21]. In order to formally extend the model of the kicked rotor in such a way, additional terms have to be included in the Hamiltonian (1). Using standard methods [22, 23], a heat bath is incorporated as a large number of harmonic oscillators, and a coupling term is specified according to the intended application: Friction can be modelled by coupling the rotor to the bath oscillators via creation and annihilation operators of quanta of the action p, with a coupling constant that depends on p. Continuous quantum measurements [24] of characteristic dynamical quantities like, e.g., the energy and the probability distribution along the action axis are modelled by coupling the corresponding operators to the so-called "pointer variable". The pointer variable is that degree of freedom of a measuring device which carries the desired information on the measured quantum system, but can be read off classically [25].

In both cases, it is an adequate approximation to assume the response of the environment to have a memory decaying sufficiently fast such that the time-evolution of the whole system can be considered Markovian on the relevant time scales [17]. Under this and some other, more technical assumptions it is possible to derive a quantum map in the form of a propagator for the density operator ρ of the central system, i.e. here, the kicked rotor:

$$\rho' = G\rho \tag{6}$$

In general, the propagator G is non-unitary and generates irreversible behavior. Indeed, it reduces to the dissipative map (5) in the classical limit if friction has been included. Furthermore, it allows to derive a stochastic map of the form

$$\begin{aligned} p' &= f(p, q') + \xi \\ q' &= g(p, q) + \varphi \end{aligned} \tag{7}$$

as a semiclassical limit of the quantum dynamics, where $f(p, q')$ and $g(p, q)$ generate the deterministic classical dynamics and the random variables ξ, φ represent quantum noise.

Like the classical maps (2) and (5) and the conservative quantum map (3), the propagator (6) and its semiclassical approximation (7) can be used to perform numerical studies of the long-time behavior. Typical results of such numerical experiments [16, 19] are shown in Figs. 4 and 5, again as plots of the time-evolution of the energy absorbed by the rotor. Figure 4a is based on a model of the quantum kicked rotor under the influence of continuous measurements of the energy, with a weak coupling to the measuring device. A slow increase of the energy, compared with the isolated quantal case, is indeed visible. It is evidence for the gradual "reduction of the wavepacket" of the rotor, caused by interactions with the macroscopic measuring apparatus: The localization of the quasienergy eigenstates is thus resolved. However, the coupling strength assumed in this case would be insufficient for a measurement with acceptable accuracy. Figure 4b shows the corresponding result for strong coupling. Here, the collapse of the wavefunction and the concomitant destruction of localization are instantaneous on the relevant time scales. The energy increase for the

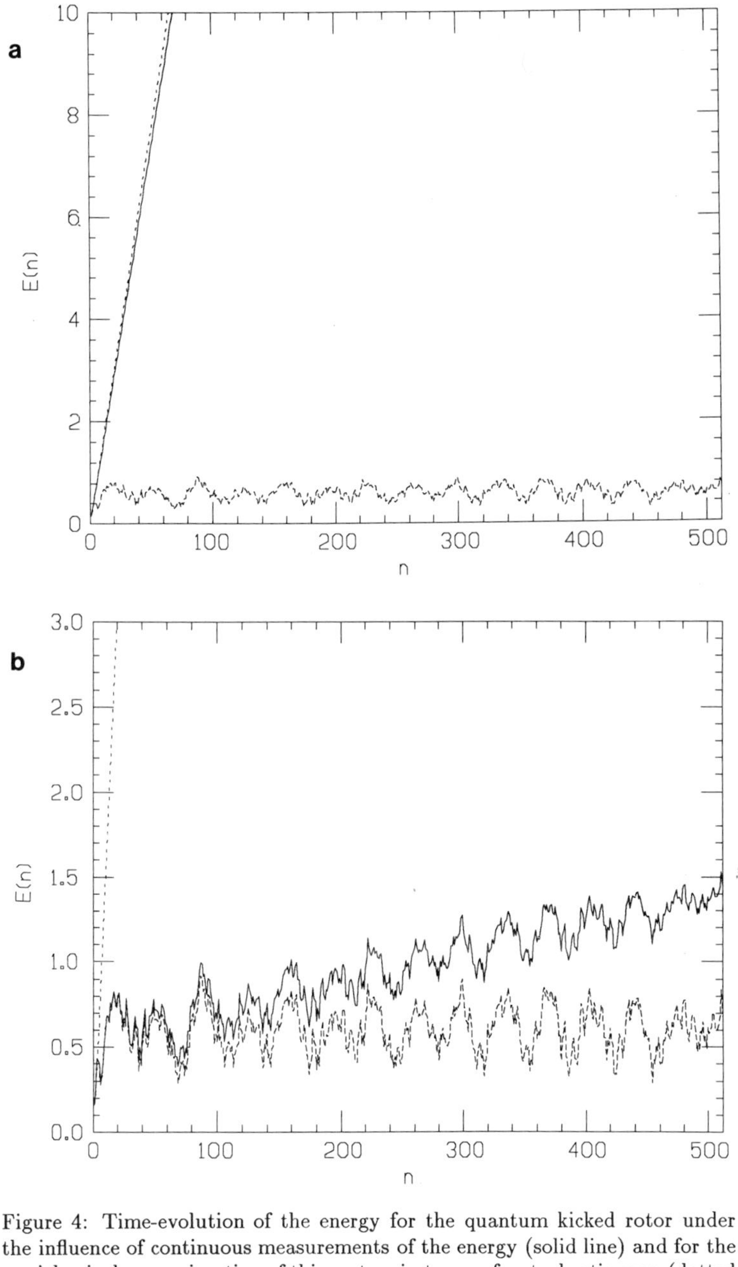

Figure 4: Time-evolution of the energy for the quantum kicked rotor under the influence of continuous measurements of the energy (solid line) and for the semiclassical approximation of this system in terms of a stochastic map (dotted line), for weak (part a) and strong coupling (part b) to the measuring device. For comparison, the corresponding data for the isolated quantum system are included (dashed line). Parameter values are $K = 5$ and $2\pi\hbar = 0.2/(\sqrt{5}-1)$.

measured quantum system reaches the classical rate, and the semiclassical approximation of the measured dynamics by a stochastic map, which neglects all coherence effects and is poor for weak coupling, becomes valid.

An analogous comparison is made in Fig.5 for a version of the quantum kicked rotor with dissipation, corresponding to the classical map (5). The results for various values of the dissipation constant λ (cf. Eq.(5)) have been combined in this plot, showing again a transition, with increasing coupling strength, from a very slow disruption of coherence to almost classical behavior. In the case of strongest dissipation shown in this figure, a saturation of the energy growth is visible. It indicates the approach to a stationary state, analogous to the relaxation, in the classical counterpart, into a steady state where diffusion is balanced by dissipation.

For even stronger dissipation, this quantal stationary state can be reached numerically and studied in detail. In order to resolve its phase space structure, which forms the quantal analogue of the classical strange attractor shown in Fig.2, phase-space representations of the density operator ρ have been calculated and plotted: Fig.6a is a plot of the Wigner function [26, 27], a representation of the density operator which bears the full information contained in the quantal state. It exhibits two significant features, a smoothing of the fine structure of the classical attractor which reconciles the quantal state with the uncertainty principle, and regular ripples in regions of phase space which are not visited in the classical steady state [28]. These waves represent remnants of quantum coherence and render the Wigner function negative in their troughs: Being equivalent to the density operator, the Wigner function is not an observable probability density. An observable phase-space density for the same state, the Husimi or Glauber function [29], is shown in Fig.6: The ripples have disappeared, and the only remaining quantum effect consists in an additional broadening of the distribution, due to the idealized measuring process to which the transformation to the Husimi function amounts [30]. Indeed, this result can be reproduced by using a stochastic map of the form of Eq.(7) instead of the exact propagator. Quantum noise thus proves sufficient to describe the observable quantum effects in the stationary state generated by the dissipative quantum map near the classical limit.

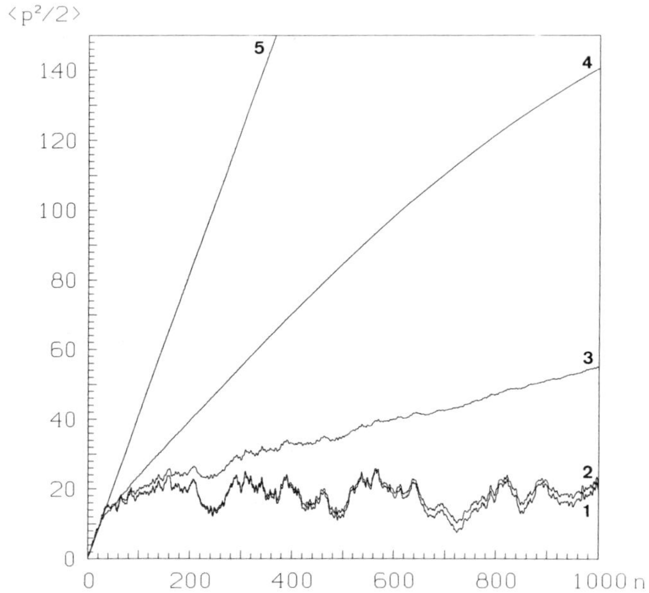

Figure 5: Time-evolution of the energy for the quantum kicked rotor with dissipation (graphs 2-4), for the conservative quantum kicked rotor (graph 1), and its classical counterpart (graph 5). For the dissipative cases the dissipation constant is $\lambda = 1 - 5 \cdot 10^{-6}$ (graph 2), $1 - 10^{-4}$ (graph 3), and $1 - 10^{-3}$ (graph 4), other parameter values are as in Fig.3.

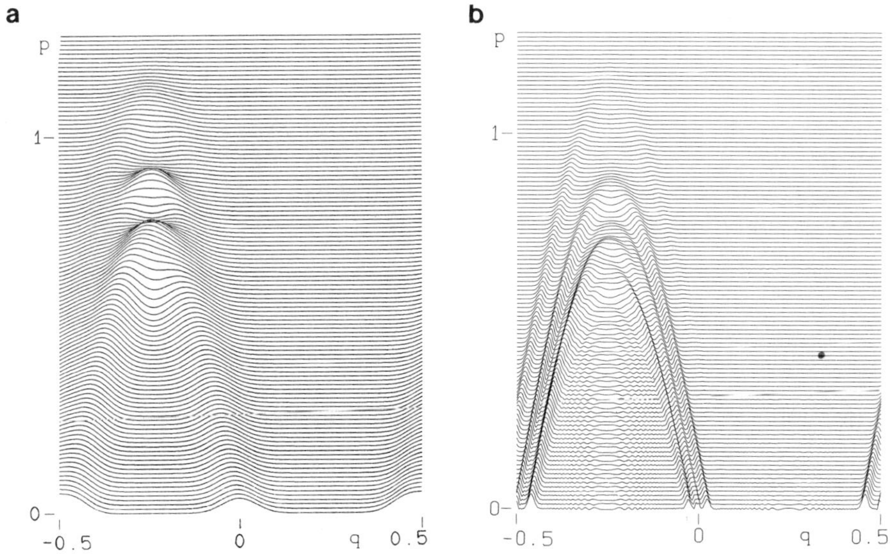

Figure 6: Steady state of the quantum kicked rotor with dissipation, corresponding to the classical strange attractor shown in Fig.2. Part a is a plot of the Wigner function for successive values of the discrete quantum number of the action p, covering the same part of phase space as Fig.2. Part b represents the corresponding continuous Husimi function, plotted as sections along lines $p = const$ at the eigenvalues forming the support for the Wigner function of part a. The relative Planck's constant is $2\pi\hbar = 0.01$, the classical parameters K and λ are as in Fig.2.

4. Conclusion

Quantum systems that assume chaotic behavior in their classical limit provide particularly rich examples to study the conceptual differences between classical and quantum physics and the subtleties of the correspondence principle. This is due to the fact that a classical chaotic system, in the course of its time-evolution, probes progressively finer scales in phase space and thus is capable of a continuous positive information production. In isolated quantum systems, on the other hand, the number of accessible states is effectively finite. They mimic the behavior of their chaotic classical counterparts for some finite time, but must eventually "run out of information" and enter into recurrent behavior. This very general scheme of the quantal suppression of chaos may take the form of intricate quantum coherence effects like a dynamical analogue of Anderson localization in the case of the kicked rotor.

It is altered, in turn, if the interaction of a quantum system with its environment is taken into account. By perturbing the system in an incoherent way, the ambient degrees of freedom degrade the coherence necessary for the quantal suppression of chaos. They form an inexhaustible external information source which renders a quantum system truly stochastic.

It thus appears that concepts of information theory are particularly well suited to understand systems which combine aspects of quantum coherence, classical deterministic chaos and the randomness of systems with many degrees of freedom. Up to now, however, their use in this context is still metaphorical, and much work remains to be done to make it quantitative and rigorous and to prove its fruitfulness.

References

1. R. Shaw. *Z. Naturforsch.* **36 a**, 80 (1981).
2. M.V. Berry. In *Chaotic Behaviour of Deterministic Systems*, Proceedings of the Les Houches Summer School, Session XXXVI, 1981. Editors G. Iooss, R.H.G. Helleman, and R. Stora. North Holland, Amsterdam (1983).
   B. Eckhardt. *Phys. Rep.* **163**, 205 (1988).
   T. Dittrich and R. Graham. *Naturwissenschaften* **76**, 401 (1989). (In German.)
3. B.V. Chirikov. *Phys. Rep.* **52**, 263 (1979).
4. A.J. Lichtenberg and M.A. Lieberman. *Regular and Stochastic Motion.* Springer, Berlin, 1983.
5. G. Casati, B.V. Chirikov, F.M. Izrailev, and J. Ford. *Lecture Notes in Physics* **93**, 334. Springer, Berlin, 1979.
6. G.M. Zaslavsky. *Phys. Lett.* **69 A**, 145 (1978).
   G.M. Zaslavsky and Kh.-R.Ya. Rachko. *Zh. Eksp. Teor. Fiz.* **76**, 2052 (1979) [*Sov. Phys. JETP* **49**, 1039 (1979)].
7. G. Schmidt and B.W. Wang. *Phys. Rev.* **A 32**, 2994 (1985).
8. T. Hogg and B.A. Huberman. *Phys. Rev.* **A 32**, 2338 (1985).
   B.A. Huberman and W.F. Wolff. *Phys. Rev.* **A 32**, 3768 (1985).
   W.F. Wolff and B.A. Huberman. *Z. Phys. B* **6 3**, 397 (1986).
9. C. Beck and G. Roepstorff. *Physica* **25 D**, 173 (1987).
10. T. Hogg and B.A. Huberman. *Phys. Rev. Lett.* **48**, 711 (1982).
    D.L. Shepelyansky. *Physica* **8 D**, 208 (1983).
    B. Dorizzi, B. Grammaticos and Y. Pomeau. *J. Stat. Phys.* **37**, 93 (1984).
11. S. Fishman, D.R. Grempel, and R.E. Prange. *Phys. Rev. Lett.* **49**, 509 (1982).
    S. Fishman, D.R. Grempel, and R.E. Prange. *Phys. Rev.* **A 29**, 1639 (1984).
12. F.M. Izrailev and D.L. Shepelyansky. *Dokl. Akad. Nauk. SSSR* **249**, 1103 (1979) [*Sov. Phys. Dokl.* **24**, 996 (1979)].
    F.M. Izrailev and D.L. Shepelyansky. *Teor. Mat. Fiz.* **43**, 417 (1980) [*Theor. Math. Phys.* **43**, 553 (1980)].
13. G. Casati and I. Guarneri. *Commun. Math. Phys.* **95**, 121 (1984).
14. G. Casati, J. Ford, I. Guarneri, and F. Vivaldi. *Phys. Rev.* **A 34**, 1413 (1986).
15. F.M. Izrailev, B.V. Chirikov, and D.L. Shepelyansky. *Sov. Sci. Rev.* **2C**, 209 (1981).
    D.L. Shepelyansky. *Phys. Rev. Lett.* **56**, 677 (1986).
16. T. Dittrich and R. Graham. *Z. Phys. B* **6 2**, 515 (1986).
    T. Dittrich and R. Graham. *Europhys. Lett.* **4**, 263 (1987).
    T. Dittrich and R. Graham. *Europhys. Lett.* **7**, 287 (1988).
    T. Dittrich and R. Graham. *Ann. Phys. (N.Y.)* **200**, 363 (1990) and references therein.
17. D. Cohen and S. Fishman. *Phys. Rev.* **A 39**, 6478 (1989).
    D. Cohen. Israel Inst. Tech. (Haifa) preprints Technion-Ph-89-52 (1989); -90-12 (1990); -90-17 (1990); -90-21 (1990).
18. S. Sarkar and J.S. Satchell. *Europhys. Lett.* **4**, 133 (1987).
    S. Sarkar and J.S. Satchell. *Physica* **29 D**, 343 (1988).

19. T. Dittrich and R. Graham. *Europhys. Lett.* **11**, 589 (1990).
T. Dittrich and R. Graham. *Phys. Rev.* **A 42**, 4647 (1990).

20. J.D. Hanson, E. Ott, and M. Antonsen, Jr. *Phys. Rev.* **A 29**, 1819 (1984).
S. Adachi, M. Toda, and K. Ikeda. *Phys. Rev. Lett.* **61**, 659 (1988).

21. A.O. Caldeira and A.J. Leggett. *Phys. Rev. Lett.* **46**, 211 (1981).
A.O. Caldeira and A.J. Leggett. *Ann. Phys. (N.Y.)* **149**, 374 (1983). Erratum *Ann. Phys. (N.Y.)* **153**, 445 (1984).
C.M. Savage and D.F. Walls. *Phys. Rev.* **A 32**, 2316 (1985).

22. H. Haken. Laser theory. In *Encyclopedia of Physics* Vol. XXV/2c. Springer, Berlin, 1970.

23. W.H. Louisell. *Quantum Statistical Properties of Radiation.* Wiley, London, 1973.

24. C.M. Caves. *Phys. Rev.* **D 33**, 1643 (1986).
C.M. Caves. *Phys. Rev.* **D 35**, 1815 (1986).
C.M. Caves and G.J. Milburn. *Phys. Rev.* **A 36**, 5543 (1987).
A. Schmid. *Ann. Phys. (N.Y.)* **173**, 103 (1987).

25. W.H. Zurek. *Phys. Rev.* **D 24**, 1516 (1981).
W.H. Zurek. *Phys. Rev.* **D 26**, 1862 (1982).

26. E.P. Wigner. *Phys. Rev.* **40**, 749 (1932).

27. M.V. Berry. *Trans. R. Soc. (London)*, Sec. A **287**, 237 (1977).

28. K. Takahashi and N. Saitô. *Phys. Rev. Lett.* **55**, 645 (1985).
S.-J. Chang and K.-J. Chi. *Phys. Rev. Lett.* **55**, 269 (1985).
S.-J. Chang and K.-J. Chi. *Phys. Rev.* **A 34**, 7 (1986).

29. K. Husimi. *Proc. Phys. Math. Soc. Jpn.* **22**, 264 (1940).
R.J. Glauber. *Phys. Rev.* **130**, 2529 (1963).
R.J. Glauber. *Phys. Rev.* **131**, 2766 (1963).
R.J. Glauber. In *Quantum Optics.* Editors S.M. Kay and A. Maitland. Academic Press, London, 1970.

30. N.D. Cartwright. *Physica* **83 A**, 210 (1976).
E. Prugovecki. *Ann. Phys. (N.Y.)* **110**, 102 (1978).
A.K. Rajagopal. *Phys. Rev.* **A 27**, 558 (1983).

A DYNAMICAL SYSTEM APPROACH TO QUANTUM MECHANICS WITH IMAGINARY TIME

CHRISTIAN BECK

Institut für Theoretische Physik
RWTH Aachen
D-5100 Aachen, FRG

1. Introduction

Within the traditional approach, quantum mechanics is formulated by means of a partial differential equation, i.e., the Schrödinger equation. In this article we will present an alternative formulation of quantum mechanics in terms of certain nonlinear chaotic mappings. In an appropriate scaling limit these maps generate a behaviour equivalent to that of the Schrödinger equation with imaginary time. The method is applicable to arbitrary potentials. There are three basic ingredients that make this approach possible:

1. the path integral representation of the Schrödinger propagator (Feynman-Kac formula) [1, 2, 3];
2. a functional central limit theorem for the iterates of certain deterministic systems with strong mixing properties [4, 5];
3. the thermodynamic formalism of dynamical systems [6, 7, 8].

In fact, we will introduce a formal inverse temperature, a partition function, and a free energy for our chaotic mappings. The interaction of the underlying spin system is a complicated long range interaction depending on the potential under consideration — quite different from the next-neighbour interaction used in the conventional formulation of Euclidean quantum mechanics on the lattice [9]. Our maps are easily implemented on a computer and can be used to calculate, e.g., ground state energies and wave functions.

2. Path Integrals and Chaotic Maps

In quantum field theories it is a useful standard procedure to rotate the time axis to imaginary values and to consider Euclidean field theories [10]. This can also be done in nonrelativistic quantum mechanics, obtaining the Euclidean Schrödinger

Information Dynamics, Edited by H. Atmanspacher and H. Scheingraber, Plenum Press, New York, 1991

equation:

$$-\hbar\frac{\partial}{\partial t}\Psi(y,t) = H\Psi(y,t) \tag{1}$$

Let us restrict to Hamiltonians H of the form:

$$H = -\frac{\hbar^2}{2m}\,\Delta + V(y) \tag{2}$$

The general solution of Eq. (1) is given by:

$$\Psi(y,t) = \int d^3y'\; G_{t,t'}(y,y')\Psi(y',t') \qquad (t \geq t') \tag{3}$$

Here $G_{t,t'}(y,y')$ is the Green's function (the propagator) of the Schrödinger problem (1). According to the Feynman-Kac formula, the propagator can be presented as a path integral:

$$G_{t,t'}(y,y') = E^{y,t}_{y',t'} \exp\left\{-\frac{1}{\hbar}\int_{t'}^{t} dt''\; V\left(\sqrt{\tfrac{\hbar}{m}}\; W(t'')\right)\right\} \tag{4}$$

Here $W(t) = (W_1(t), W_2(t), W_3(t))$ is the normalized 3-dimensional Wiener process, and $E^{y,t}_{y',t'}$ is the expectation with respect to all realizations of the Wiener process that are at position y' at time t' and at position y at time t. Unfortunately, path integrals cannot be calculated in most cases. Hence it is necessary to introduce an appropriate regularization. Usually a path integral is approximated by a large number of ordinary integrals. For nonrelativistic quantum mechanics one defines

$$\begin{aligned} I_N := & \frac{1}{(2\pi D\epsilon)^{\frac{3}{2}(N+1)}} \int d^3x_1 \cdots \int d^3x_N \exp\left\{-\frac{(y-x_1)^2}{2D\epsilon}\right\}\cdots \\ & \cdots \exp\left\{-\frac{(x_N-y')^2}{2D\epsilon}\right\} \exp\left\{-\frac{\epsilon}{\hbar}\sum_{j=1}^{N} V(x_j)\right\} \\ & \epsilon = \frac{t-t'}{N+1} \qquad D = \frac{\hbar}{m} \end{aligned} \tag{5}$$

obtaining the Green's function as $G_{t,t'}(y,y') = \lim_{N\to\infty} I_N$. In this article we want to describe a different kind of regularization. As has been shown in [4, 5], there exists a class of rescaled *deterministic* dynamical systems $T : X \to X$ that generate the Wiener process. Defining

$$S_\tau(t) = \sigma^{-1}\tau^{1/2} \sum_{n=1}^{\lfloor t/\tau \rfloor} x_n \tag{6}$$

($\lfloor\ \rfloor$: integer part) with $x_{n+1} = Tx_n$ and $\sigma^2 = \langle x_0^2 \rangle + 2\sum_{n=1}^{\infty} \langle x_0 x_n \rangle$, one has:

$$\lim_{\tau\to 0} S_\tau(t) = W(t) \tag{7}$$

Here $\langle \cdots \rangle = \int_X \cdots d\mu$ denotes the expectation with respect to the natural invariant measure $d\mu$ of T. As a consequence, any path integral $J = E(g[W(t)])$ of an observable g can be approximated by a single integral

$$J_\tau := \int_X d\mu(x_0) g[S_\tau(t)] = \langle g[S_\tau(t)] \rangle \tag{8}$$

with $J = \lim_{\tau\to 0} J_\tau$.

It may come as a surprise that there exist simple deterministic maps T that can generate Brownian motion, i.e., a process that is commonly thought to be the standard example of random motion. In fact, the statement (7) is highly nontrivial, as the iterates x_n are *dependent* random variables, i.e., there are non-vanishing higher order correlations [11] and the standard central limit theorem cannot be used. The simplest example of a map T for which the rescaled sum of iterates converges to Brownian motion is the piecewise linear map $Tx = 2x \bmod 1$ with subtracted mean [12]. It is also possible to extend the discussion to processes of Langevin type [5, 13]. For applications in hydrodynamics see [14].

3. Euclidean Quantum Maps

To simulate quantum mechanics, let us define the following 7-dimensional dynamical system:

$$Q: \qquad \begin{aligned} x_{n+1} &= Tx_n \\ y_{n+1} &= y_n + \sigma^{-1}\sqrt{\frac{\hbar\tau}{m}}\, x_{n+1} \\ z_{n+1} &= z_n \exp\left\{-\frac{\tau}{\hbar}V(y_{n+1})\right\} \end{aligned} \tag{9}$$

$x_n = (x_n^{(1)}, x_n^{(2)}, x_n^{(3)})$ and $y_n = (y_n^{(1)}, y_n^{(2)}, y_n^{(3)})$ are vectors with 3 components,[1] z_n is a scalar variable. τ is a small time constant and V an arbitrary potential. The initial values $x_0^{(i)}, i = 1, 2, 3$, are randomly chosen from X, whereas y_0 is an arbitrary constant initial vector and $z_0 = 1$. The 1-dimensional map T is assumed to possess the so called φ-mixing property [5]. It acts onto every component of x_n, i.e., $Tx_n = (Tx_n^{(1)}, Tx_n^{(2)}, Tx_n^{(3)})$. A possible choice is

$$Tx = 1 - 2x^2, \quad X = [-1, 1], \quad d\mu(x_0) = \frac{dx_0}{\pi\sqrt{1 - x_0^2}}, \quad \sigma^2 = \frac{1}{2} \tag{10}$$

or any other map conjugated to the Bernoulli shift with average 0.

We will argue now that for $\tau \to 0$ the dynamical system Q simulates a quantum mechanical system with potential V. For that purpose we define conditioned expectations $< z_n >_{y,y_0}$ by

$$< z_n >_{y,y_0} := \Delta^{-3} \int_{X^3} d\mu(x_0)\chi_y^{\Delta}(y_n) z_n(x_0, y_0) \tag{11}$$

where:

$$\chi_y^{\Delta}(y_n) = \begin{cases} 1 & \text{if} \quad |y_n - y| \leq \frac{\Delta}{2} \\ 0 & \text{otherwise} \end{cases} \tag{12}$$

Here y is an arbitrary vector in $\mathbf{R}^3$. Δ is a parameter that measures how precisely y_n shall coincide with y. The quantity $< z_n >_{y,y_0}$ contains the factor Δ^{-3} to make the expectation independent of the box size Δ for $\Delta \to 0$. We now claim

$$\lim_{\Delta \to 0} \lim_{\tau \to 0} < z_n >_{y,y_0} = G_{t,0}(y, y_0) \qquad (n\tau = t \text{ finite}) \tag{13}$$

---

[1]For a d-dimensional quantum mechanical system, x_n and y_n have d components and Q is a $2d + 1$-dimensional map.

i.e., the propagator of the Schrödinger problem (1) is given as the conditioned expectation of the z-variable of the dynamical system Q.

Proof: Iteration of Q yields

$$\begin{aligned} z_n &= \exp\left\{-\frac{\tau}{\hbar}\sum_{j=1}^{n} V(\sigma^{-1}\sqrt{\frac{\hbar\tau}{m}}\sum_{k=1}^{j} x_k)\right\} \\ &= \exp\left\{-\frac{1}{\hbar}\int_0^t dt'\, V\left(\sqrt{\frac{\hbar}{m}}\, S_\tau(t')\right)\right\} \qquad (t = n\tau) \end{aligned} \tag{14}$$

(for simplicity we have chosen $y_0 = 0$). To obtain the conditioned expectation (11) we have to choose all those trajectories that start at y_0 and end at $y_n = y$ with a precision of measurement Δ. This is guaranteed by the choice of the initial value y_0 and the measure $d\mu(x_0)\chi_y^\Delta(y_n)$. For $\tau \to 0$ we have $S_\tau(t') \to W(t')$. Thus, according to the Feynman-Kac formula (4), $\lim_{\tau\to 0} < z_n >_{y,y_0}$ is the propagator $G_{t,0}(y, y_0)$ averaged over a region of size Δ^3. In the limit $\Delta \to 0$ we obtain the propagator itself, q.e.d.

We will call the map Q "Euclidean quantum map" corresponding to the potential V. With the choice $Tx = 1-2x^2$ it is very easy to implement this map on a computer. Instead of a mixing map T it is also possible to choose independent random variables $x_n = \pm 1$, as approximately produced by a pseudo-random generator ($\sigma^2 = 1$). In this case the convergence to the Wiener process is guaranteed by the standard functional central limit theorem. One might ask: What is the difference between Euclidian quantum maps and the conventional Monte-Carlo simulations of Euclidean quantum mechanics on the lattice [9]? The map Q generates Brownian motion on a very small time scale τ, whereas the conventional methods, based on Eq. (5), generate a stroboscopic picture of Brownian motion on a larger time scale ϵ. Notice that in Eq. (5) $x_1, x_2, \ldots$ take continuous values in $\mathbf{R}^3$, whereas in the dynamical system appraoch the x_n take values in a compact phase space X. If we choose $x_n = \pm 1$ there are just two discrete alternatives for the microscopic random variables x_n. From the aspect of information dynamics, this corresponds to the transfer of one bit information per time unit τ [15]. If we choose for x_n the iterates of a chaotic map T, we still can associate with every time step τ a production of information (or loss, respectively) given by the positive Ljapunov exponent of the map T.

4. Determination of Ground State Energies

We now describe a simple method how to determine ground state energies with the help of Euclidean quantum maps. According to Eqs. (11) and (13) we have for small τ and Δ:

$$\int d\mu(x_0)\chi_y^\Delta(y_n) z_n(x_0, y_0) \approx \Delta^3 G_{t,0}(y, y_0) \tag{15}$$

Summing over all possible endpoints $y = (j_1\Delta, j_2\Delta, j_3\Delta), j_k \in \mathbf{Z}$, we obtain in the limit $\tau \to 0, \Delta \to 0$:

$$< z_n > = \int d\mu(x_0) z_n(x_0, y_0) = \int d^3y\, G_{t,0}(y, y_0) \tag{16}$$

Let us assume that the following series representation of the Green's function exists:

$$G_{t,0}(y, y_0) = \sum_{j=0}^{\infty} \Psi_j(y)\Psi_j^*(y_0) e^{\frac{E_j t}{\hbar}} \tag{17}$$

Here Ψ_j are eigenfunctions of the Hamiltonian and $E_0 \leq E_1 \leq E_2 \leq \cdots$ are discrete energy eigenvalues. Combining (16) and (17) we obtain for large $t = n\tau$

$$< z_n > = \Psi_0^*(y_0) \int d^3y \; \Psi_0(y) e^{\frac{E_0 t}{\hbar}} + O(e^{-\frac{E_1 t}{\hbar}}) \tag{18}$$

where we have assumed integrability of the wave functions. Thus:

$$\begin{gathered} \log < z_n > = -E_0 \frac{\tau}{\hbar} \cdot n + const + O(e^{-(E_1 - E_0)\frac{\tau}{\hbar} \cdot n}) \\ (\tau \to 0, \; t = n\tau \text{ large}) \end{gathered} \tag{19}$$

Hence the ground state energy E_0 can be determined as the asymptotic slope in a $(\log < z_n >, n)$-plot provided τ is small enough and $t = n\tau$ large. Eq. (19) is valid for both, positive and negative energy eigenvalues. Varying the initial value y_0, the constant in Eq. (19) can be used to determine the ground state wave function $\Psi_0^*(y_0)$.

As an example we have determined ground state energies for the 3-dimensional harmonic oscillator and for the hydrogen atom. The Euclidean quantum map has been iterated n_{max} times for a large number n_{path} of initial values x_0. Averaging the variable z_n over the various trajectories we obtained the expectation $< z_n >$. For the harmonic oscillator we have chosen the coupling constant $g = 10^{-4}$, the time contant $\tau = \hbar/(mc^2)$, $Tx = 1 - 2x^2$, $n_{max} = 500$, and $n_{path} = 5000$. The result $E_0 = \frac{3}{2}\hbar\omega$ was reproduced with an error smaller than 1%. For the hydrogen atom we have chosen the coupling constant $g = 1/137$, the time constant $\tau = \hbar/(mc^2)$, $Tx = \pm 1$, $n_{max} = 300000$, and $n_{path} = 10000$. This reproduced the Schrödinger result $E_0 = \frac{1}{2}g^2mc^2$ with a statistical error of about 5%. Numerical investigations of the Coulomb potential are much more difficult than those of an oscillator potential. The reason for this will become clear in Section 6.

5. Thermodynamic Formalism

The dynamical system approach can also be described in the language of statistical mechanics. The partition function is:

$$\begin{gathered} < z_n > = < \exp\left\{ -\frac{\tau}{\hbar} \sum_{j=1}^{n} V\left(\sqrt{\tfrac{\hbar\tau}{m}}\, S_j\right) \right\} > \\ S_j := \sum_{k=1}^{j} x_k \end{gathered} \tag{20}$$

Notice that for $x_k = (\pm 1, \pm 1, \pm 1)$ or for a map T conjugated to a $(1/2, 1/2)$-Bernoulli shift the underlying spin system is an Ising system. It is, however, an Ising system with an exotic long range interaction, which depends on the potential V. As has been shown in [8], the partition function $< z_n >$ can be obtained if the usual thermodynamic formalism of dynamical systems [7] is applied to the 6-dimensional map $(x_{n+1}, y_{n+1}) = (Tx_n, y_n + \sigma^{-1}\sqrt{\frac{\hbar\tau}{m}} x_n)$ with the test function $-\frac{\tau}{\hbar}V(y)$. Even magnetic fields can be treated in that way, resulting in a test function that also contains the vector potential [8].

The quantity $\frac{\tau}{\hbar}$ may be interpreted as a formal inverse temperature $\beta = (kT)^{-1}$. The free energy per spin is then defined as $F(t) = \lim_{n\to\infty} -\frac{1}{n}\frac{\hbar}{\tau} \log < z_n >$. To reproduce the continuum Schrödinger theory, we have to let $\tau \to 0$ keeping the time $t = n\tau$ constant. $\tau \to 0$ implies the thermodynamic limit $n \to \infty$ and vice versa. According to Eq. (19), we obtain for large $t = n\tau$ the ground state energy E_0 as free energy per spin:

$$E_0 = \lim_{t \to \infty} \lim_{\tau \to 0} -\frac{1}{n} \frac{\hbar}{\tau} \log < z_n > \tag{21}$$

In principle, the map Q can be used to calculate expectations of arbitrary observables A as long as A can be written as a functional of the trajectory $\{S_j\}$. The usual formalism of statistical mechanics yields:

$$E(A) = \lim_{t \to \infty} \lim_{\tau \to 0} \frac{< A(S_1, \ldots, S_n) z_n >}{< z_n >} \tag{22}$$

6. Estimation of the Necessary Lattice Size

How much computing time do we need to obtain reasonable values for ground state energies? As the interaction of our spin model depends on the potential V, the necessary number of iterations (= the necessary lattice size) is also a function of the potential. Let us restrict to potentials of the form $V(y) = b\,|y|^w, w \in \mathbf{R}, w > -2$. b has the dimension energy/(length)^w. It is useful to introduce a dimensionless coupling constant $g \in \mathbf{R}$ by setting:

$$b = \frac{(mc^2)^{w+1}}{(\hbar c)^w} \cdot g \tag{23}$$

For $V(y) = b\,|y|^w$ Eq. (20) reduces to:

$$< z_n > = < \exp\left\{ -f \sum_{j=1}^{n} |S_j|^w \right\} > \tag{24}$$

$$f := g \left(\frac{mc^2 \tau}{\hbar} \right)^{\frac{w+2}{2}} \tag{25}$$

We will call the dimensionless parameter f *bare coupling*. For potentials of the form $V(y) = b\,|y|^w$ we recognize the following fact: The partition function does not know the physical coupling g and the time constant τ solely, but only the product $f = g\,(mc^2\tau\hbar^{-1})^{\frac{w+2}{2}}$ of both. The bare coupling f is the only relevant parameter of the system.

A priori it is not known how the function $\log < z_n >$ depends on f. However, we can make a statement on the limit case $\tau \to 0 \Longrightarrow f \to 0$. From dimensional considerations it is easy to derive that the energy eigenvalues of the Schrödinger operator scale with the coupling g as follows:

$$\frac{E_j}{mc^2} \sim |g|^{\frac{2}{w+2}} \qquad j = 0, 1, 2, \ldots \tag{26}$$

Eqs. (25) and (26) imply $E_j \sim |f|^{\frac{2}{w+2}} \frac{\hbar}{\tau}$. Inserting this into Eq. (19) we obtain:

$$\log < z_n > = const\,|f|^{\frac{2}{w+2}} \cdot n + const + O(e^{-const\,|f|^{\frac{2}{w+2}} \cdot n}) \qquad (f \to 0,\ n \text{ large}) \tag{27}$$

Eq. (27) is very useful if one wants to determine ground state energies on a computer. Then for the iteration of the quantum map Q a very small f has to be chosen in order to be near to the Schrödinger theory, which is reproduced for $\tau \to 0 \Longrightarrow f \to 0$ only. On the other hand, the relaxation time n^* of the exponential function in Eq. (27) is given by:

$$n^* \sim |f|^{-\frac{2}{w+2}} \qquad (w > -2) \tag{28}$$

Hence, to reach an asymptotic linear behaviour in a $(\log < z_n >, n)$-plot, we have to iterate the quantum map Q the longer the better we want to approximate Schrödinger's theory, i.e., the smaller values f takes. In the language of statistical mechanics n^* is the minimum lattice size for senseful results. As an example let us choose $f = 10^{-4}$. For the harmonic oscillator ($w = 2$) we obtain $n^* \sim 10^2$. For the hydrogen atom ($w = -1$), however, the same f yields $n^* \sim 10^8$. Hence, for numerical investigations of the hydrogen atom we need much more computing time than for the harmonic oscillator.

References

1. L.S. Schulman. *Techniques and Applications of Path Integration.* Wiley, New York, 1981.
2. Z. Schuss. *Theory and Applications of Stochastic Differential Equations.* Wiley, New York, 1980.
3. R.P. Feynman and A.R. Hibbs. *Quantum Mechanics and Path Integrals.* McGraw - Hill, New York, 1965.
4. P. Billingsley. *Convergence of Probability Measures.* Wiley, New York, 1968.
5. C. Beck and G. Roepstorff. *Physica* **145 A**, 1 (1987).
6. D. Ruelle. *Thermodynamic Formalism.* Addison-Wesley, Reading, 1978.
7. T. Bohr and T. Tél. *The Thermodynamics of Fractals.* In *Directions in Chaos*, Vol.2. Ed. B.-L. Hao. World Scientific, Singapore, 1988.
8. C. Beck. Thermodynamic formalism for quantum mechanical systems. Preprint RWTH Aachen, 1990.
9. M. Creutz and B. Freedman. *Ann. Phys. (N.Y.)* **132**, 427 (1981).
10. P. Ramond. *Field Theory: A Modern Primer.* Benjamin/Cummings, London, 1981.
11. C. Beck. Higher correlation functions of chaotic dynamical systems: A graph theoretical approach. Preprint RWTH Aachen, 1990.
12. C. Beck. *Physica* **A**. To be published.
13. C. Beck. *Commun. Math. Phys.* **130**, 51 (1990).
14. C. Beck and G. Roepstorff. *Physica* **165 A**, 403 (1990).
15. H. Atmanspacher. *Found. Phys.* **19**, 553 (1989).

DO QUANTUM PARTICLES TRAVEL IN REAL SPACE TIME? EXPERIMENTAL EVIDENCE AND THEORETICAL IMPLICATIONS

JEAN PIERRE VIGIER

Institut Henri Poincaré
Université Pierre et Marie Curie
Laboratoire de Physique Théorique
F-75231 Paris Cedex 05, France

1. Introduction

A new situation is developing in the Bohr–Einstein discussion on the nature of the wave – particle dualism in individual microobjects, a controversy which has never stopped despite the majority position occupied by Bohr's Copenhagen interpretation. This situation is not difficult to understand. Until a very recent period one could only observe statistical distributions of microscopic particles which all turned out to be correctly described by the mathematical formalism of quantum theory, so that one could not distinguish between the alternative possible conflicting interpretations such as the Heisenberg–Born–Pauli or de Broglie–Bohm points of view developed by their respective followers. This situation has recently changed for two main reasons:

1. The expansion of modern technology has now made possible individual observation/detection of individual particles associated with point–like exchange of quanta. One has thus recently observed the interference with themselves of individual neutrons, photons, and electrons in double slit experiments. All quantum experiments boil down to the analysis of individual localized particle impacts (quantum exchanges) whose statistical distribution always (up to now) confirms the validity of the quantum mechanical formalism. We live in a particle world. One only observes particles directly: the physical reality of all fields (such as gravitational, electromagnetic, and ψ fields) being indirectly inferred from their action on the observed sets of "particle" distributions.
2. The second reason is that this technical breakthrough (i.e., the possibility to observe individual particles) has opened the possibility of realizing crucial experiments on individual objects (only formally discussed on the "Gedanken"level in Bohr's and Einstein's time) in which the conflicting interpretations predict different results. We can thus pass from collective observations (which satisfy

Information Dynamics, Edited by H. Atmanspacher and H. Scheingraber, Plenum Press, New York, 1991

the quantum mechanical formalism) to individual observations which can test the validity of specific interpretations.

To clarify the theoretical implications of these experiments we briefly recall here some well known elements of the Bohr–Einstein controversy, i.e., the points of agreement and disagreement between Bohr and Heisenberg (BH) and Einstein and de Broglie (EdB) on the wave – particle dualism in double slit experiments.

- For BH and EdB all quantum experiments can be reduced to the direct observation of individual localized particles. Experimentally we live in a particle world. The physical reality of all fields is only established through the observation of their action on individual test particles [1]. One never sees the quantum mechanical ψ – waves directly since they only represent observable particle distributions and energy is exchanged by contact between localized particles.
- For BH and EdB the statistical distributions of observed individual localized particles is correctly given by the present quantum mechanical formalism (ψ – waves, wave equations, brackets, etc.) at least in the present range of energy – momenta or wavelengths 10^{-19} cm $< \lambda < 10^{-7}$ cm.
- For BH and their followers (Pauli, Born, Fock, etc.), since this quantum formalism is a complete final unsurpassable description, one has a wave – particle dualism. The observed particle aspect does not travel in space and time (since such paths would correspond to "hidden" parameters) *and does not pass through one slit* (in a double slit experiment) when one particle (neutron, photon, electron, etc.) interferes with itself. Individual microobjects are waves *or* particles, never the two simultaneously.
- For EdB and their followers (Bohm, etc.), this quantum formalism is incomplete. Individual microobjects are waves *and* particles simultaneously. The waves which "control" or "pilot" the particle's real stochastic trajectories go through both slits (and interfere) while particles go through one slit only.

Of course both BH and EdB agreed that this problem should be experimentally settled within the frame of the existing quantum mechanical formalism. We shall not recall here (since they are already part of the history of physics) Einstein's various "Gedanken" experimental proposals to show that particles indeed go through one slit only, and Bohr's corresponding no–go rebuttals. Evidently, to support Einstein's view one should distinguish two types of experiments:

(I) "Einweg" experiments which should explicitly show that individual particles indeed go through one slit only (in double slit interference setups) without information on which slit.

(II) "Welcher Weg" experiments which would tell through which specific slit each individual particle has effectively gone. Of course (II) implies (I) but not *vice versa.* Clearly experiments of type (II) are more delicate since it is difficult to observe (i.e., measure with an energy exchange related to an eigenfunction change) in quantum mechanics a particle's passage on one path without destroying the two interfering waves' phase coherence (i.e., the final interference pattern) as a consequence of Heisenberg's inequalities.

In this contribution we shall briefly discuss:

- in Section 2 the "Einweg" experiment recently proposed by Rauch and myself [2];

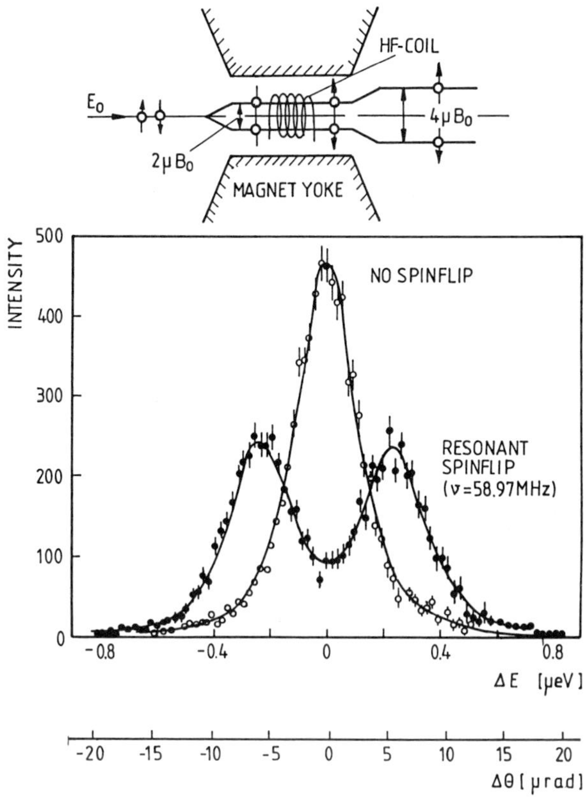

Figure 1: Principle of the energy transfer between the neutron system and a resonance spin flipper device for an unpolarized beam and characteristical results for the energy separation.

- in Section 3 a possible "welcher Weg" experiment based on the technology recently utilized by Tonomura *et al.* [3] to prove the reality of the Aharonov – Bohm effect;
- in Sections 4 and 5 some new theoretical implications opened by the (possible) existence of real particle trajectories in real physical space time, providing the equivalence of Einstein's, de Broglie's, and Feynman's views on the physical meaning and interpretation of quantum mechanics [4].

2. The Rauch – Vigier "Einweg" Experiment

The neutron is a well suited system for fundamental investigations of the basic principles of quantum mechanics because it exhibits well defined particle properties (mass, magnetic moment, decay, internal structure, confinement radius, etc.) and wave properties as well. In the last decade two different neutron optics experiments have been performed which lead – if energy–momentum is absolutely conserved – to the conception that neutrons inside a perfect neutron interferometer choose a certain beam path although the decision which path has been chosen remains unknown. These two experiments deal with the coherent energy shift due to a magnetic spin flipper [5] (see Fig.1) and the persistance of interference properties in the double coil interference experiment [6] (see Fig.2).

In the Rauch – Vigier (RV) proposal [2] the "Einweg" argument goes as follows:

1. There is only one neutron at a time in the setup associated with incoming packets of polarized neutrons with spin up.
2. The split packets go through two spin flippers (coils with resonance field rf) which act as harmonic oscillators and invert the spin direction of each separated wave packet (since the first slab has split them into two non overlapping packets).
3. Each neutron flips its spin with the exchange of one photon, i.e., an energy loss $\Delta E = \hbar\omega_{rf}$ of one photon, ω_{rf} being confined to the harmonic oscillator's internal quantum jumps.
4. The frequency distributions of the two flippers can be identical, in which case one sees an ordinary interference pattern of neutrons which have all lost a quantum $\hbar\omega$ since they all have spin down. If both flip coils have been operated at different frequencies, one observes a time beat of spin down neutrons with frequency $(\omega_{r1} - \omega_{r2})/2\pi$.
5. If one assumes absolute energy–momentum conservation at each possible variation of particle energy, this variation must have occured in coil I or in coil II since there is no photon with half the energy in the process.

Indeed, with assumption (5) in this double coil experiment the situation exists that a photon with the Larmor frequency is exchanged between the neutron and

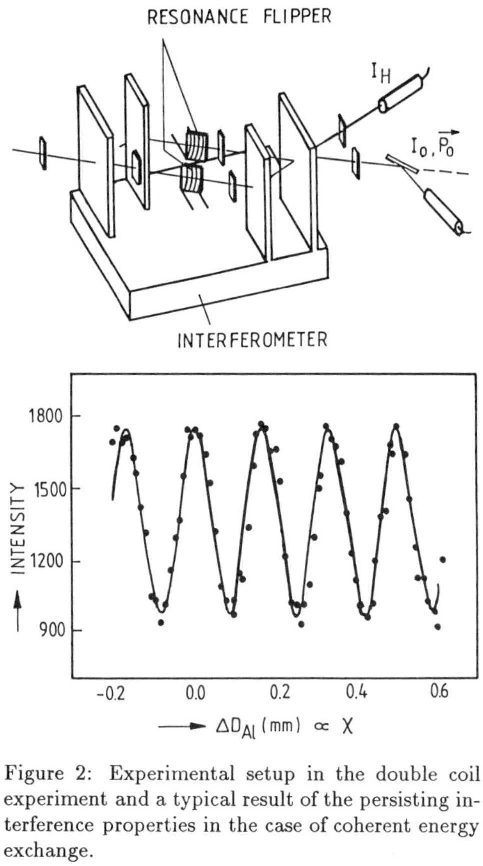

Figure 2: Experimental setup in the double coil experiment and a typical result of the persisting interference properties in the case of coherent energy exchange.

the resonators. Because there are only photons with those Larmor frequencies, it is reasonable that the exchange occured only in one of both coils which suggests the existence of a trajectory although which of both possible trajectories has been chosen remains undetermined. Here the energy exchange is not a measuring process: it rather shifts the system in a preparatory stage permitting a forthcoming measurement of this energy exchange. This argument, however, is evidently neither complete nor absolutely convincing from Bohr's point of view due to the fact that the existence of a one photon energy exchange has only been verified in those two separate experiments with different setups. Since for Bohr's Copenhagen interpretation [7] "phenomena exist only if they are measured phenomena" one is justified to claim that point (5) is only valid if one can

- measure the energy shift $\Delta E = \hbar\omega_{rf}$ in the same experiment where one observes interference,
- prove that this has occured in one coil only by some process (i.e., modified setup) which does not destroy this interference.

A setup is therefore necessary which satisfies both conditions. To describe such a setup let us recall the following points:

1. In a first experiment it has been demonstrated that an energy shift ΔE_{hf} larger than the energy width ΔE can be achieved if a rather strong magnetic field B_o is applied [5],

$$\Delta E_{hf} = 2|\mu|B_o > \Delta E \tag{1}$$

   where μ is the magnetic moment of the neutron. In this case the neutron has absorbed (or emitted) a photon from the resonance field with an energy $\Delta E_{hf} = 2|\mu|B_o = \hbar\omega_{rf}$. The energy width ΔE_{ph} of the photon resonance field has to be narrow due to the resonance condition

$$P\left(\frac{1}{2}, -\frac{1}{2}, t\right) = \frac{\sin^2\left[\frac{\mu B_1 t}{2\hbar}\sqrt{1+\left(\frac{2(B_o-\hbar\omega/2\mu)}{B_1}\right)^2}\right]}{1+\left(\frac{2(B_o-\hbar\omega/2\mu)}{B_1}\right)^2} \tag{2}$$

   where B_1 is the amplitude of the resonance field. We thus have the resonance conditions

$$\begin{aligned} \hbar\omega &= \hbar\omega_{rf} = 2|\mu|B_o \\ |\mu|B_1 t &= \hbar\pi \end{aligned} \tag{3}$$

   where t is the interaction time which is given by the length of the coil and by the velocity of the neutrons ($t = l/v$). In order to satisfy Eq. (1), rather strong magnetic fields B_o have to be used which gives a very narrow resonance with $B_1/B_o \sim \Delta E_{ph}/E_{ph} \sim 10^{-6}$ which indicates that the photon band is very narrow. Then the energy exchange is a clear indication that the neutron spin has been flipped and that the neutron has passed through that coil (Fig.1). Flipping probabilities very near to 1 can be achieved.

2. In the second experiment [6] which was performed after a proposal published earlier [8] it has been shown that interference still persists in the case where a spin reversal and a related energy exchange for neutrons exists in both beam paths (Fig.2). In this experiment unfortunately the energy exchange ΔE_{hf} was

considerably smaller than the energy width of the beam, and therefore could not be observed simultaneously with the interference pattern. The magnetic field strength was 20 G, providing an energy shift to the neutrons of 0.24 neV only, where the energy width of the beam was 0.6 meV. The interference pattern is in this case determined by the related change of the overlapping wave functions:

$$\psi \rightarrow e^{i(\omega-\omega_r)t} \mid -z > + e^{i\chi} e^{i(\omega-\omega_r)t} \mid -z > \tag{4}$$

which yields the interference pattern

$$I \propto 1 + \cos\chi \tag{5}$$

where χ is caused by an auxiliary nuclear phase shift which is used to make the interference properties visible ($\chi = -Nb_c\lambda D$, N is the particle density, b_c is the coherent scattering length, λ is the neutron wavelength, D is the thickness of the phase shifter). The interference phenomena are shown in Fig.2.

This experiment has been discussed to see whether it can be considered or extended to a "welcher Weg" ("which path") experiment [8]. In this respect the resonator coil was treated as a coherent state system which preserves the neutron phase but does not allow an extraction of information out of the photon field due to the photon–number phase uncertainty relation $\Delta N_{ph}\Delta\Phi_{ph} > 1$, or it was treated as a number state system which randomizes the neutron phase, so that no neutron interference would occur in this case. This suggests that a "welcher Weg" experiment is not feasible under these conditions. An analysis of the double coil experiment in the sense of the Copenhagen interpretation appeared recently [9]. Here we shall show that it can also be interpreted in terms of an "Einweg" experiment in the sense of Einstein's view [7]. In this context we now propose an experimental situation which makes the interpretation in terms of beam trajectories more obvious: always being aware of the fact that different interpretations of quantum mechanics are experimentally equivalent as long as the outcome of an experiment is described by the same formalism.

After this reminder one can evidently add to the setup of Ref. [6] a device to measure ΔE_{hf}. There is no limitation on the energy transfer which is determined by the strength of the magnetic field only. Therefore it can also be made larger than the energy width ΔE of the beam, which permits a simultaneous measurement of the interference pattern *and* of the individual energy transfer to the neutrons. Such an apparatus is shown schematically in Fig.3 where the energy analysis is performed by means of a high resolution perfect crystal camera. In this case the initial momentum distribution $|A(k)|^2$ is shifted to $|A(k+\Delta k)|^2$, where $\Delta k = \mu B_o m/\hbar^2 k$ and the beam intensity behind the interferometer reads

$$I(k) \propto |A(k+\Delta k)|^2(1+\cos\chi) \tag{6}$$

whereby the momentum shift Δk and the interference appearing as a phase shift χ can be measured simultaneously.

Probably the guide field has to consist of a superconductor to achieve about 3 T which corresponds to a Larmor resonance frequency of about 86 MHz. Because the neutron – photon energy exchange can happen inside one resonator coil only, it can be assumed that the neutron has passed through one of these coils even if one does not know through which one it went. The absorption of half the photon energy in

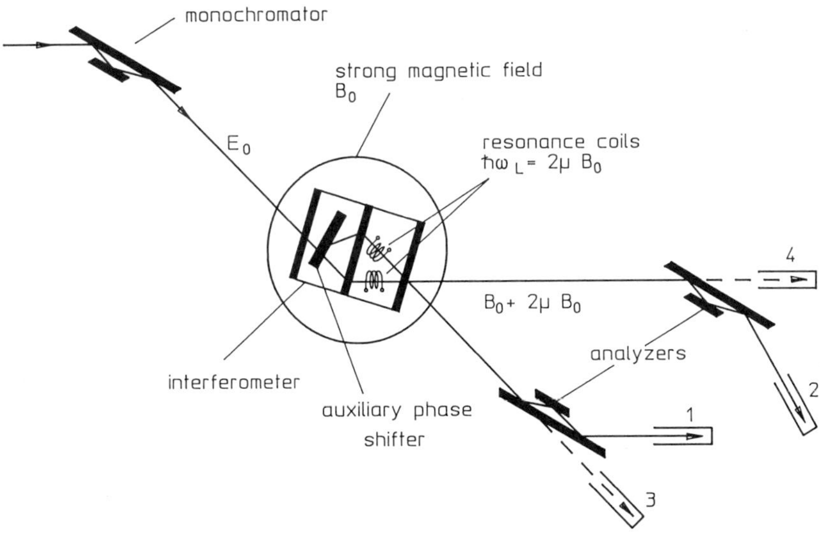

Figure 3: Proposed experimental setup for a simultaneous detection of interference and energy exchange.

each coil is physically impossible [10] because photons of that energy are not excited in either coil. We therefore conclude that every neutron went through one coil only which is equivalent to the statement of Einstein that every particle has a trajectory but we do not know distinctly which coil was passed.

Neutron interferometry evidently contains supplementary possibilities: to prepare different initial quantum states and to make new types of delayed choice or unsharp measurements feasible. Here we like to mention one more modification (see Fig.4) which makes the "Einweg" statement even stronger:

1. A stroboscopic chopping of the initial wave packets to small dimensions (1 cm long, 1 mm radius) is possible and has been tested in the past [10].
2. The coils are spatially separated and therefore the separated parts of the wave packet cross through coil I or coil II at different times.
3. The use of an alternative switcher operated in such a way that coil I works when coil II is interpreted and *vice versa*, so that there is only one coil working at a time during the performance of the experiment. Clearly the frequency of this alternative switching can be chosen in such a way that when one packet goes through slit (coil) I (II) no packet goes through slit (coil) II (I).

In this case the wave function of the whole system at a time interval $t_1 = l_1/v < t < t_2 = l_2/v$, i.e., when the packet in beam path I has passed through coil I, and the packet in beam path II has not yet reached coil II, can be written as

$$\psi \;\rightarrow\; A(r^I, k+\Delta k)\, e^{i(\omega-\omega_r)t}\; |-z> + \,A(r^{II}, k)\; e^{i\omega t} e^{i\chi}\; |z> \tag{7}$$

Behind coil II the second part of this wave function changes accordingly and the final result of the superposed wave functions is again given by expression (6). In this time–resolved situation it can be shown explicitly, that for the time interval mentioned ($t_1 < t < t_2$) the energy exchange occured in beam path I only. With a special arrangement it is also possible to measure this behavior directly by using a gated crystal for the central plate of the interferometer. Such crystals can switch the beam between the interferometer direction and the energy (beam path) measuring direction by applying a proper magnetic field [11].

All components of the proposed setup shown in Fig.4 have been tested separately and therefore the realization of such an experiment seems to be feasible. Nevertheless it might be quite difficult due to the necessity of strong magnetic fields, a pulsed neutron beam, stroboscopically pulsed flipper fields, high energy resolution spectrometers, and (occasionally) a magnetically gated central crystal.

Within the frame of the usual quantum mechanical formalism the experiment simultaneously provides

- interference,
- energy shift,
- one way coil operation at any given time,
- delayed choice option.

To our knowledge this is the first time that Einstein's "Einweg" assumption is shown to be deducible from conventional quantum mechanics. If experiments confirm (as believed by the author) the validity of the presently known quantum mechanical formalism in the preceding experimental proposals, this will raise crucial new theoretical problems for all interpretations of quantum theory. We only mention two of them:

1. Since we know that the observed neutron's loss (or gain) of energy ΔE cannot be divided between the two coils (since each one cannot absorb (emit) photons of energy $\Delta E/2$), the correct quantum mechanical formalism tells us that its

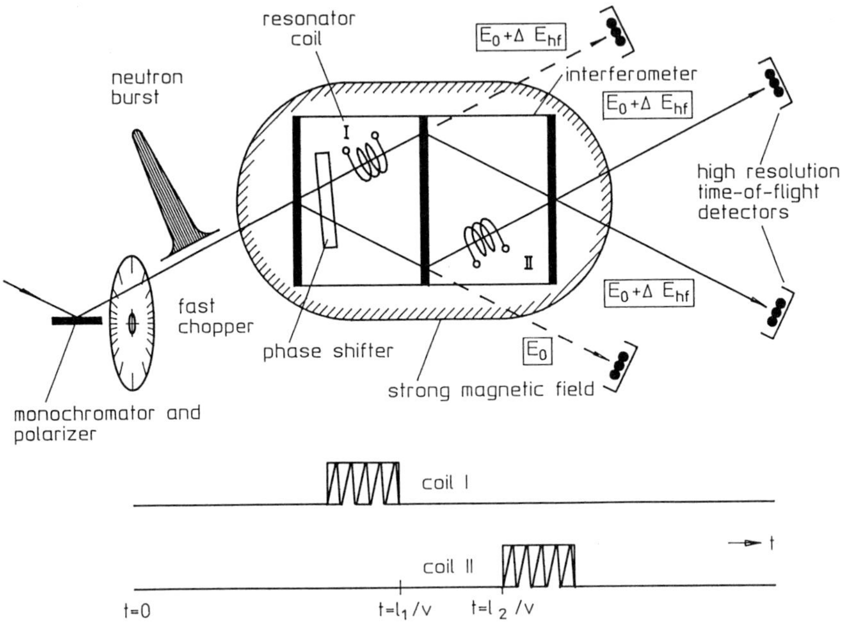

Figure 4: The stroboscope limits the incoming wave packet to dimensions shorter than the lengths of the coils and shorter than the distance between the coils. The coils are arranged spatially separated and operated at time intervals only when the wave packet passes through. The energy and the spin direction of the neutrons between the coils can be measured in the course of a delayed choice mode.

final state is *in a coherent superposition* of states: coil I is in a state where it absorbed this photon *and* coil II is in a state where it also absorbed it. In such a situation the Copenhagen interpretation tells us that both absorption processes *must coexist* with the same degree of reality (i.e., probability), so that (since) an energy loss ΔE has been effectively observed for each individual neutron process, each individual neutron has lost one quantum ΔE only: a process which evidently contradicts the principle (postulate) of absolute energy–momentum conservation in all (observed or unobserved) physical interaction processes.[1]

2. If one observes the simultaneous existence of an individual energy loss ΔE and neutron interference patterns and interprets this as evidence (as suggested by EdB) that individual microprocesses are waves and particles simultaneously (so that their particle aspect follows real space time paths) then the present quantum mechanical formalism (whose validity has been tested in so many experiments) tells us (as shown theoretically by John Bell and confirmed by Aspect's EPR–type experiments) that they must be correlated by non–local interactions. This shows that the EdB interpretation of quantum mechanics, far from being a regression into classical views, requires the introduction into physics of new non–local types of forces (tied to the existence of non–local quantum potentials) which have been shown to be compatible with Einstein's causality [12].

3. Possible "welcher Weg" Experiments

Possible "welcher Weg" experiments (experiments of type II mentioned in the Introduction) are:

1. the proposal to utilize recent photon cloning techniques to tell the photon's path in Mach–Zehnder interferometry [13],
2. the proposal by Scully and Walther [9] to use quantum beat micromaser "welcher Weg" atomic experiments.

Here we shall discuss another promising setup based on a simple modification of the technological breakthrough (microscopic superconducting toroidal magnets and electron holograms) which allowed Tonomura *et al.* [3] to provide conclusive evidence of the reality of the Aharonov–Bohm (AB) effect. The relative AB phase shifts were measured (one electron at a time) between two electron waves passing through spaces inside and outside a tiny toroidal ferromagnet covered completely with a superconductor layer and a Cu layer. The AB fringes and fringe shifts were clearly observed. As we shall now show, a similar, slightly modified version of this setup (which evidently corresponds topologically to a double slit experiment with the sources at infinity) can detect whether the electrons have passed (or not) inside the toroid: a result which corresponds to a "welcher Weg" quantum non–demolition measurement (see Fig.5).

The proposed modification enlarges a suggestion of Kuper [14], using a superconducting torus which confines the magnetic flux of a superconducting current (see Fig.6). By applying a temperature above the superconducting critical temperature T_c and then cooling to well below T_c we can "freeze in" any desired number of flux

---

[1] Of course, such a principle (postulate) cannot be proven by experiment since there is no way in which one can show that it holds for all space points and for all (past, present, or future) possible times.

quanta $n\Phi_o$ with $\Phi_o = \pi(\hbar c/e)$. Different fringe patterns have been predicted (and confirmed experimentally) when n is even or odd. For $\lambda = 2$nm electron wavelength one finds $r \approx 3\mu$m. The proposed modification is to introduce a constant magnetic field H (see Fig.5) which obliges the magnetic moment of the incoming electrons (observed one by one in 10^{-5} sec) to be parallel to their incoming velocity. This does not modify the AB effect since the passage of such an electron's magnetic moment in zone P in Fig.6 (but not in zone Q) induces a pulse of the superconducting current in T, which (according to the quantum mechanical formalism) only reduces its velocity, i.e., adds a constant phase shift to the associated wave. The essential point is that one can in principle detect such an individual pulse (with a SQID for example) so that, if these electrons distribute themselves on the fringe pattern, this will tell that each electron passing in P "knows" of the simultaneous existence of zones P and Q: an evident "welcher Weg" measurement. The existence of such real space time motions (paths) of the particle aspect of individual microobjects implies important theoretical fallouts.

4. Theoretical Implications of Real Particle Trajectories

The coordinates of real paths represent "hidden parameters". They cannot satisfy the classical relativistic laws of motion since their average motion must be influenced by the surrounding "pilot wave" propagation and boundary conditions represented by the quantum potential. To obtain the observed average quantum distributions one must (as shown by Bohm and Vigier [15]) utilize the quantum potential (and quantum torques) deduced from the ψ – field random stochastic path – fluctuations. This reintroduces some subquantal covariant form of stochastic ether of the Dirac – type which carries the average ψ – waves and generates the mentioned fluctuations.

If these stochastic paths are time–like (as they should to satisfy relativity theory) then (as first suggested by Maddox [16]) they should be associated with real positive probability weights: so that the observed quantum probabilities at a point B should

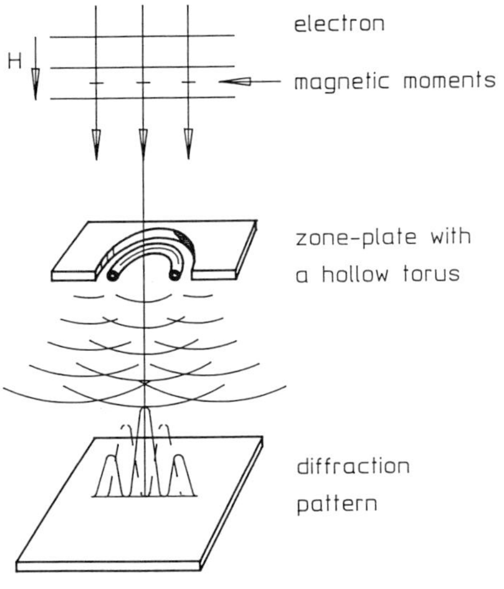

Figure 5: "Welcher Weg" quantum nondemolition experiment as proposed in the text.

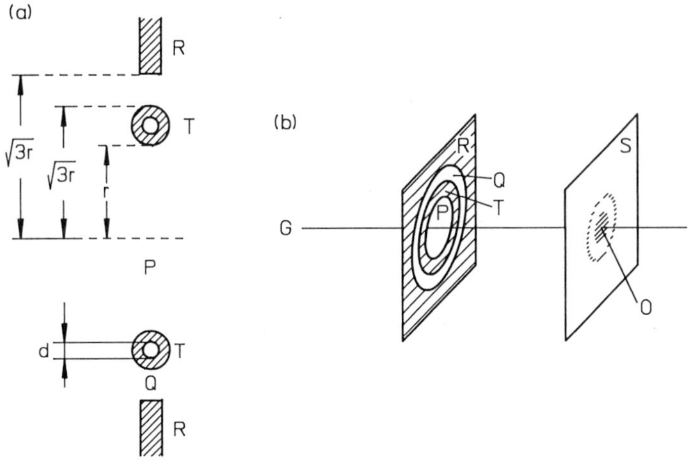

Figure 6: Layout of the experiment sketched in Fig. 5. a) Sectional view of torus T and screen R. b) Schematical diagram of the electron optics.

result from the summation of the probabilities of all stochastic paths joining B to all possible initial space time points A. The introduction into quantum mechanics of such stochastic paths was first done by Feynman [17] who only discussed the laws of associated amplitude superpositions so that he needed complex weights on each such path to reproduce quantum mechanics correctly. Since then some progress has been made. At least in the non-relativistic case, one can associate to each Feynman path a real positive probability if one applies Feynman's path integral formalism to the Fokker–Planck equations associated with all quantum wave equations [17]. This corresponds to the causal stochastic interpretation of quantum mechanics first proposed by Bohm and Vigier [15] and developed later by Nelson [18], de Broglie [19], Guerra and Marra [20], Cetto and de la Pena [21], and Cufaro – Petroni *et al.* [22]. This interpretation considers the ψ – field as describing the average behavior of a stochastic fluctuating field: an assumption which amounts to saying that the elements of the Madelung fluid undergo Feynman–like random stochastic motions, which also corresponds to a Markovian process which combines (average) drift motions $v_i = (1/m)\partial_i S$ with Einstein's osmotic motions $u_i = D\frac{\partial_i \rho}{\rho}\frac{\hbar}{2m}\partial_i P$ within the stochastic differential equations introduced by Nelson [18], i.e.,

$$
\begin{aligned}
d\mathbf{x}(t) &= \mathbf{v}_+(\mathbf{x}, t)dt + d\omega \qquad (dt > 0) \\
d\mathbf{x}(t) &= \mathbf{v}_-(\mathbf{x}, t)dt + d\omega^* \qquad (dt < 0)
\end{aligned}
\tag{8}
$$

with $\mathbf{v}_\pm = \mathbf{v} \pm \mathbf{u}$.

The causal stochastic interpretation considers the quantum potential in terms of the corresponding osmotic energy distribution which reflects the vacuum's interaction with the propagating ψ – wave or Madelung fluid. Indeed, introducing the Hamilton–Jacobi–Bohm equation as a constraint on the motion, one obtains (with $v = 0$) the Lagrangian:

$$
\mathcal{L} = \frac{\hbar^2}{2}(\Delta P + 2\partial_k P \partial^k P) = \frac{\hbar}{4m}\Delta e^{2P} \tag{9}
$$

Since one has $H = -\rho mc^2\sqrt{1 - v_{total}^2/c^2}$ and $\partial_t H = 0$, with $P_i = \partial_i S$ one can write $H_{part} = (P_i P^i/2m) + Q' + V = const$ for the fluid elements with $Q' = (m/2)u_i u^i$ so that the quantum potential just corresponds to the osmotic energy $(m/2)u_i u^i$ introduced by Einstein in Brownian motion theory. In other terms, the Madelung – Takabayasi pilot fluid can be considered as associated with a superfluid, non–dissipative, frictionless wave propagating on Dirac's chaotic ether [15]. In that case the osmotic energy compensates exactly any variation in the drift energy.

Describing the particle's (or soliton's) behavior as also controlled by the fluctuating ψ–field (wave) [23], we see as a consequence of the phase synchronism (locking) of de Broglie, that they must follow the same Feynman–like trajectories in E_4 as the controlling fluid's elements. From the conservation and Liouville equations of both fluid and particles, Bohm and Vigier [15] have deduced that the real physical particle probability $\rho(\mathbf{x}, t)$ associated with the same ψ (pilot) wave $\psi = \sqrt{\rho} \exp iS/\hbar$ can be different from ρ, but always goes back to ρ (which thus represents an equilibrium state) after a certain initial relaxation time.

This implies that the corresponding particles' physical E_4 motions, usually described by a Fokker–Planck equation, coincide on the average with the Fokker–Planck equation which can be deduced from the Schrödinger equation. This equation (equivalent to the conservation equation $\partial_t \rho + \partial_i(\rho/m\ \partial_i S) = 0$) can be written as:

$$\frac{\partial \rho}{\partial t} + \partial_i(\rho v_+^i) - \frac{\hbar}{2m}\Delta\rho = 0 \tag{10}$$

where v_+ represents the Bohm–Vigier–Nelson forward–in–time particle stochastic velocity.

We are now going to compare Feynman's interpretation of quantum mechanics with the causal stochastic interpretation. Before we do so, however, we must emphasize that, following an idea developed by de Broglie [19], Yukawa [25] and their followers [26], all microobjects (i.e., both the piloting fluid elements and their guided particle aspect) which move in space time E_4 are considered as extended spatial structures. They are described by hypertube–like structures in relativistic space time which contain real internal oscillating bilocal structures (clocks) beating with a frequency ν_o in their rest frames where $E_o = h\nu_o$.

The preceding analysis suggests two evident questions which the BH and the EdB interpretations answer in completely different, incompatible, and antagonistic ways.

1. Is it true (or not) that the particle (and wave) aspect of individual microobjects really travels in E_4? Or, in other terms, do these particles exist objectively (or not) even when they are not observed directly? Is it true, for example, that, as assumed by EdB, the individual neutrons observed one by one in Rauch's experiments [5, 6] travel (or not) through the silicium interferometer? This is an experimental question which should and will be answered experimentally [2].
2. If this particle aspect really travels in E_4,

- is it possible (or not) to attribute a positive probability weight to each stochastic space time path between $(\mathbf{x}_i, t_i)$ and $(\mathbf{x}_f, t_f)$?
- is it legitimate to add these probabilities to evaluate the total probability to pass from $(\mathbf{x}_i, t_i)$ to $(\mathbf{x}_f, t_f)$, i.e., to calculate by addition of all possible path contributions the value of the density $\rho(\mathbf{x}, t)$ at any given space time point if one knows the initial particle distribution?

A positive answer to question 2 clearly conditions the possible validity of interpretations of quantum mechanics, and the possibility of giving a realistic interpretation to Feynman's formalism (or not). Indeed, a positive answer to question 2 would imply that *each arbitrarily chosen Feynman path could also be endowed with a positive probability weight*, i.e., considered (as once suggested by Feynman himself to the author) as corresponding to real motions in E_4.

5. Stochastic Feynman–Like Paths Endowed with Positive Probabilities

The answer to question 1 (related to Rauch's recent experiments [5, 6] is beyond the scope of the this paper. The aim of the present contribution is to show on the basis of recent publications of Cufaro–Petroni *et al.* and Raffi–Tabar [27] that one can (but of course not necessarily should) answer the second question positively. To show this let us analyze how the presence of a fluctuating ψ–field carrying any number of particle–like structures (solitons) influences (through the associated quantum potential and guiding theorem) the behavior (i.e., the probability weight) of the particle's possible random path between two arbitrary points in the EdB interpretation.

Clearly, in such a model the transition probability $P(\mathbf{x}, t|\mathbf{x}', t')$ (i.e., the conditional probability of being in $(\mathbf{x}, t)$ starting from $(\mathbf{x}', t')$) will be ruled exactly by the Fokker–Planck equation (10). The random path within the ψ–field is thus controlled by the field itself, so that we can derive the particle's behavior from stochastic control theory. The system can thus be described by means of a Markovian process $\mathbf{q}(t)$ taking values on the configuration space and satisfying an Ito stochastic differential equation:

$$d\mathbf{q}(t) = \mathbf{v}_+[\mathbf{q}(t), t]dt + d\mathbf{w}(t) \tag{11}$$

where $\mathbf{v}_+$ is the control field and $\mathbf{w}(t)$ is a Brownian motion.

Now, as shown by Guerra and Marra [20] and by Weigel [28] in that specific case, our problem is reduced to the evaluation of a stochastic action functional:

$$A = \int_{t_i}^{t_f} E\{\mathcal{L}[\mathbf{q}(t), t]\}dt \tag{12}$$

where $E\{.\}$ denotes expectation values. Guerra and Marra [20] and Cufaro–Petroni *et al.* [22] have discovered that if one starts from the stochastic Lagrangian field $\mathcal{L}$, i.e.,

$$\mathcal{L}(\mathbf{x}, t) = \frac{m}{2}\mathbf{v}_+(\mathbf{x}, t)\mathbf{v}_-(\mathbf{x}, t) - V(\mathbf{x}(t)) \tag{13}$$

where

$$\mathbf{v}_\pm(\mathbf{x}, t) = \mathbf{v}_+(\mathbf{x}, t) \pm \frac{\hbar}{m}\frac{\nabla\rho(\mathbf{x}, t)}{\rho(\mathbf{x}, t)}$$

and which now describes both the Madelung fluid elements and the set of all associated particle motions, one recovers[2] Schrödinger's equation from $\delta A = 0$ [20].

---

[2]The demonstration is evident. If, starting from a dynamical system in configuration space R^n with the classical Lagrangian $L = (m/2)\dot{q}^2 - V(q)$ we introduce stochastic motions characterized by the density $\rho(\mathbf{x}, t)$ and the given forward and backward velocities $v_\pm$, then the average action $A(t_i, t_f; q) = \int_{t_i}^{t_f} \lim E\{(m/2)(\Delta q/\Delta t)^2 - V(q(t))\}dt$ is equal to $A(t_i, t_f; q) = \int_{t_i}^{t_f} dt \int ((m/2)v_+v_- - V(x))\,\rho(\mathbf{x}, t)dx$. With the help of the forward and backward transport operators $D_\pm = \partial_t + v_\pm\nabla \pm \nu\Delta$ and the Hamilton–Jacobi principal function S with $D_+S = (m/2)v_+^2 + m\rho\nabla v_+$ one sees that an arbitrary small change $\delta\rho$ and $\delta v_\pm$ in the process yields

This implies that, if one starts from $\mathcal{L}$ and the second order Fokker–Planck equation for $P \to \rho$ (instead of the complex function ψ), one can follow Feynman's idea of calculating the Green kernel function for the positive real density and express it in terms of path integrals: thus attributing a real positive probability weight to each Feynman path introduced in Feynman's formalism. This has been done independently by Cufaro–Petroni, Raffi–Tabar, and Vigier [27] and Wang [29].

Following Feynman, starting from (10) and the initial conditions

$$\lim_{t \to t'} P(\mathbf{x}, t, \mathbf{x}', t') = \delta^3(x - x')$$

one can first calculate the infinitesimal transition probabilities which satisfy it and thus obtain a probabilistic ponderation for any finite path between $(\mathbf{x}_i, t_i)$ and $(\mathbf{x}_f, t_f)$ by iteration. By this method one finds the probabilistic density $\rho[q(.)]$ in the space of possible paths with respect to a formal infinite–dimensional Lebesgue measure $D[\mathbf{q}(.)]$ in the standard form

$$P[\mathbf{q}(.)] = \exp\left\{-\frac{q}{4D}\int_{t_i}^{t_f}[\dot{\mathbf{q}}(\tau) - \mathbf{v}_+(\mathbf{q}(\tau), t)]^2 d\tau\right\} \geq 0 \tag{14}$$

where $d\tau$ denotes the interval on each path, and $D = (\hbar/2m)$.

The Green function for the Fokker–Planck equation can be written

$$\begin{aligned} G_{fi} = \lim_{\epsilon \to 0, n \to \infty} (4\pi\epsilon D)^{\frac{-(n+1)}{2}} \\ \int \ldots \int d\mathbf{x}_1 \ldots d\mathbf{x}_n \, \exp\left\{\frac{\epsilon}{4D}\sum_{k=1}^{n+1}\left[\frac{\mathbf{x}_k - \mathbf{x}_{k-1}}{\epsilon} - \mathbf{v}_\pm(\mathbf{x}_{k-1} t_{k-1})\right]\right\} \end{aligned} \tag{15}$$

if we write $(n+1)\epsilon = t_f - t_i$, with $\mathbf{x}_{n+1} = x_f$, $\mathbf{x}_o = x_i$, and the interval $[t_i, t_f]$ is split into n subintervals of equal length $\epsilon = t_{i+1} - t_i$. The expression $P[.]$ can be shown to define a non–pathological suitable probability measure $P[q(.)]D[q(.)]$ in the space of all possible random paths.

In this model *the stochastic random paths followed by particles and Madelung fluid elements (since microobjects are waves and particles) are both characterized by relations (14) and (15). This overcomes* [4] *the objections raised against the Nelson formalism which only attributed such paths to particles. Moreover, the set of particle motions is only influenced in time by its own surrounding Madelung fluid element motions as a consequence of de Broglie's (phase locking) guiding theorem. The identity of this model with Feynman's formalism becomes evident if one remarks that, for any real given stochastic path between fixed end points, one can calculate both Feynman's amplitudes and the preceding real positive probabilities by moving successively on small finite time steps*, i.e., deducing step by step the $\psi(\mathbf{x}, t+\epsilon)$ distribution. Of course, the procedure is somewhat different from and more complicated than Feynman's in our case. To calculate Feynman's amplitudes it is enough to know the initial distribution $\psi(\mathbf{x}, 0)$ and to use the values of the classical Lagrangian $L = (m/2)\dot{q}^2 - V(q)$ in the expression

$$\psi(\mathbf{x}, t) = \int d\mathbf{x}_o \; K(\mathbf{x}, t|\mathbf{x}_o, t_o)\psi(\mathbf{x}_o, t_o) \tag{16}$$

$\delta A = \int S(\mathbf{x}_f, t_f)\delta\rho(\mathbf{x}_f, t_f)dx_f \ldots - \int S(\mathbf{x}_i, t_i)\delta\rho(\mathbf{x}_i, t_i)dx_i + \int_{t_i}^{t_f} E\{m\mathbf{v} - \nabla\}\delta v_+(q(t), t)dt$. For limit variations $\delta\rho = \delta v_\pm = 0$ this implies $m\mathbf{v} = \nabla S$. Combined with the preceding value of $D_+S(\mathbf{x}, t)$ we obtain relations which are equivalent to Schrödinger's equation.

with:

$$K(\mathbf{x},t|\mathbf{x}_o,t_o) = \int_{\mathbf{q}(t_o)=\mathbf{x}_o}^{\mathbf{q}(t)=\mathbf{x}} dq\ \exp\left\{\frac{i}{\hbar}\int_{t_o}^{t} dt\ L(\dot{q},q,t)\right\} \tag{17}$$

This simplicity explains the success of Feynman's model. In our stochastic model, relations (14) and (15) imply that one should also know at each point the value of $\mathbf{v}_+(\mathbf{x})$. To follow Feynman's procedure [17], one must thus be able to derive from the initial knowledge of $\rho(\mathbf{x},t)$ and $S(\mathbf{x},t)$ at any given time t their values at a later time $t+\epsilon$ in order to calculate the value $\mathbf{v}_+(\mathbf{x},t+\epsilon)$. This can be done by noting that in our model the initial conditions on the set ψ imply the knowledge of S, P, $\mathbf{v}$, $\dot{S}$, and $\mathbf{u}$ at any point $(\mathbf{x},0)$ which yields their subsequent values at all subsequent points. For example, we can follow step by step the evolution of ρ and S as a consequence of the relations $d\rho/dt = dE/dt = 0$, i.e., $(\partial_t + \nabla\mathbf{v})\rho = 0$ and $(\partial_t + \mathbf{v}\nabla)E = 0 = \partial_t S$, and this determines the set $\mathbf{v}_+$ from which one can calculate all subsequent real, positive path probabilities from relation (15) in any given finite time interval.

Since we have now explicitly shown that an interpretation of quantum mechanics in terms of random particle paths moving within an underlying stochastic medium (such as our Madelung fluid) is possible, and deduced positive probabilities to weigh these Feynman–like motions, one can (following standard classical stochastic definitions) distinguish:

1. *The most probable paths* which connect any pair of fixed points in space time. They are defined by the average of the real trajectories defined by the Green functions

$$G^+(\mathbf{x}_f,t_f|\mathbf{x}_i,t_i) = \lim_{\epsilon\to 0, N\to\infty} (4\pi\epsilon D)^{-N/2} \int\ldots\int d\mathbf{x}_1\ldots d\mathbf{x}_n\ \exp\left\{-\epsilon\sum_{j=0}^{N-1} L^+(\mathbf{x}_{j+1},\mathbf{x}_j,\epsilon)\right\} \tag{18}$$

associated with the stochastic Lagrangian

$$L^+(\mathbf{x}_{j+1},\mathbf{x}_j,\epsilon) = \frac{1}{4D}\left[\frac{1}{2}\mathbf{v}_+(\mathbf{x}_{j+1},t_{j+1}) + \frac{1}{2}\mathbf{v}_+(\mathbf{x}_j,t_j) - \frac{\mathbf{x}_{j+1}-\mathbf{x}_j}{\epsilon}\right]^2 + \frac{1}{2}\partial_{x_{j+1}}\mathbf{v}_+(\mathbf{x}_{j+1},t_{j+1}) \tag{19}$$

and similar terms for motions backward in time.

2. *The most probable transition paths* which correspond to a path (chain) of the most probable small time ϵ transitions with one end point fixed and the other one free. To derive their equations, we consider (15) and (19) with $\mathbf{x}_j$ fixed (for both forward and backward trajectories) and determine the likeliest values of $\mathbf{x}_{j+1}$ (for the forward paths) and $\mathbf{x}_{j-1}$ (for the backward paths). For example, for forward motions we define the stochastic action variation by

$$\begin{aligned}\frac{\delta S^+}{\delta\mathbf{x}_{j+1}} &= \epsilon\frac{\partial L^+(\mathbf{x}_{j+1},\mathbf{x}_j,\epsilon)}{\partial\mathbf{x}_{j+1}} \\ &= \frac{1}{2D}\left[\frac{1}{2}\mathbf{v}_+(\mathbf{x}_{j+1},t_{j+1}) + \frac{1}{2}\mathbf{v}_+(\mathbf{x}_j,t_j) - \frac{\mathbf{x}_{j+1}-\mathbf{x}_j}{\epsilon}\right] \\ &\times\left[1+\frac{\epsilon}{2}\partial_{x_{j+1}}\mathbf{v}_+(\mathbf{x}_{j+1},t_{j+1})\right] = 0\end{aligned} \tag{20}$$

and a simple calculation leads for both, guided particles and fluid elements, to the first order differential equations

$$\begin{aligned}\dot{\mathbf{x}}_+(t) &= \mathbf{v}_+(\mathbf{x},t) = \frac{1}{m}(\nabla S) + D\frac{\nabla\rho}{\rho}\\ \dot{\mathbf{x}}_-(t) &= \mathbf{v}_-(\mathbf{x},t) = \frac{1}{m}(\nabla S) - D\frac{\nabla\rho}{\rho}\end{aligned} \tag{21}$$

which shows that these paths (for both guided particles and fluid elements) correspond to our stochastic Feynman motions.

3. *The most probable average paths* are those paths followed by the quantum particles (or their surrounding fluid elements) with an average velocity defined at any fixed point in space time within a small surrounding space time volume. They evidently correspond to the relation

$$\dot{\mathbf{x}}(t) = \frac{1}{2}\left[\dot{\mathbf{x}}_+(t) + \dot{\mathbf{x}}_-(t)\right] = \frac{\mathbf{x}_{j+1} - \mathbf{x}_{j-1}}{2\epsilon} = \mathbf{v}(\mathbf{x}_j) = \frac{1}{m}\nabla S(\mathbf{x}_j, t_j) \tag{22}$$

so that the corresponding average motions are just the de Broglie–Bohm paths.

6. Conclusions

We conclude with three remarks:

1. The existence of real particle paths in quantum correlated particle systems implies as results from experiments on EPR type situations the reality of multi-body non–local (faster than light) quantum interactions necessarily carried by the many body quantum potential deduced from the corresponding many–body wave equations. This implies that quantum mechanics contains new types of physically real, non–local interactions never observed or discussed before in macroscopic relativistic mechanics. This, however, has been shown not to conflict with Einstein's definition of causality [12]. Indeed, these non–local interactions cannot carry superluminal signals and satisfy Einstein's causality criteria. The sequence of events on any time–like quantum path cannot be modified by the subquantal ether [30]. This interpretation is thus absolutely deterministic.
2. The existence of causal, non–local interactions justifies the results of Bose–Einstein and Fermi–Dirac statistics in terms of distinguishable (by their paths) particle motions [31].
3. The simultaneous existence in general relativity and in quantum mechanics of real particle space time motions suggests a new approach to unify the two theories. All attempts to quantize relativity theories have failed despite the known fact that de Broglie [32] first discovered wave mechanics (i.e., the relation $E = h\nu$ and the idea that particles behave as oscillators beating in phase with the surrounding wavefield) with the help of special relativistic ideas and justified the Bohr–Sommerfeld quantization laws (i.e., $\int \frac{B}{c} dl/\sqrt{1-\beta^2} = nT_o$) on the basis of general relativity. One can show for example that the de Broglie–Bohm average trajectories with $\mathbf{v} = (1/m)\nabla S$ (with $\psi = \sqrt{\rho}\,\exp(iS/\hbar)$) correspond to geodesics of a conformally distorted $g_{\mu\nu}$ metric [23, 33] so that, if (as suggested by Einstein himself [7]) one introduces fluctuating $g_{\mu\nu}$'s at a microscopic level in general relativity, one might associate the stochastic paths of the stochastic interpretation of quantum mechanics with a particular set of microscopic

geodesic paths around a particle–like singularity associated with a particular set of Einstein's field equations.

References

1. J.P. Vigier. Matter waves interferometry. *Physica* **151 B**, 386 (1986).
2. H. Rauch and J.P. Vigier. Proposed neutron interferometry test of Einstein's Einweg assumption in the Bohr–Einstein controversy. *Phys. Lett. A*. To be published.
3. N. Osakabe, T. Matsuda, T. Kawasaki, J. Endo, A. Tonomura, S. Yano, and H. Yamada. *Phys. Rev.* **A 34**, 815 (1986).
4. J.P. Vigier. Real physical paths in quantum mechanics – equivalence of the Einstein–de Broglie and Feynman points of view on quantum particle behavior. *Proc. 3rd Symp. Found. Quant. Mech.* Tokyo, 1989, p. 140.
5. B. Alefeld, G. Badurek, and H. Rauch. *Z. Phys. B* **4 1**, 231 (1981).
   H. Weinfurter, G. Badurek, H. Rauch, D. Schwahn. *Z. Phys. B* **7 2**, 195 (1988).
6. G. Badurek, H. Rauch, and D. Tuppinger. *Phys. Rev.* **A 34**, 260 (1986).
7. N. Bohr. Discussion with Einstein. In *Albert Einstein: Philosopher – Scientist.* Ed. P.A.Schilpp. Open Court, LaSalle, 1970, p.199.
   A. Einstein. *Sitz. Preuss. Akad. Wiss.* 1917, p.606.
   A. Einstein. *Verh. Dtsch. Phys. Ges.* **19**, 82 (1917).
   W. Heisenberg. *Physics and Philosophy.* Benjamin, New York, 1960.
8. J.P. Vigier. *Pramana J. Phys.* **25**, 495 (1985).
9. M.O. Scully and H. Walther. *Phys. Rev.* **A 39**, 5229 (1989).
10. T. Unnerstal. *Phys. Lett. A*. To be published.
11. M. Heinrich, H. Rauch, and H. Wölwitsch. *Physica* **156/157 B**, 588 (1989).
12. J.P. Vigier. *Astr. Nach.* **303**, 55 (1982).
13. A. Garuccio, K.R. Popper, and J.P. Vigier. *Phys. Lett.* **86 A**, 397 (1981).
    See also H.P. Yuen. Quantum amplification and detection of empty de Broglie waves. Northwestern University Evanston Preprint.
14. C.G. Kuper. *Phys. Lett.* **79 A**, 413 (1980).
15. D. Bohm and J.P. Vigier. *Phys. Rev.* **96**, 208 (1954).
    P.A.M. Dirac. *Nature* **168**, 906 (1951).
    K.P. Sinha, E.C.G. Sudarshan, and J.P. Vigier. *Phys. Lett.* **114 A**, 298 (1986).
16. J. Maddox. *Nature* **331**, 101 (1984).
17. R. Feynman. *Rev. Mod. Phys.* **20**, 367 (1946).
    R. Feynman and A.R. Hibbs. *Quantum Mechanics and Path Integrals.* McGraw - Hill, New York, 1965.
18. E. Nelson. *Phys. Rev.* **150**, 1079 (1966).
19. L. de Broglie. *Nonlinear Wave Mechanics.* Elsevier, Amsterdam, 1960.
20. F. Guerra and R. Marra. *Phys. Rev.* **D 28**, 1916 (1983).
21. A.M. Cetto and E. de la Pena. *Found. Phys.* **19**, 419 (1989).
22. N. Cufaro–Petroni, C. Dewdney, P. Holland, A. Kyprianidis, and J.P. Vigier. *Phys. Rev.* **D 32**, 1375 (1985).

23. J.P. Vigier. *Phys. Lett.* **135 A**, 99 (1989).

24. D. Bohm and J.P. Vigier. *Phys. Rev.* **96**, 208 (1954).
For references see J.P. Vigier. *Astr. Nach.* **303**, 55 (1982).

25. H. Yukawa. *Phys. Rev.* **91**, 415 (1953).
See also his scientific works. I Wanami Shoten Publishers, Japan, 1979.

26. D. Bohm and J.P. Vigier. *Phys. Rev.* **109**, 1882 (1958).
F. Halbwachs, J.M. Souriau, and J.P. Vigier. *Journal de Physique et le Radium* **22**, 393 (1961).
F. Halbwachs. *Théorie Relativiste des Fluids a Spin.* Gauthier Villars, Paris, 1960.

27. N. Cufaro–Petroni, H. Raffi–Tabar, and J.P. Vigier. Brownian and quantum motion based on the reality of random paths in physical space and time. IHP Preprint 1988.
H. Raffi–Tabar, *Phys. Lett.* **138 A**, 353 (1989).

28. F.W. Weigel. *Physica* **37**, 105 (1967).
H. Haken. *Z. Phys. B* **2 4**, 321 (1976).
H. Dekker. *Phys. Lett.* **67 A**, 90 (1978).

29. M.S. Wang. *Phys. Lett.* **137 A**, 437 (1989).

30. K.P. Sinha, E.C.G. Sudarshan, and J.P. Vigier. *Phys. Lett.* **114 A**, 298 (1986).

31. A. Kyprianidis, S. Roy, and J.P. Vigier. *Phys. Lett.* **119 A**, 333 (1987).

32. L. de Broglie. *C. R. Acad. Sci.* 1959, p.1426.

33. J.P. Vigier. *C. R. Acad. Sci.* **266**, 598 (1968).

WHAT HAPPENS DURING A QUANTUM TRANSITION?

C. Dewdney and M.M. Lam

Department of Applied Physics
Portsmouth Polytechnic
Portsmouth PO1 2DZ
United Kingdom

Abstract

We show that it is possible to give a causal, continuous and deterministic description of the motion of an individual system undergoing a transition between states using the de Broglie – Bohm interpretation of quantum mechanics.

1. Introduction

The de Broglie – Bohm interpretation of quantum mechanics has now been applied in detail to most of the fundamental phenomena of nonrelativistic quantum mechanics. Explicit calculations of the motion of individual quantum systems, in various situations, have been carried out. These calculations have shown just how it is possible to explain the statistical predictions of quantum mechanics in terms of a precisely defined conceptual scheme in which individual particles have well defined coordinates and are associated with an objective wave. The associated wave determines the values of the particle's dynamical attributes (momentum, spin etc.), these are evaluated from the wave actual particle position. Once an initial position is given (along with the initial form of the associated wave) the particle velocity is determined from the gradient of the phase of the wave at that point. Hence the time development of the wave determines the particle trajectory and controls the evolution of the particle's dynamical attributes.

The de Broglie – Bohm approach is deterministic; exact (but uncontrollable) and unique initial conditions for individual events lead to unique individual outcomes. It is because the particle always has a well defined actual set of attributes that there is no need for wave packet collapse. Interference effects [2, 3], tunneling [4], spin superposition [5], spin measurement [6], quantum statistics [7], EPR spin correlations [8], and correlated–particle interferometry [9] have all been discussed in terms of individual particle trajectories and spin vectors. Clear and intuitive descriptions have been given which demonstrate exactly how the de Broglie – Bohm approach accounts

Information Dynamics, Edited by H. Atmanspacher and H. Scheingraber, Plenum Press, New York, 1991

for the the usual results of quantum mechanics in a consistent manner. Indeed in the case of tunneling, interference and spin superposition (in neutron interferometers) a computer generated motion picture has been developed [10] which demonstrates how the de Broglie – Bohm approach works in an even more striking manner.

Work is currently in progress extending this type of detailed analysis to the interpretation of quantum field theory proposed by Bohm [11] and the relativistic regime.

In the de Broglie – Bohm point of view quantum systems have both wave and particle attributes simultaneously. The various particle attributes can be simultaneously well defined and their equations of motion deduced from the appropriate wave equation. There is a formal similarity between these equations of motion and the Hamilton – Jacobi formulation of classical mechanics. A comparison of the de Broglie – Bohm interpretation of quantum mechanics and the Hamilton – Jacobi formulation of classical mechanics enables an identification of the essentially new features of matter that the quantum theory has introduced. The equations of motion deduced in the de Broglie – Bohm approach demonstrate that, although the particle attributes in one sense can be thought of as belonging to the particle alone, their evolution is determined by the development of the wave function of the complete quantum system of which the particle is a part. The evolution of the wave function of the quantum system depends on the whole context in which the quantum system is placed and so the motion of an individual particle in the system depends on the whole context.

It is really this feature that constitutes the most radical departure from classical mechanics. In classical mechanics the motion of an individual particle in the system is determined simply by specifying the initial coordinates and (independently) the momenta of all the particles along with the forces acting. One can solve the associated Hamilton – Jacobi equation for Hamilton's function and calculate the system trajectory from a given initial position in configuration space. Individual particle trajectories can be obtained from the projection of the system trajectory on the individual coordinate axes. Alternatively one can integrate Newton's equation for each particle. In quantum mechanics one must calculate the evolution of an initial wave function in configuration space. The gradient of the phase of this wave function determines the velocity of the system representative point in configuration space. Hence the evolution of this wave function controls the motion of the particles. This feature of the approach prompted de Broglie to describe the wave as a "pilot" wave. In any problem the initial conditions which must be specified are thus the particle coordinates (determining the initial system point in configuration space) and the configuration space wave function.

The de Broglie – Bohm interpretation of quantum mechanics makes it clear that it is necessary to introduce a completely new type of determination in the description of the motion of physical systems: the control of the particles' motion by the pilot wave. The novel character of this description of motion can be seen in the examples already referred to and in those given below.

It is perhaps worth pointing out that there are significant differences in the nature of the control in the single and in the many particle case. As the evolution of the particle's associated wave function is determined, in the single particle case, simply by the structure of the particle's environment the evolution of the particle attributes can come to depend on features of the environment that are distant from a given individual particle trajectory, but these features can not influence the particle's motion if they are far from the region in which the associated wave function is finite. In terms of the usual approach this simply means that distant features of the environment can not affect the particle's motion if there is zero probability of the particle ever being at the distant region. There must be some probability that a particle will interact with the distant features if they are to affect the wave function! Further, any influence on an individual particle caused by a change in the environment where the wave is not zero must be transmitted locally to the particle position through the propagation of the wave according to the time dependent Schrödinger (or other appropriate) equation. Changes made in the environment do not instantly alter the wave over all space. So there is no nonlocality in the sense that spacelike separated regions can influence each other. In this respect we have a wave theory similar to any other in which the wave propagates locally and causally in real space.

In the many particle case the wave propagates locally and causally in configuration space but this can result in truly nonlocal effects in the real space in which the individual particles move. That is the motion of one particle may depend on conditions pertaining in regions where other particles in the system are that are spacelike separated. The quantum system has a single trajectory in configuration space, projection of this trajectory on the individual coordinate axes yields the individual particle motions and, when the wave function is entangled (not factorisable into a product of functions each dependent on a single particle coordinate), an individual trajectory may become correlated with the motion of the other particles in the system. This is shown clearly in references [8, 9].

2. Transitions in Two–Level Systems

In order to complete the set of calculations for the non–relativistic case, we discuss here, in terms of specific examples, an initial approach to transitions between non–degenerate eigenstates induced by resonant perturbations in two–level systems [12]. Of course a complete description must quantise both the material two–level system and the transition inducing field and in a further paper such a description (from the de Broglie – Bohm point of view) will be given. In this paper we will simply discuss the motion undergone by the two–level system, treating the field classically and ignoring the radiative reactions. The main point is to demonstrate that the usual description of transitions between states in terms of individual discontinuous jumps does not necessarily follow from the quantum formalism itself, but is an additional interpretive description invoked to explain what happens in an individual transition and to show what lies beyond the statistical predictions of the quantum theory. We will show that it is possible to give a description of transitions in terms of a continuous evolution of the system coordinates.

To discuss transitions it is usually necessary to use time dependent perturbation theory. However it is well known that there is one case in which it is possible to solve the Schrödinger equation exactly, this is the case in which the system has just two energy levels, E_1 and E_2 with the associated eigenfunctions ψ_1 and ψ_2. This type of system starts at $t = 0$ in one of the states (say ψ_1), under the influence of an appropriate perturbation of frequency:

$$\omega = (E_2 - E_1)/\hbar$$

It thereafter oscillates with period τ between ψ_1 and ψ_2 according to

$$\Psi = a_1(t)\psi_1 \exp(-iE_1t/\hbar) + a_2(t)\psi_2 \exp(-iE_2t/\hbar)$$

where

$$a_1(t) = \cos \Omega t \quad \text{and} \quad a_2(t) = -i \sin \Omega t$$

and Ω is related to the strength of the perturbation. In time the system passes from one state to the other and no measurement need be carried out.

According to the usual approach to quantum theory the state (1) means that if the energy of the system is measured at a time t then the result E_1 will be found with a probability $|a_1(t)|^2$ and the result E_2 will be found with a probability $|a_2(t)|^2$. In the absence of measurement since (1) is a pure state nothing can be said about the energy of a single system except of course for integer multiples of the period. Between times the energy should be described indeterminate. If we consider a statistical ensemble of systems then, adopting a relative frequency approach to probability, we may say that if a measurement of the energy is carried out at time t then the result E_1 will be found in a proportion approximately $|a_1(t)|^2$ of the cases and the result E_2 in a proportion approximately $|a_2(t)|^2$ of the cases. This does not mean however that *before* the measurement the fraction $|a_2(t)|^2$ of systems had made a quantum jump from state ψ_1 (the definite state at $t = 0$) with energy E_1 to the state ψ_2 with energy E_2 whilst the fraction $|a_1(t)|^2$ remained in the state ψ_1. Again, in the usual approach, in the absence of a measurement, nothing can be said about the energy of an individual system at time t. This is because (1) is a pure state. Even if the system is interacted with in some way so that the interference terms in $\Psi^*\Psi$ are made negligible then in the usual approach it is still not possible to say that an individual system has a definite energy. This is because the wave function of an individual system is still a pure state, and not a mixture, but now with the parts containing ψ_1 and ψ_2 orthogonal. These two parts could however be brought to interfere once more, which would not be the case if (1) actually was a mixture of the two components and each individual system associated with either ψ_1 or ψ_2.

In the de Broglie – Bohm interpretation the interpretation of (1) goes further since the wave function does not exhaust the possibilities for the description of individual systems. The wave function does however exhaust the possible information that can be obtained about individual systems through measurement. In addition to its wave function each system has a well defined set of particle attributes and once the coordinate of the particle is specified these attributes are determined from the associated wave function. In a superposition state such as (1) the energy evolves in a continuous manner. It is not confined to the eigenvalues E_1 or E_2 (except when the state is definitely ψ_1 or ψ_2) but evolves continuously between them. This is the basis for a continuous and causal description of transitions. If at some arbitrary time the system is interacted with in such a way that the interference terms are made negligible then the energy evolves during this interaction to become equal to one of the appropriate eigenvalues. The energy in the de Broglie – Bohm approach is given by $-\frac{dS}{dt}$. When the interference terms are made small the energy tends to either E_1 or E_2, as is clear from

$$E = -\frac{dS}{dt} = \frac{1}{\rho}\left[{R_1}^2 E_1 - {R_2}^2 E_2 + R_1 R_2 (E_1 + E_2) \cos (E_2 - E_1)\, t/\hbar\right]$$

where $\rho = {R_1}^2 - {R_2}^2 + 2R_1R_2 \cos (E_2 - E_1)\, t/\hbar$, $R_1 = \cos \Omega t \psi_1$, $R_2 = \sin \Omega t \psi_2$, and we assume that ψ_1 and ψ_2 are real in these expressions.

Which of these values is adopted in an individual case depends on the precise values of the initial coordinates as we shall see below in specific examples.

As far as the particle attributes in the de Broglie – Bohm approach are concerned the pure state (1) with negligible interference terms in the probability density gives the same result as a statistical mixture. (In the description of Bohm and Hiley [11] the information contained in the orthogonal component of the wave function is inactive.) However since each individual particle is in fact still associated with the whole wave function if the two parts are brought back to interfere the particle attributes will once more contain contributions from the interference terms. (The information in the once orthogonal part of the wave function has become active again.) Evidently this is not possible with a statistical mixture.

3. Transitions in a Square Well

The first example we consider is that of a particle bound in a one–dimensional infinite square potential well. The particle is initially in its ground state and is then subjected to a perturbation which has a frequency given by $\omega = (E_4 - E_1)/\hbar$, where E_4 is the energy of the third exited state u_4 and E_1 the energy of the ground state u_1. Ignoring transitions to any other energy levels we treat this system approximately as a two–level system. It is easily shown that if the system starts in the state u_1 at $t = 0$ then the state evolves thereafter according to

$$\Psi = \cos \Omega t \; u_1 \exp(-iE_1 t/\hbar) - i \sin \Omega t \; u_4 \exp(-iE_4 t/\hbar)$$

where Ω is dependent on the strength of the interaction. According to the usual approach to quantum mechanics this wave function is interpreted as giving the probability of finding an individual system in the state u_1 at time t as $(\cos \Omega t)^2$ whilst the probability that the system is in state u_4 at time t is $-(\sin \Omega t)^2$. Only when the evolution of the state is such that it is simply u_1 or u_4 can the energy be said to have a definite value, between times it must be said to be indefinite. If, on measurement, an individual particle is found to have the energy E_4 then the system is often described as having made a quantum jump from u_1 to u_4. There is nothing in the formalism of quantum mechanics to justify this statement, it is an *ad hoc*, but often repeated, interpretative assumption. No such additional assumption is required in the de Broglie – Bohm approach.

The de Broglie – Bohm interpretation of the single particle Schrödinger equation is arrived at in a straight forward manner by substituting

$$\psi = R \; \exp(iS/\hbar)$$

in the Schrödinger equation

$$i\hbar \frac{\partial \psi}{\partial t} = -\frac{\hbar^2}{2m} \nabla^2 \psi + V\psi$$

to yield

$$\frac{\partial \rho}{\partial t} = \nabla(\rho v)$$

the continuity equation for the probability density $\rho = \psi^* \psi$, and

$$-\frac{dS}{dt} = -\frac{\hbar^2 \nabla^2 R}{2mR} + \frac{(\nabla S)^2}{2m} + V$$

a Hamilton–Jacobi type of equation with the particle momentum given by:

$$\mathbf{p} = \nabla S$$

This equation can be integrated to give the particle trajectory providing ∇S is known and the initial position of the particle is given.

In Figure 1 we show a set of possible individual particle trajectories using the wave function (4) describing the transitions of the two–level system. The particle's motion depends on its initial position. At $t = 0$ and $t = \tau$ the particle is stationary corresponding to the system being momentarily in one of the two possible stationary states. When the system is in one of the stationary states u_1 or u_4 the particle energy is E_1 or E_4, respectively: although its kinetic energy $(\nabla S)^2/2m$ is zero its quantum potential energy is exactly equal to E_1 or E_4 as a simple calculation shows. In between the particle oscillates back and forth with the frequency ω, whilst its energy is continuously variable. In this way each individual system executes a continuous motion during the transitions between the two states. Under these conditions the particle will make the transition from one state to the other through a continuous evolution and without the intervention of a measurement.

If at some time t' we interact with the system to destroy the interference between the two superposed components the particle's motion evolves so that the energy becomes E_1 or E_4 depending on its initial position. One way of achieving this in a rather approximate manner is to switch off the perturbation and the confining potential at the time t'. The wave then runs out in either direction. Two wave packets emerge in each direction travelling at different average velocities corresponding to the two energies E_1 and E_4. We consider the potential to be switched off at $t = \pi/4\Omega$. The probability density at this time and for a series of times after the potentials are switched off is shown in Figure 2. (We use a numerical integration of the time–dependent Schrödinger equation.) Evidently the observation of the particle position at this later time can be considered to be a measurement of the energy of the particle to the extent that the wave packets corresponding with u_1 and u_4 separate. A set of trajectories for the situation after the potential has been turned off is shown in Fig.3, clearly each particle must enter one packet and which one it enters depends on its position when the potentials are removed.

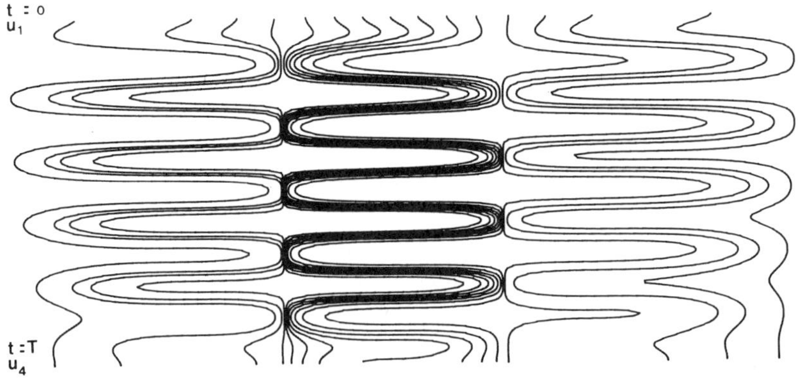

Figure 1: A set of possible individual trajectories for a particle undergoing a transition from the ground state to the fourth excited state in an infinite square well. Time evolves down the page.

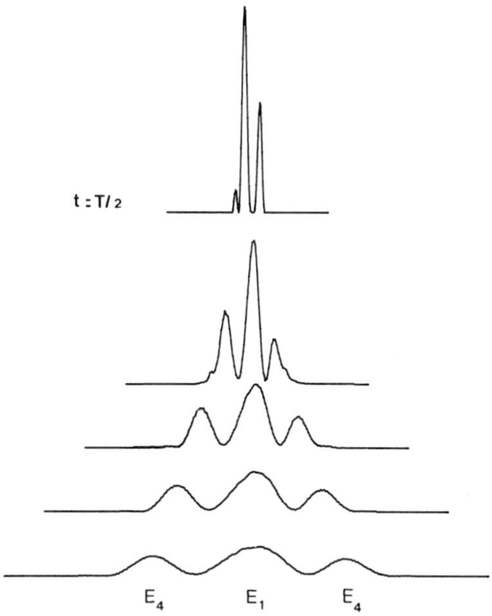

Figure 2: The evolution of the probability density after the confining potential is removed at the time when the particle's wave function is an equal superposition of the ground and the fourth state.

4. Spin Flips

As our second specific example of a two–level system we consider a massive spin one–half particle with magnetic moment μ in a static magnetic field aligned with the z–direction. This system is exactly a two–level system but the interpretation is more complex. The theory of this situation, both classical [13] and quantum [14] is well known and we will only outline it here.

Classically a spinning charged sphere with an angular momentum s and magnetic moment μ will precess about an applied magnetic field under the influence of the torque $\mathbf{T} = \mu \times \mathbf{B}$. In order to alter the angle between the applied field and the magnetic moment it is necessary to exert a torque about the direction of the applied static field. If the static field is applied in the z–direction then another field must be applied in the xy–plane. The most efficient way to apply the field is so that it is always at right angles to the precessing moment, hence it must rotate. Fields applied in this way will cause the magnetic moment to rotate so that it passes the z–axis in the positive and negative senses in a periodic manner. During this process the energy varies in a continuous manner.

The quantum mechanical treatment of this situation reveals a similar behaviour. The magnetic moment precesses about the applied static field with the Larmor frequency $\mu B_z/\hbar$. If a field rotating with the Larmor frequency is applied as described above, transitions between the two possible eigenstates of the z–component of μ can be obtained. The difference from the classical description being of course that in quantum mechanics only one component of μ can be well defined and each component can adopt one of two possible eigenvalues. Quantum mechanically this is a two–level system, μ_z may either be aligned with the field or opposed to the field, the two states have different energies separated by $2\mu_z B$ whose populations vary with time.

In the usual description μ_z cannot be thought of as changing continuously. The common description is that the z–component of the spin changes discontinuously, it is said to "flip" from being "up" to "down". Such statements are not derived from the formalism of quantum theory, they are *ad hoc* interpretative assumptions designed to explain what happens at the level of the individual system. An individual system in the configuration of fields described here is said to have flipped at a given time with a certain probability. In the de Broglie – Bohm approach the description is somewhat different and not at all ambiguous. We first outline the theory of spin in the de Broglie – Bohm approach.

In order to develop an approach to the spin in the context of the de Broglie – Bohm interpretation of quantum mechanics, and there are several related possibilities, it is necessary first to be able to interpret the spinor. We use the well known geometric interpretation of the spinor in which it is interpreted as defining a state of rotation by writing

$$X = \begin{pmatrix} \cos\Theta/2 & \exp i(\Phi+\psi)/2 \\ i\sin\Theta/2 & \exp -i(\Phi-\psi)/2 \end{pmatrix}$$

where Θ, Φ, ψ are the Euler angles. In this representation of the spinor it is possible to define a spin vector by

$$\mathbf{s} = X^{+}\sigma X$$

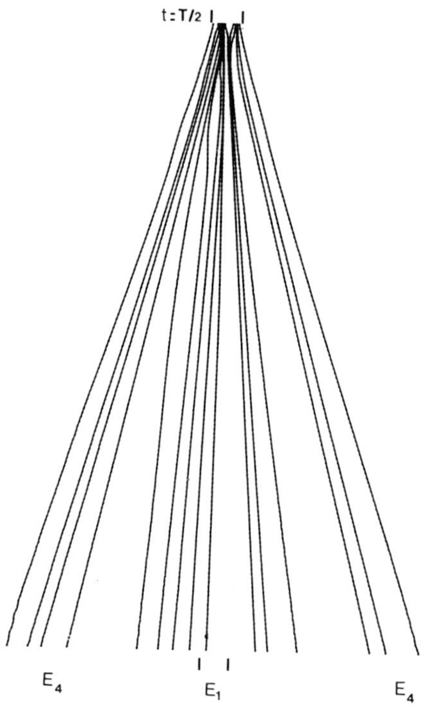

Figure 3: A set of possible individual particle trajectories corresponding with the evolution of the state whose probability density is shown in Figure 2.

or in terms of components and the Euler angles:

$$\mathbf{s} = \hbar/2\ (\sin\Theta\sin\Phi, \sin\Theta\cos\Phi, \cos\Theta)$$

We adopt the causal interpretation of spin outlined by Bohm *et al.* [15] and applied in detailed calculations by Dewdney *et al.* [5, 6, 8] in which all components of the spin are well defined simultaneously. One substitutes

$$\Psi = RX$$

where R is a real amplitude function and X a unit spinor in the Pauli equation

$$i\hbar\frac{\partial\Psi}{\partial t} = H\Psi$$

where H is the Pauli Hamiltonian:

$$H = \frac{-\hbar^2}{2m}\left[\left(\nabla - \frac{ie}{\hbar c}\mathbf{A}\right)^2 + eA_o + V\right] + \mu\sigma\cdot\mathbf{B}$$

Carrying out this procedure one finds a set of equations of motion which determine the evolution of the particle coordinates and spin vector. The equations are

$$\frac{\partial\rho}{\partial t} + \nabla(\rho\mathbf{v}) = 0$$

a continuity equation for the probability density ρ, where the particle velocity $\mathbf{v}$ is given by

$$\mathbf{v} = \frac{\hbar}{2m}(\nabla\psi + \cos\Theta\ \nabla\Phi)$$

and a generalized Hamilton–Jacobi type of equation

$$\frac{\hbar}{2}\left(\frac{\partial\psi}{\partial t} + \cos\Theta\frac{\partial\Phi}{\partial t}\right) + \frac{mv^2}{2} + Q + Q_s + \frac{2\mu}{\hbar}\mathbf{B}\cdot\mathbf{s} + eA_o + V = 0$$

where

$$Q = -\frac{\hbar^2\nabla^2R}{2mR}$$

the usual quantum potential and

$$Q_s = \frac{\hbar^2}{8m}\left((\nabla\Theta)^2 + \sin^2\Theta(\nabla\Theta)^2\right)$$

is a spin–dependent addition.

The evolution of the spin vector is best described by the equation

$$\frac{d\mathbf{s}}{dt} = \mathbf{T} + \frac{2\mu}{\hbar}\mathbf{B}\times\mathbf{s}$$

where:

$$\mathbf{T} = \frac{1}{m\rho}\mathbf{s}\times\sum_i\frac{\partial}{\partial x_i}\left(\rho\frac{\partial\mathbf{s}}{\partial x_i}\right)$$

For the system under consideration the spinor has no spatial variation and so there is no quantum torque acting on the spin vector which executes the same notion as a classical magnetic moment subjected to a combination of fields with magnitude $B_\perp$ in the xy–plane and $B_{||}$ in the z–direction. If the initial spinor wave function is given by

$$X = \begin{pmatrix} 1 \\ 2 \end{pmatrix}$$

with spin vector $\mathbf{s} = \hbar/2(0,0,1)$ at time t when the driving frequency is the Larmor frequency, then the spinor wave function will be given by [13]

$$X = \begin{pmatrix} \cos(\Omega_\perp t) & \exp(i\Omega_{||} t) \\ i\sin(\Omega_\perp t) & \exp(-i\Omega_{||} t) \end{pmatrix}$$

yielding

$$\mathbf{s} = \hbar/2 \left(\sin(2\Omega_\perp t)\sin(2\Omega_{||} t), \sin(2\Omega_\perp t)\cos(2\Omega_{||} t), \cos(2\Omega_\perp t)\right)$$

where $\Omega_\perp = \mu B_\perp/2\hbar$ and $\Omega_{||} = \mu B_{||}/2\hbar$. At $t = 0$ the spin vector lies along the z–axis and the system is in the state ψ_1. After a time $t = \pi/2\Omega_\perp$ the spin vector lies in the opposite direction and the state is ψ_2. In between these times the spin vector has evolved in a continuous manner as the spin makes the transition from one state to the other. If the rotating field is applied for a time $t = \pi/4\Omega_\perp$ then the spin vector will lie in the xy–plane. If the spin component in the z–direction is now measured, then the result $\hbar/2$ will be found with a probability of one half and the result $-\hbar/2$ with a probability of one half. This does not mean however that before a measurement is carried out the spin component in the z–direction is $\pm\hbar/2$. In particular it does not mean that in a statistical ensemble of systems one half of them have undergone a spin flip. According to our approach to spin the z–component of the spin is actually zero in this state. In order to see how this value is consistent with the statistical

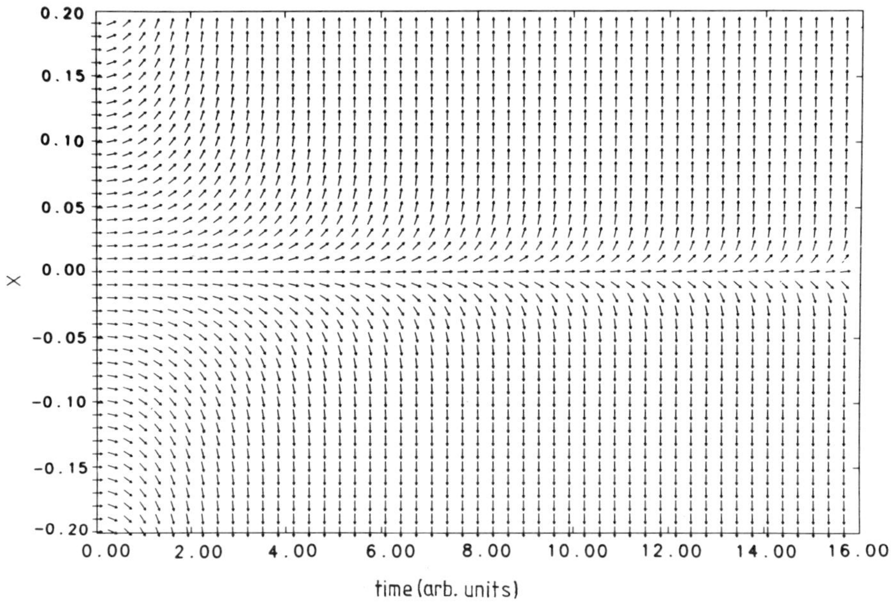

Figure 4: A set of possible trajectories for a spin one–half particle, whose spin initially lay in the xy–plane ($\Theta = \pi/2$). After interacting with a Stern–Gerlach field it is oriented in the z–direction.

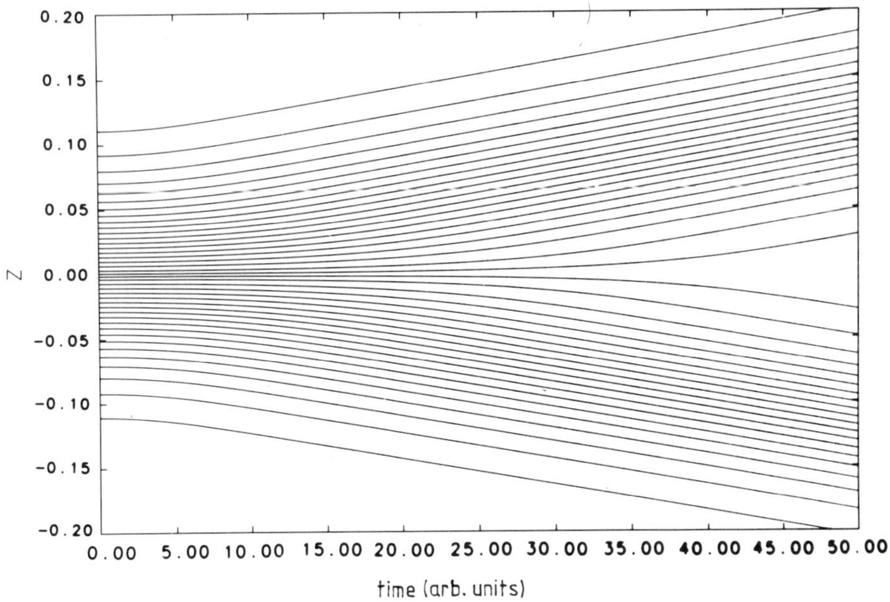

Figure 5: The spatial variation of the polar angle of the spin after the interaction with the Stern–Gerlach field. Initially $\Theta = \pi/2$, as the state evolves it tends to either 0 or π.

predictions of quantum theory it is necessary to discuss the process by which the spin is measured.

A discussion of spin measurement using a Stern–Gerlach device has already been given from our point of view [6], here we just present the results to illuminate the present discussion. The interaction between the spin and the inhomogeneous field is taken to be impulsive and the beam is assumed to have a Gaussian profile in the direction of the analyzing field. Other parameters are chosen so that the two packets formed in the interaction, corresponding to a positive and a negative value of the z–component of the spin, separate along the z–direction faster than they spread. This allows an unambiguous deduction of the spin component from the observation of the position of the particle. The apparatus coordinate in this case is the particle position (by observing it we deduce the value of the spin) and the function of the inhomogeneous field is to introduce a correlation between the system coordinate (the spin component) and the apparatus coordinate (the particle position). As shown in Fig.4, under the influence of the (spin dependent) quantum potential the set of trajectories splits into two along the center of the packet as two separating packets are formed. As can be seen in Fig.5, the separation of the two packets introduces a spatial dependence of the spin, the spin vector precesses, even in the absence of any externally applied fields, under the influence of the quantum torque. The torque acts in such a way that any particle following a trajectory entering the upper (lower) packet has its spin vector twisted to lie along (opposed to) the z–axis. In our description no system makes a jump from one state to another, the spin vector evolves continuously, as does the particle energy, under the influence of the applied fields. It is clear in this description just how the interaction with the system brings about the eigenvalues of the operator being measured on the system and the appropriate correlated state of the apparatus variable through a continuous and deterministic evolution. There is no need to talk about quantum jumps.

5. Conclusion

The wave function of a system undergoing a transition between the eigenstates of some operator is, in general, a superposition of states. According to the de Broglie – Bohm interpretation when a system is in such a superposition the value of the particle's corresponding attribute will not be stationary. Instead the value of the attribute will oscillate continuously. The system cannot be thought of as being in any definite single component state. Certain interactions with a two–level system can cause it to oscillate between the two states and we have seen how this process can be described without quantum jumps. Other interactions (which may or may not constitute a measurement) can destroy the interference between the component terms in the superposition. In such a case the particle's attribute will evolve in a continuous manner to become equal to those values associated with each component eigenstate alone. During such a measurement–like interaction the system coordinate evolves in conjunction with the apparatus coordinate so that each individual system adopts one of the possible eigenvalues of the observable being measured, whilst the apparatus coordinate adopts a correlated and unambiguous value.

In this way the whole process of transitions between states is described in a continuous fashion without the need for quantum jumps. We have seen how and in what sense interactions such as we have discussed bring about the eigenvalues of the observables concerned. We have seen how the results of measurement are created by the interaction between the system and the measuring device and understand exactly the nature of the "disturbance" of quantum systems undergoing measurements. Since the particle attributes always have definite values, even in superposition states, we can discuss how they change when they are measured. The statistical predictions concerning the results of measurement are recovered *and* a deterministic continuous and causal description is given of individual transitions.

References

1. L. de Broglie. *C. R. Acad. Sci. Paris* **183** 447 (1926).
   L. de Broglie. *C. R. Acad. Sci. Paris* **185**, 380 (1927).
   L. de Broglie. *Nonlinear Wave Mechanics*. Elsevier, Amsterdam, 1960.
   D. Bohm. *Phys. Rev.* **85**, 166 (1952).
   D. Bohm. *Phys. Rev.* **85**, 180 (1952).
2. C. Philippidis, C. Dewdney, and B. Hiley. *Nuovo Cim.* **52 B**, 15 (1979).
3. C. Dewdney. *Phys. Lett.* **109 A**, 377 (1985).
4. C. Dewdney and B. Hiley. *Found. Phys.* **12**, 27 (1982).
5. C. Dewdney, P.R. Holland, and A. Kyprianidis. *Phys. Lett.* **121 A**, 105 (1987).
6. C. Dewdney, P.R. Holland, and A. Kyprianidis. *Phys. Lett.* **119 A**, 259 (1986).
7. C. Dewdney, A. Kyprianidis, and J.P. Vigier. *J. Phys. A* **17**, L741 (1984).
8. C. Dewdney, P.R. Holland, and A. Kyprianidis. *J. Phys. A* **20**, 4717 (1987).
9. M.M. Lam and C. Dewdney. *Phys. Lett.* **150 A**, 127 (1990).
10. C. Dewdney. Copies on video available from the author.
11. D. Bohm, B.J. Hiley, and P.N. Kaloyerou. *Phys. Rep.* **144**, 321 (1987).

12. A. Yarif. *Theory and Applications of Quantum Mechanics.* Wiley, New York, 1982.
13. H. Goldstein. *Classical Mechanics.* Addison – Wesley, Cambridge, 1980.
14. R.L. Liboff. *Introductory Quantum Mechanics.* Addison – Wesley, Cambridge, 1980.
15. D. Bohm, R. Schiller, and J. Tiomno. *Nuovo Cim. Suppl.* **1**, 48 (1955).

EPILOGUE: DUALITIES, CONTEXT, AND MEANING

HARALD ATMANSPACHER

Max Planck Institut für Physik und Astrophysik
Institut für extraterrestrische Physik
D-8046 Garching, FRG

EDWARD D. WEINBERGER

Max Planck Institut für biophysikalische Chemie
Am Fassberg
D-3400 Göttingen, FRG

Much has been written about the conceptual crisis that necessitated the development of quantum mechanics. The backdrop for this workshop is a similar crisis, which may be called the "crisis of complexity". The conventional scientific wisdom of previous centuries expected First Principles to be "intuitively obvious", and thus as sure a foundation for natural law as Euclid's postulates were for geometry. It was also expected that such First Principles would ultimately lead to a "theory of everything" because quantitative discussions of *all* natural phenomena could be deduced from them *simply*. The first blow against this simplistic view was the counter-intuitiveness of quantum phenomena, but the postulates of quantum mechanics, if counter-intuitive, were still simple to state (in the appropriate mathematical language). It was "chaos theory" that showed an *essential* difficulty with the above assertions: a simple system can exhibit such complex dynamics as to be essentially unpredictable over a reasonable time scale. Chaotic dynamics thus emphasizes the need for a "science of complex systems", the study of systems in which "the whole is more than the sum of its parts," in contrast to earlier science, which emphasized the study of isolated systems, or "things in themselves". (It is interesting to note that the Poincaré-Bendixson Theorem [1] rules out the existence of chaotic attractors in systems with fewer than three degrees of freedom. Haken's slaving principle, which states that system dynamics is essentially determined by the slowest acting degrees of freedom, thus rules out chaos in autonomous systems in which fewer than three degrees of freedom vary significantly on the same time scale, thus proving rigorously that the study of chaos is necessarily the study of interactions.)

Our thesis in this Epilogue is that science can confront the crisis of complexity with the same conceptual tool that proved so useful in the formulation of quantum

Information Dynamics, Edited by H. Atmanspacher and
H. Scheingraber, Plenum Press, New York, 1991

mechanics; namely, the notion of duality. Indeed, many of the articles presented in this volume can be embedded in recent developments in the philosophy of science, proposing a "synthesis" of realism and relativism in the sense of Elkana's "anthropology of knowledge" [2], or of Putnam's viewpoint of an "internal realism" [3]. This synthesis offers space for a huge variety of kinds of knowledge representation. It is relativistic since there is an unavoidable context–dependence ("subjectivity") of any representation of knowledge. It is realistic since within a given representation of knowledge (within a given context) "objective" criteria for rationality and truth can be developed and used. Relativity is to be understood in the sense of different possible levels of reality. Realism characterizes an attitude resulting from the restriction to a particular one of these levels.

The apparent dichotomy of objective realism and subjective relativism can be extended to further pairs of terms. While the realistic attitude is related to the level of facts or events as observable phenomena, relativism stresses the notion of ideas, concepts, models, theories, and laws (ordered by an increasing degree of "objectivity"). Facts and models can again be considered as a two level matter of affairs. Facts are always theoretically contaminated, and models can equivalently be considered as facts on a higher order level, i.e., as facts for meta–models. This view stresses the relativistic aspect mentioned above. The realistic aspect relates to the application of logical rules pertaining to each particular one of those levels. In this regard, the general relevance of non–Boolean logics has repeatedly been indicated in different contexts, e.g., in the contributions of Euler, Krueger, and Rosenblum.

In physics, the dualities of realism/relativism and of facts/models appear as generalizations of the well known incommensurabilities of position/momentum and time/energy. It has been suggested by Atmanspacher that all of these incommensurabilities might be understood as dualities in the same strict mathematical sense (which may be expressed in a lattice theoretical manner). Relating position and momentum to spatial properties (structure) of a system, and relating time and energy to its temporal properties (dynamics, function, behavior), an additional duality appears: that of space and time. What we observe and measure are spatial components of events, whereas their temporal component is the subject of dynamical modelling and prediction, issues to which the articles by Farmer, Fraser, Jensen, and Smith are devoted.

Several contributions included here stress the duality between an "internal" and an "external" description of a physical system. Correspondingly, one can list the pairs internal (proper) time *versus* external (parameter) time, intrinsic irreversibility *versus* environment induced irreversibility, potential information *versus* actual information, statistics *versus* causality, (Boltzmann) "probability" distribution for single events *versus* (Gibbs) probability distribution for event ensembles, etc. For instance, internal time is one of the key concepts in the approach of the Brussels school, here represented by the articles of Antoniou and Courbage. This approach relates to an intrinsic irreversibility of classical and quantum systems arising from the system's internal information processing capabilities, which can, in turn, be related to a positive dynamical entropy. Various additional aspects, problems, and extensions of such an approach are described by Beck, Hudetz, and Wehrl.

Zurek (conceptually similar to Dittrich & Graham) proposes the complementary view: that irreversibility might result from interactions with the system's environment. The main emphasis in Zurek's article concerns a unification of actual and potential information via the decomposition of "physical entropy" into the sum of the algorithmic information content of data (actual component) and its statistical entropy (potential component). (Muller discusses problems with specific experimental setups related to this context.) Fundamental deliberations about the relation among the concept of causal law and statistical correlations between events are the subject of Lloyd's article. Further general aspects, especially concerning non–local correlations in complex systems, are addressed by von Lucadou.

Klir's contribution deserves specific emphasis since it describes novel mathematical theories that go beyond the capabilities of traditional probability theory in dealing with uncertainty and information. His notion of nonspecificity jibes perfectly with Boltzmann's concept of uncertainty for single particles, where traditional probability (if applied rigorously) is ill–defined. Finally, the subject of discrete *versus* continuous transitions and paths in quantum mechanics (see the articles of Dewdney & Lam and Vigier) may be mentioned here.

Last but not least we have the *duality of context–dependence and universality.* The importance of the context in which information is gathered or used has been stressed by several contributors to this volume. In particular, context–dependence has been discussed in approaches toward a reasonable definition of complexity. In this regard a number of notions have shown up: algorithmic complexity, graph complexity, complexity as logical or thermodynamic depth, complexity as difficulty of performing a task, complexity and scaling properties, regular language complexity, etc. These subjects are discussed in the contributions of Badii *et al.*, Crutchfield, and Grassberger. Many of the corresponding measures of complexity are differential measures, relating a prior state to a later state of a specific system under specific constraints. (Note that such measures can only be applied in retrospect, i.e., when the later state is known already.) Context–dependence also turns out to be essential with respect to the problem of qualifying and quantifying *meaning.* Some of the similarities between the notions of meaning and complexity are best demonstrated via an example due to Haken [4]:

Consider the behavior of a laser system when it is driven over the lasing threshold. Below threshold the system emits basically incoherent light quanta corresponding to spontaneous emission processes in the laser gain medium. Beyond threshold, it is well known that the laser starts to emit coherent light, generated by stimulated emission in the gain medium. In this case, the gain medium and the radiation field behave cooperatively — they influence each other in a self–referent and self–organizing manner. Everybody agrees that this sort of behavior is far more complex than simple spontaneous emission.

The self–organized appearance of complexity is in a certain sense associated with a self–generation of meaning. Namely, both the radiation field and the gain medium can be said to *understand* that the modified behavior of their respective counterpart *means* a corresponding change in their own behavior. (Fortunately, correlation times below threshold are exceedingly small compared with those beyond threshold — otherwise it would be harder to make the system *forget* its former behavior.) Moreover, meaning, once it is understood, is immediately transferred into *action*, here an increase in efficiency (output power with respect to input power) of the laser. In less anthropomorphic terms, the knowledge that the gain medium has of the radiation field (and *vice versa*) is active in the sense that it alters the dynamical behavior of the system.

This example emphasizes that terms usually applied to cognitive systems may have relevance for very basic (and quite well understood) physical systems. Moreover, the notions of understanding and usage are strongly reminiscent of semantics and pragmatics. Together with syntactics, these notions constitute the field of *semiotics*, founded late in the last century [5]. While the syntactic level is relevant for the interrelation of symbols (denoting objects), semantics deals with the relation between symbols and what they symbolize, and pragmatics addresses the relation between symbols and their users. It is clear that issues like meaning and action, understanding and usage have to be included in any semiotic discussion of a situation as described above. The contribution by Nicolis stresses the relevance of such a semiotic approach in relation to information processing systems.

The denotation of an object by a symbol also gives rise to a fundamental conceptual problem: the identification of an object with a symbol separates the symbolized object from everything else, effectively creating a conceptual barrier between that object and every other object. Symbolic denotation therefore assumes that objects exist independently in an objective world (*noumena*), in contrast to the viewpoint of objects as contextual entities (*phenomena*). Noumena are relevant only in a situation without any coupling (e.g., observational coupling). Because the existence of such a coupling automatically generates phenomena, noumena are artifical limiting cases of phenomena. If symbols are to represent phenomena, with their attendant plurality of contextual meanings, rather than context independent noumena, the symbols must consequently be defined very carefully.

Within this framework, the semiotic disciplines of syntactics, semantics, and pragmatics are independent only in the limiting case of noumena, because each phenomenon can to a certain extent be regarded as created by the use of symbols, which in turn implies that usage is not independent from meaning. In terms of physics the inseparability of syntactics, semantics, and pragmatics is a consequence of the interaction between the observing and observed system, which might roughly be illustrated by the unity of a data string (syntactics), its measurement (pragmatics), and its interpretation (semantics). Although this interaction is inevitably present in any system, its relevance has most strikingly been recognized in the quantum domain. (Classical mechanics where actions on the order of Planck's action are negligible can be considered as a limiting case of observational decoupling.) This argument shows the necessity of including the observing/predicting system into any measurement/modelling of phenomena.

This necessity expresses the significance of interactions, or — in other, perhaps trendier, words — of the holistic interconnectedness of the world. With respect to its information theoretical formulation, a surprisingly clear historical development of corresponding concepts can be reconstructed. The aspect of syntactic information has been extensively studied in the influential book of Shannon and Weaver [6]. Shannon–type information is purely syntactic in omitting any reference to contextual or meaning–related issues. It vanishes in case of a message which leaves the receiver in a given state of "syntactic" ignorance (i.e., its syntax is redundant with the prior state of its receiver), and it is greater than zero if it reduces the amount of ignorance of the receiver (i.e., reduces uncertainties).

Shortly after the publication of Ref.[6], Bar Hillel and Carnap [7] proposed a quantification of information based on a receiver's ability to draw logical inferences from received messages, thus addressing the semantic aspect of information. What happens if a message contains a huge amount of syntactic information which is not or cannot be understood by its receiver? For instance, imagine a "Babylonian library", containing all books which can ever be written. (For simplicity, the library may be assumed to contain all possible sequences of a finite number of symbols, up to a finite length.) Obviously, most such books are meaningless to most (standard) readers because they offer mainly random sequences of letters (which might not even be distinguishable from foreign language texts, at least with regard to their meaning). The Babylonian library thus contains an incredible amount of syntactic information, but only a small fraction of it is (understandable) semantic information for standard readers.

As indicated in the example discussed above, semantic information can be qualified by its usage. If semantic information (meaning) is understood, then it triggers action, e.g., changes efficiency, or leaves some other imprint on the behavior of its receiver. As a starting point toward a quantification, E. v. Weizsäcker [8] has introduced the concept of pragmatic information. Taking into account the context of the reader of a text (education, socialization, native language and other pre-knowledge) this concept provides vanishing information measures for completely redundant (Shannonian limit) as well as for completely novel messages. Pragmatic information (or pragmatic value) includes all semiotic levels of information, and it clearly points toward a more complete, and thus more adequate description of an interacting world.

The need for a theory of pragmatic value has long been felt in theoretical studies of biological evolution. From the Shannonian point of view, every possible genome of the same length has the same amount of information, so that the Shannon theory gives no clue as to the reasons why evolution has preferred some sequences over others, and thus has nothing to say about the accumulation of *biologically relevant* information. As Eigen [9] pointed out, "We need further specification of what we call knowledge or unknowledge. We need 'valuation' in order to characterize the degree of self-organization of a functional order and to define a gradient for evolution." This quotation was taken from a seminal paper in which Eigen proposes that the net growth rate be taken as this value parameter in the study of the sequence specific growth rate of RNA replicating *in vitro*. Weinberger's contribution to the present volume elaborates on the notion of *value landscapes*, which are mappings from some regular graph (e.g., the Boolean hypercube) to the real numbers. His paper discusses the characterization of such landscapes via analysis of the statistical properties of the sequence of values obtained by a random walk on the underlying graph. In an impromptu discussion at the workshop, Weinberger proposed a framework for thinking about pragmatic value that was influenced by this work: he postulated that a message has value because it changes the probability distribution $\mathbf{p}$ of possible outcomes of some activity whose value scores are the components of some vector $\mathbf{v}$. In this context, the only definition of pragmatic value that is consistent with the laws of large numbers is $\Delta V = <\mathbf{v}>_{\mathbf{p}'} - <\mathbf{v}>_{\mathbf{p}}$, where $<\mathbf{v}>_{\mathbf{p}}$ is the expected value of the outcomes *before* the message is received, and $<\mathbf{v}>_{\mathbf{p}'}$ is the expected value *afterwards*, which may be less than $<\mathbf{v}>_{\mathbf{p}}$. A message can be viewed as a partial specification of a particular vertex in a value landscape, and the outcomes can be viewed as the completion of the specification. Such partial specifications will lead to completions that are significantly different than the average over all landscape values *only* if the associated message has a *certain* minimal length, thus reflecting context dependencies in the message.

In contrast to the view of value as a fixed, "built-in" property of a particular system, Packard presented a model considering the *emergence* of value from the interactions of a model eco-system. Here, messages take the form of "genetically" prescribed behaviors that are adopted in a wide variety of contexts and evaluated as to their average contribution to the adopting organism's survival value. The remarkable thing about this model is that the lookup table in which the behaviors are coded is almost trivially simple, so that complexity (and behavior valuation) can only arise because of interactions. Indeed, these interactions are so complex that they merely allow to identify appropriate "macrovariables". Such a model leads us back to a perspective, in which the goal of science is the description of a system in terms of system specific quantities (e.g., "order parameters"). Complementary to the universal generality ("First Principles") that we seek to achieve in theoretical abstract models aiming toward a unity of nature, such a contextual viewpoint emphasizes the enormous diversity of nature, simply appearing in its concrete experience. In this sense, context–dependence might turn out to provide deep and essential relationships between physics and other sciences, allowing us to consider much more complex systems than are usually studied in physics.

A famous quotation by Wittgenstein [10] says: "If all scientific questions one can imagine were answered, then the problems of our daily life would still remain completely untouched." If we consider subjective context–dependence no less fundamental than the traditional scientific *credo* of objective universality, then any separation of science and life is ultimately superficial. In concrete terms, this abstract formulation might indicate a shift not just in the philosophy of science, but also in the way that science is practiced on a day–to–day basis.

References

1. M.W. Hirsch and S. Smale. *Differential Equations, Dynamical Systems, and Linear Algebra.* Academic Press, New York, 1974, p.248.
2. Y. Elkana. In *Sciences and Cultures. Sociology of the Sciences*, Vol.5. Editors E. Mendelsohn and Y. Elkana. Reidel, Dordrecht, 1981, pp. 1-76.
3. H. Putnam. *Reason, Truth, and History.* Cambridge University Press, Cambridge, 1980.
4. H. Haken. *Information and Self–Organization.* Springer, Berlin, 1988. See also H. Atmanspacher and H. Scheingraber. *Can. J. Phys.* **68**, 728 (1990).
5. C.S. Pierce. *Collected Papers.* Harvard University Press, Cambridge, 1931–1958.
6. C. Shannon and W. Weaver. *The Mathematical Theory of Communication.* University Press, Urbana, 1949.
7. Y. Bar Hillel and R. Carnap. *Brit. J. Phil. Science* **4**, 144 (1953). R. Carnap and Y. Bar Hillel. In *Language and Information.* Editor Y. Bar Hillel. Massachusetts 1964.
8. E. v. Weizsäcker. In *Offene Systeme I: Beiträge zur Zeitstruktur von Information, Entropie und Evolution.* Klett, Stuttgart, 1974, pp. 82–113. See also K. Kornwachs and W. v. Lucadou. *Cognitive Systems* **1**, 79 (1985).
9. M. Eigen. *Die Naturwissenschaften* **58**, 465 (1971).
10. L. Wittgenstein. *Tractatus Logico–Philosophicus (6.52).* Annalen der Naturphilosophie (1921).

PARTICIPANTS

Ioannis E. Antoniou
Service de Chimie Physique
Campus Plaine CP 231
Université Libre de Bruxelles
B - 1050 Bruxelles
Belgium

Harald Atmanspacher
Max Planck Institut
für extraterrestrische Physik
D - 8046 Garching
FRG

Christian Beck
Institut für Theoretische Physik
Sommerfeldstr.
RWTH Aachen
D - 5100 Aachen
FRG

Jörg Becker
Institut für Physik
Werner Heisenberg Universität
D - 8014 Neubiberg
FRG

Maurice Courbage
Laboratoire de Probabilités
Université de Paris
4 place Jussieu
F - 75252 Paris cedex 05
France

James P. Crutchfield
Department of Physics
University of California
Berkeley, CA 94720
USA

Christopher Dewdney
Department of Applied Physics
Portsmouth Polytechnic
King Henry I Street
Portsmouth, PO1 2DZ
United Kingdom

Thomas Dittrich
Department of Physics
Universität Augsburg
D - 8900 Augsburg
FRG

Stefan Döbereiner
Max Planck Institut
für extraterrestrische Physik
D - 8046 Garching
FRG

Nuretdin Eren
Physics Department
Selcuk University
TR - 42049 Kampus, Konya
Turkey

Manfred Euler
Fachbereich Elektrotechnik
FH Hannover
Richlinger Stadtweg 120
D - 3000 Hannover 91
FRG

J. Doyne Farmer
Los Alamos National Laboratory
Theoretical Division
MS B213
Los Alamos, NM 87545
USA

Valentin Demmel
Max Planck Institut
für extraterrestrische Physik
D - 8046 Garching
FRG

Marco Finardi
Paul Scherrer Institut
CH - 5232 Villigen PSI
Switzerland

Andrew Fraser
Systems Science Program
Portland State University
Portland, Oregon 97207 - 0751
USA

Thomas Hudetz
Institut für Theoretische Physik
Universität Wien
Boltzmanngasse 5
A - 1090 Wien
Austria

Peter Grassberger
Physics Department
Universität Wuppertal
D - 5600 Wuppertal 1
FRG

Mogens Jensen
NORDITA
Blegdamsvej 17
DK - 2100 Copenhagen
Denmark

Nihal Fatma Güler
Mühendislik Fakültesi
Erciyes University
TR - 38090 Kayseri
Turkey

Tassos Katsikas
Service de Chimie Physique
Campus Plaine CP 231
Université Libre de Bruxelles
B - 1050 Bruxelles
Belgium

Gülsen Gürbüz
Department of Mechanics
Istanbul Technical University
TR - 80626 Maslak – Istanbul
Turkey

George J. Klir
Department of Systems Science
Thomas J. Watson School
State University New York
Binghamton, NY 13901
USA

Avadis Hacinliyan
Department of Physics
Istanbul Technical University
TR - 80626 Maslak - Istanbul
Turkey

Franz R. Krueger
Messeler Str. 24
D - 6100 Darmstadt 12
FRG

Wolfgang Haidegger
Institut für Theoretische Physik
Boltzmanngasse 5
A - 1090 Wien
Austria

Jürgen Kurths
Zentralinstitut für Astrophysik
Akademie der Wissenschaften der DDR
DDR - 1501 Tremsdorf/Potsdam
GDR

Hermann Haken
Institut für theoretische Physik
und Synergetik
Universität Stuttgart
D - 7000 Stuttgart 80
FRG

Mai Mai Lam
Department of Applied Physics
Portsmouth Polytechnic
King Henry I Street
Portsmouth, PO1 2DZ
United Kingdom

Daniel Lavallee
Department of Physics
McGill University
3600 University Street
Montreal, Quebec H3A 2T8
Canada

Hans Leydolt
Atominstitut der österr. Universitäten
Schüttelstr. 115
A - 1020 Wien
Austria

Seth Lloyd
Department of Physics
California Institute of Technology
Pasadena, CA 91125
USA

Walter von Lucadou
WGFP
Franziskanerstr. 9
D - 7800 Freiburg i. Br.
FRG

Théo Muller
Centre de Recherches Nucleaires
Université Louis Pasteur
B.P. 20
F - 67037 Strasbourg Cedex
France

John S. Nicolis
School of Engineering
University of Patras
Patras
Greece

Riza Ogul
Physics Department
Selcuk University
TR - 42049 Kampus, Konya
Turkey

Norman Packard
Center for Complex Systems Research
Department of Physics
University of Illinois
405 N. Mathews Ave.
Urbana, IL 61801
USA

George V. Palassopoulos
Military Academy of Greece
41 Blessa Street
Papagos, Athens 15669
Greece

Bernd Pompe
Sektion Physik
Universität Greifswald
Jahnstr. 16
DDR - 2200 Greifswald
GDR

Hubert Preißl
Max Planck Institut
für biologische Kybernetik
Spemannstr. 38
D - 7400 Tübingen
FRG

Fred Richards
Center for Complex Systems Research
Department of Physics
University of Illinois
405 N. Mathews Ave.
Urbana IL 61801
USA

Ute Riemann - Kurtz
Fachbereich Informatik
Universität Bremen
Postfach 330440
D - 2800 Bremen 33
FRG

Arnold Rosenblum
Int. Inst. for Theoretical Physics
Utah State University
Logan, UT 84322
USA

Herbert Scheingraber
Max Planck Institut
für extraterrestrische Physik
D - 8046 Garching
FRG

Rob Shaw
Box 8218
Santa Cruz, CA 95061
USA

Leonard Smith
Department of Engineering
University of Warwick
Coventry CV4 7AL
United Kingdom

Charles P. Sonett
Department of Planetary Sciences
University of Arizona
Tucson, AZ 85721
USA

Maria de Sousa Vieira
The James Franck Institute
University of Chicago
5640 S. Ellis Ave.
Chicago, IL 60637
USA

Jose A. Tenreiro Machado
Departamento des Engenarhia
Electrotecnica e de Computadores
Universidade do Porto
P - 4099 Porto
Portugal

Rudolf A. Treumann
Max Planck Institut
für extraterrestrische Physik
D - 8046 Garching
FRG

Inga Tschiersch
Institut für kybernetische Verfahren
der Ingenieurwissenschaften
RWTH Aachen
Rolandstr. 7-9
D - 5100 Aachen
FRG

Dimitris Vassiliadis
Physics Department
University of Maryland
College Park, MD 20742
USA

Jean Pierre Vigier
Laboratoire de Physique Theorique
Institut Henri Poincaré
11 Rue Pierre et Marie Curie
F - 75231 Paris Cedex 05
France

Alfred Wehrl
Institut für theoretische Physik
Universität Wien
Boltzmanngasse 5
A - 1090 Wien
Austria

Edward D. Weinberger
Max Planck Institut
für biophysikalische Chemie
Am Fassberg
D - 3400 Göttingen
FRG

Wojciech H. Zurek
Los Alamos National Laboratory
Theoretical Division
MS B288
Los Alamos, NM 87545
USA

INDEX